Possible Knowledge

Possible Knowledge

The Literary Forms of Early Modern Science

Debapriya Sarkar

PUBLISHED IN COOPERATION WITH FOLGER
SHAKESPEARE LIBRARY

PENN

UNIVERSITY OF PENNSYLVANIA PRESS

PHILADELPHIA

Published by
University of Pennsylvania Press
Philadelphia, Pennsylvania 19104-4112
www.upenn.edu/pennpress

Printed in the United States of America on acid-free paper
10 9 8 7 6 5 4 3 2 1

Hardcover ISBN: 978-1-5128-2335-6
eBook ISBN: 978-1-5128-2336-3

A Cataloging-in-Publication record is available from the Library of Congress

For Ma

&

To the memory of Baba

CONTENTS

Possible Knowledge

INTRODUCTION

> Right poets . . . borrow nothing of what is, hath been, or shall be; but range, only reined with learned discretion, into the divine consideration of what may be and should be.
>
> —Philip Sidney, *The Defence of Poesy* (ca. 1581)

Living through the slow collapse of the Aristotelian cosmos, early modern thinkers felt long-prevailing ideas of truth and reality shifting under their feet.[1] The expressions of skepticism, doubt, and even bewilderment that pervade Renaissance writing (in discourses ranging from natural philosophy to theology to political theory) convey the unease of a culture struggling to cope with this widespread sense of incertitude. The stakes, after all, were immense. Not only were novel theories like Copernicus's heliocentrism aggravating feelings of uncertainty by remodeling the cosmos, they were also displacing both earth and humans from the center of the universe. Even as early modern thinkers gradually developed techniques to reckon with a world that might remain essentially unknowable (English naturalists, one familiar argument goes, thrived amid this period of epistemic uncertainty by embracing probabilistic and empirical methods to manifest the workings of the physical world), they had to confront an even more fundamental—and profoundly more disturbing—prospect: that their own existential status in the cosmos was, in consequence, precarious. This dual crisis of epistemological uncertainty and ontological precarity could not be resolved merely by studying what was manifest, perceivable, or even probable. Instead, as this book shows, writers and thinkers became fascinated by the capacious domain of the "possible." *Possible Knowledge* studies the radical ingenuity of sixteenth- and seventeenth-century English writers who mobilized acts of literary worldmaking to forge new theories of physical and metaphysical reality during this period of intellectual ferment. By treating poesy—a general early modern term for literature—as the engine of a dynamic intellectual paradigm,[2] this book explores how the imaginative allure of the "possible" reshaped central problems of Renaissance thought, including relations between words and things, between form and matter, even between self and world.

Possible Knowledge approaches the "possible" not as a singular keyword but as a pluralistic concept that incorporates notions of contingency, open-endedness, suspension, incompletion, and futurity. Simultaneously ubiquitous and ungrounded, everywhere and nowhere, the term "possible" is one of those untheorized concepts that pervades Renaissance discourse.[3] Period linguists defined "possible" simply as "that may be."[4] The term, however, adumbrates non-actualized states of being ranging from the potential to the hypothetical, and from the counterfactual to the predictive and the prophetic. The "possible" is an assemblage of imaginative habits of thought and action that enabled writers to grapple with the challenges of constructing knowledge in and about an incomprehensible world. While there was no unified early modern theory of the possible, my phrase "possible knowledge" evokes the shared epistemology that is crafted in and through literary form. Writers including Edmund Spenser, William Shakespeare, Francis Bacon, Margaret Cavendish, and John Milton negotiate the loss of philosophical certainty that pervades early modern intellectual culture by marshaling the knowledge-making potential of poesy, an enterprise that claimed this mosaic of the "possible"—in all its variety—as its ontological compass.

Philip Sidney's *The Defence of Poesy* (ca. 1581) is a foundational text for theorizing possible knowledge. Sidney's writing and figure loomed large in the English cultural imaginary in the years after his death in 1586, and his treatise outlined an influential series of claims about poetic existence, truth, and knowledge that offer a provocative intervention in the problem of early modern uncertainty. In the epigraph that opens this introduction, Sidney describes the labors of *poiesis* by carving out a special ontological status for poetry: the poetic world is opposed to what exists or will be proven to exist, since poets "borrow nothing of what is, hath been, or shall be" and deal, rather, with the "consideration of what may be and should be" (26). In this formulation, poesy does not merely represent or explain reality. Instead, at its most ambitious, it envisions a reality that *could be otherwise*, suggesting a literary epistemology that understands knowledge as conceivable rather than verifiable, practical rather than theorized, ongoing rather than perfected, particular rather than universal. The *Defence* further insists that such possible and ideal spheres of existence cannot be disentangled from *how* they come into being. It is the poet's act of "freely ranging only within the zodiac of his own wit" (24) that enables them to "grow in effect another nature" (23). The poet's peculiar deed of "rang[ing]" beyond what "is," "hath been," and "shall be" is therefore instrumental to "figuring forth" (25) entities that have an asymptotic relation to actuality. Yoking together the activities of poetic making and the worlds they produce, Sidney hints that the ontology of poesy is inseparable from its methods of becoming.

Written at the cusp of the so-called Scientific Revolution,[5] Sidney's "pitiful defence of poor poetry" (18) defines literature as an intellectual pursuit by comparing the poet's scope and practices to the work of other knowledge producers. This method of definition by comparison prompts Sidney to claim the domain of the "possible" as the space in which literature asserts its authority. The *Defence* compares the work of the poet to that of the historian, moral philosopher, musician, arithmetician, geometrician, astronomer, logician, rhetorician, lawyer, grammarian, metaphysic, and physician, and as scholars have demonstrated, Sidney's poet is interested in topics and methods of inquiry that also occupy contemporary natural philosophers.[6] He compares physical and poetic realms, and he argues that the poet's ability to exceed the "brazen" world of "Nature" to "deliver a golden" (24) one liberates them from the limits within which all other artists and philosophers have to work.[7] While figures such as the astronomer, the geometrician, the arithmetician, the musician, the natural and moral philosopher, and the physician are "tied" to the "subjection [of nature]," the poet, "lifted up with the vigour of his own invention, doth grow in effect another nature, in making things either better than nature bringeth forth, or, quite anew, forms such as never were in nature" (23).[8] This capacity to produce "another nature" characterizes the compass of an art that professes a practical ethics—poesy operates "with the end of well-doing and not of well-knowing only" (29). Aiming toward *praxis* ("well-doing") and not merely *gnosis* ("well-knowing"),[9] poesy's conjuration of what "may be" and "should be" is, in Sidney's formulation, directed toward reshaping the actual world. At a moment when the epistemic value of poesy was at best unsettled, and at worst summarily dismissed,[10] the *Defence* flaunts the art's audacious intellectual ambitions by indicating that while disciplines like natural philosophy might explain the physical world, imaginative writing could remake reality.

My book constructs the paradigm of "possible knowledge" by turning to works of English poesy that are as unabashed about their ambitions as Sidney's *Defence*; writers from Spenser to Milton mobilized the ontological affordances of literary form to generate methods of learning that were coextensive with the kinds of worlds they engendered.[11] The "knowledge" of my title is not *one* static concept that exists across, or a singular idea to be extracted from, different forms of literary writing. Instead, "possible knowledge" encompasses the mobile and unpredictable energies that constitute varied practices of literary worldbuilding. "Knowledge," in other words, refers to methods of *poiesis* rather than to a perfected artifact. I excavate this polyvalent paradigm from notions and practices that are embedded in, and that morph across, the generic forms of romance, tragedy, utopia, lyric, and epic. The abstract definition of "possible knowledge" thus emerges from the diverse structures and techniques (stylistic, rhetorical, formal, and discursive) intrinsic to literary texts themselves. I treat imaginative intellectual modes—such as hypothesis, prophecy, prediction, probability, conjecture, conditionals, and counterfactuals—as instruments of possible knowledge that span these genres: epic events in *Paradise Lost* serve as units of learning that expose the limitations of probabilistic experiments, for instance, while the certainty of prophecy in *Macbeth* remains contingent on the teleological unfolding of tragic form. In particular, I illustrate how formal techniques typical of different genres of literary writing undergird the imaginative dimensions of scientific experiment, induction, and theories of probability. For instance, I show how the error and endlessness that govern Spenserian romance are at the heart of Baconian induction, and I trace how Cavendish's utopian thought experiment rewrites premodern physics. The phrase "possible knowledge" constellates the ways in which literary writing generated forms of thinking vital to the exchange of ideas about natural and imaginary worlds at a moment when astronomers and natural philosophers were grappling with new accounts of the cosmos. By arguing that we cannot separate the ontology of literature—what literature *is*—from the ways of thinking that govern poetic production, this book prompts literary scholars to reclaim poesy as a philosophical mode of being and knowing. And by documenting the deeply literary life of early modern uncertainty, a topic that has long fascinated scholars across disciplines, *Possible Knowledge* offers a defense of *poiesis* as a vibrant philosophical endeavor.

"Poets were the first philosophers": Literary Studies and the Methods of Poesy

Possible Knowledge demonstrates what is gained when literature and the critical labors of literary studies occupy a central place in our approach to the history of ideas. Like Sidney's poet, who acts as a "moderator" (31) between history and philosophy, literary scholarship often seems to careen between the opposing pulls of theory (formed by the impetus to abstract and generalize) and historicism (governed by the constraints of the archive). Yet, because literary criticism attends to texts that detail modes of existence untethered to actuality, it is uniquely suited to recover the polyvalent ontology of the "possible," which neither makes itself manifest to empirical (and positivist) methods nor lends itself to abstraction. To this end, this book approaches literary works both as objects of analysis and as theoretical sources that shaped the intellectual culture of early modernity. Furthermore, it turns to the methods of literary studies to revisit paradigmatic concerns in the history and philosophy of science, including the origins of modern scientific methods and debates about competing theories of physics. In doing so, my study of "possible knowledge" reevaluates formative tensions in intellectual history, including those between knowing (*gnosis*) and doing (*praxis*), process and product, action and prescription, form and matter.

In taking science—the set of practices early moderns termed "natural philosophy" and "natural history"—as the point of comparison in its study of literary knowledge-making, *Possible Knowledge* contributes to scholarship in the rich field of early modern literature/science studies.[12] Moving beyond early and mid-twentieth-century studies that focused primarily on scientific content and imagery in literary works,[13] in the past three decades scholars have explored the co-constitutive nature of "imaginative literature" and "science," recovering what Mary Baine Campbell describes as the "history of their mutually determining emergence."[14] By arguing that the works of poets and dramatists are not mutually exclusive from the labors of artisans, alchemists, mapmakers, and mathematicians, this scholarship has recovered, in Elizabeth Spiller's words, "shared aesthetics of knowledge" across premodern disciplines.[15] In the last decade, literary critics have further highlighted, as Frédérique Aït-Touati notes, the "heuristic role" of fiction (in Aït-Touati's case, through juxtapositions with cosmological discourse).[16] In the process, several studies have confronted directly what Claire Preston identifies as "a troublesome idea": the notion "that the literary might affect or influence, or even originate, the scientific."[17]

At its boldest, this recent scholarship claims that by reading acts of representation—the use of literary tropes, figurative language, generic conventions—in the writings of natural philosophers like Francis Bacon, Galileo Galilei, Johannes Kepler, and Robert Boyle as vital components of scientific practice, we recognize "the literary *as* science."[18] *Possible Knowledge* cuts against the grain of this work by reclaiming the imaginative dimensions of emergent scientific methods as components of a *literary* epistemology, rather than rebranding the apparatus of early modern *poiesis* as scientific. Instead of focusing on the content or rhetoric of what Renaissance naturalists studied (in the fields of metallurgy, optics, physics, alchemy, or astronomy), I attend to the processes of inquiry that often underpin such specialized endeavors. In other words, this intellectual history centers ways of knowing rather than objects of knowledge. As such, my work aligns with Tita Chico's formulation of "literary knowledge," which understands "literariness as a form of epistemology,"[19] and I share the commitment in literature/science studies (both early modern and beyond) to avoid what Chico identifies as the critical "belatedness" that assumes the "literature follows the scientific."[20] I am less interested, however, in expanding what constitutes *scientific* knowledge—whether by tracking rhetorical tropes that abound in works of natural philosophy, for instance, or by cataloging the expanding genres in the oeuvre of what we taxonomize today as early modern science. Instead, by arguing that possible knowledge is a literary epistemology, I link the critical concerns of literature/science studies to broader discussions of poetics and intellectual history. Thus, I situate this project alongside recent scholarship by Colleen Ruth Rosenfeld, Jenny C. Mann, Wendy Beth Hyman, and Andrea Gadberry that reveals how early modern writers mobilized the formal, rhetorical, and stylistic apparatus of imaginative discourse as intellectual and philosophical tools to propound ways of knowing that we might term "poetic."[21]

At the heart of my emphasis on the epistemological force of the literary is the hunch that in order to fully account for the profound incertitude governing early modern intellectual discourse in the decades following Andreas Vesalius's and Nicolaus Copernicus's novel theories of microcosm and macrocosm, respectively,[22] we need to turn to the art that dealt most forcefully with the unverifiable, the intangible, and the nonexistent: poesy. Early modern thinkers inherited the notion that philosophical inquiry (*scientia* or *episteme*) addressed universal, necessary, and unchanging things; it revealed truth about causes and produced certain knowledge through logical demonstration. The loss of philosophical certainty in the Renaissance, scholars have demonstrated, prompted the elevation of the epistemic purchase of *praxis*,[23] and literary scholars in particular have foregrounded the intimate connections of *poiesis* to *praxis*.[24] While *praxis* signified "practical knowledge" in which "action contains its end in itself," *poiesis* referred to "productive knowledge" where "making finds its ends in its object."[25] Read within this framework of the entangled histories of literary and scientific knowledge, *Possible Knowledge* is first and foremost a study of the peculiar workings of *poiesis*: by insisting on the inextricability of "making" and "object," this book recovers how the *kind* of worldmaking determines the epistemic scope of an act of poetic production. Dealing with what "may be and should be," literary works served as a distinctive knowledge-making enterprise, not despite but because of their tendencies toward errancy, profligacy, and rhetorical extravagance. The mobility and variety of these literary strategies enabled writers to test out questions of truth and existence in myriad ways without the constraints faced by those in many other disciplines.

To recover how the imaginative labors, procedures, and strategies of poesy shaped early modern intellectual discourse, I start from the formal and conceptual elements of literary texts themselves, rather than beginning from external rubrics of disciplinary divisions, institutional frameworks, or classical philosophical paradigms. To define "possible knowledge," I could have treated "possible" as a keyword and traced its varied usages across a century of early modern literary writing. Or, I could have traced the homologies of poetic invention and the "New Science" over a long historical period by focusing on literature's relation to particular disciplines—physics, chemistry, astronomy, geometry, or even logic and rhetoric. Or, I could have studied the materialistic and philosophical traces—of Lucretian atomism, for instance, or of Pythagorean transmigration of souls—that abound in early modern imaginative writing and documented how poets and dramatists engaged with ancient schools of thought, ranging from Platonic forms to Aristotelian hylomorphism, from classical skepticism to Galenic medicine. I have pursued a different strategy. *Possible Knowledge* is a thought experiment that enacts how close readings of imaginative methods and formal techniques can help us recover the kinds of knowledges that early modern literary forms were themselves producing.

The political and ethical import of close reading has been a subject of longstanding debate in literary studies.[26] Early modern scholars have emphasized the particular value of this practice to historicist, feminist, and queer studies in what Corey McEleney identifies as "an intellectual era marked by archive fervor, thematic criticism, countless materialisms, distant 'reading,' and quantitative approaches to literature."[27] My methodological commitments align with such approaches, which treat literary texts as aesthetic phenomena that simultaneously enact and theorize ideas—or more accurately, that theorize ideas through enactment. Thus, like J. K. Barret, I "engage imaginative literature in its capacity to act like philosophy."[28] But, by basing its own modes of thinking on the intellectual forms that are also its objects of study, *Possible Knowledge* further emphasizes that close reading performs a kind of philosophical inquiry.

This book emulates the expansive ways of thinking that literary writing performs—thinking that unfurls through the mechanics of worldmaking, or thinking as a kind of *poiesis*. To capture the myriad forms of epistemological risk-taking in early modern texts, my interpretative mode is, by necessity, eclectic. My readings move across scales—from the micro level of grammatical mood to the macro level of theories of the cosmos. They range across different early modern discourses—logic, physics, and recipe culture, for instance. They list conceptual and formal patterns to highlight a text's recursive energies (such as *Macbeth*'s obsessive language of futurity) and dally at pivotal moments (such as Eve's fall in *Paradise Lost*). I use formal elements of literary texts (such as the techniques of romance that constitute Baconian induction) to reconcile heated scholarly debates and, by extension, engage with varied scholarly conversations and methodologies in history, literature, and philosophy. This list only begins to indicate how close reading is a practice that traverses multiple registers and scales; it does not demand a priori commitments to particular historical, theoretical, or scholarly paradigms. Instead, my reading practices mimic—and continue from—the myriad ways of thinking that literary texts themselves enact through their eccentric strategies of worldbuilding.

In performing the modes of thinking scripted by literary works, this approach models a way of knowing—a method—that is not extractable (or abstractable) from my objects of inquiry. Early modern poesy's propensity to scramble notions of chronological time, to transport readers to different places, to mesh together contradictory theories and practices, as well as its refusal to limit itself to one philosophical idea or historical event, invites us to—indeed insists that we—move across contexts, forms, and scales. As a result, the methods of the authors I study—such as the use of conjecture to invite readerly engagement, or the enactment of hypothesis through utopian worldbuilding—are also central to my own methods of investigation. This form of criticism unfolds, to borrow Andrew H. Miller's words, "through the performance or display of thinking."[29] I call such formalist scholarship a thought experiment because it requires the critic to take an imaginative leap—a remaking of our critical orientations, a commitment to follow the text by suspending the belief that the course of research must lead from local evidence to a generalizable conclusion authorized by extant categories of thought, scholarly standards, or disciplines. This methodology revels in a form of knowing that early modern poesy repeatedly performs—the ability to locate an entire universe in a single phrase, "what may be and should be."[30] The practice of close reading might seem familiar, even mundane, to literary scholars who rehearse it regularly in their research and teaching. At its most ambitious, however, it asks us to think like early modern poets, urging us to consider how criticism could be otherwise.

The book's emphasis on the philosophical import of close reading has implications beyond literary studies, as it hints that a *literary* history of Renaissance literature and science can connect issues that literary critics, philosophers, and historians of science often tackle separately. Take, for example, current narratives about histories of scientific probability. It is a scholarly commonplace that sixteenth- and seventeenth-century thinkers countered the loss of philosophical certainty by turning to myriad theories of probability (from manifestations in premodern law, religion, and logic to mathematics).[31] By foregrounding imaginative modes like prophecy, conjecture, and hypothesis, I show that the dichotomy between certain and probable knowledge is insufficient to capture the volatile contingency of early modern thought. By elaborating the concept of possible knowledge, I also hope to foster conversations among scholars who have turned to possible-worlds philosophy to grapple with theories of truth, existence, and the ontological status of imagined entities.[32] Modern theorists of possible worlds sometimes trace their origins back to Aristotle's *Poetics*, an influential intertext for early modern theorists of poesy.[33] However, they typically bypass Renaissance literature as a site of inquiry, even as they echo the language of plural worlds and wrestle with questions that fascinated early modern writers.[34] The peculiar worlds of sixteenth- and seventeenth-century literature—deeply imbricated in considerations of matter, nature, and physics, as Sidney's words imply and as we will see most prominently echoed in Cavendish's physicalist worldmaking—open up new questions for literary scholars and philosophers alike. Focusing on the pivotal role of poesy in the transformation of early modern epistemology, *Possible Knowledge* invites scholars across a range of disciplines to embrace the methods of literary studies in order to historicize and retheorize concerns about the nature of being, materiality, and existence that pervade premodern and modern discourse.

Worlds of Words: Rhetoric and Natural Philosophy

Implicit in my argument about the epistemic purchase of poesy is a further claim: the transformations in—and the institutionalization of—natural philosophy in the period were intimately connected to the shifting status of language in the broader early modern intellectual landscape. Even as research on early modern epistemology has delved into the status and function of language, typically by turning to debates about *res* and *verba*, or by attending to the fluctuating fortunes of the discipline of rhetoric itself, poesy (and literary form) has not been central to these intellectual histories. This book shows, however, that the misgivings about rhetorical extravagance and the promiscuity of language that preoccupied early modern natural philosophers were inextricable from their concerns about the labors of poets and dramatists, who, trained in the rhetorical tradition, flagrantly pushed the boundaries of decorum and plausibility. *Possible Knowledge*, then, reveals that what we have long understood as criticisms of the materiality of language and the art of rhetoric are equally anxieties about the ontology of fiction.

Rhetoric was part one of the liberal arts of the trivium (along with grammar and dialectic), which, along with the mathematical arts of the quadrivium (arithmetic, geometry, music, and astronomy), formed the curriculum at Oxford and Cambridge. By the end of the fifteenth century, however, dialectic had come to "dominat[e] the trivium program."[35] This shift, as is well-documented by scholars including Lisa Jardine and Walter J. Ong, is part of broader alterations in humanist education (by reformers such as Rudolph Agricola, Lorenzo Valla, Philip Melanchthon, and Johannes Caesarius) that challenged medieval legacies of Aristotelian logic and key tenets of scholasticism. These reforms not only impacted the university curriculum but also reverberated beyond institutional bounds. The reformed, "pupil-oriented teaching" of dialectic promoted by thinkers like Agricola—whose *De Inventione Dialectica libri tres* (1479) became the standard text in Cambridge in the sixteenth century—was geared toward non-specialists.[36] In providing grounds for "rational discourse" that were widely applicable outside the university, dialectic became foundational to all processes of thinking.[37] Central to the elevation of dialectic were debates about the appropriate role of language—and especially of rhetoric as the source of linguistic extravagance—in the production of knowledge. Thinkers like Peter Ramus limited rhetoric to the domain of *verba*, or words, severing it from the realm of *res*, or things[38], such separation of *res* and *verba* in the reformed curriculum ultimately resulting in a corresponding denigration of rhetoric as a source of knowledge.

Beyond the university curriculum—and beyond the domains of logic and rhetoric—the uncoupling of *res* and *verba* also becomes of paramount importance to English naturalists who are carving out a separate domain for their work and who are also apprehensive about the role of rhetoric and extravagant language in natural philosophy. Seventeenth-century naturalists variously transport arguments questioning the knowledge-making potential of rhetoric from the sphere of the language arts to studies of the physical world. Even Bacon (who, unlike Ramus, demotes dialectic to be on par with rhetoric) separates words and things, in the process performing what Jenny C. Mann and I identify as a "conceptual transfer": "after Bacon, things, or matter, that is, what the rhetorical manuals of Cicero, Quintilian, and Erasmus term *res*, steadily lose their rhetorical sense as 'things to be discussed' or 'subject matter' and begin to refer instead to 'empirical things' or 'physical matter.'"[39] By blaming rhetoric for misdirecting one "to hunt more after words than matter" in *The Advancement of Learning* (1605),[40] Bacon formalizes a concern that English natural philosophers had already been circulating: William Gilbert, writing at the turn of the seventeenth century, rejects "graces of rhetoric" and "verbal ornateness" in *De Magnete* (1600), implying that rhetoric's disciplinary apparatus is unsuited to explain theories about the natural realm and the practices undertaken to understand it.[41] This crisis of confidence in the intellectual merit of ornate language, the common critical argument goes, leads to debates in praise of the plain style and fosters the descriptive turn in prose genres like travel writing and natural history.[42] Natural philosophy, in particular, authorizes its legitimacy by marshaling and revising controversies of how artful language could illuminate—or obscure—understandings of truth and actuality. Later seventeenth-century natural philosophers are especially attuned, as Thomas Sprat argues in *The History of the Royal-Society* (1667), to the need for a "naked, natural way of speaking" that would erase the gap between the physical world and its semiotic representations.[43] As we will see in Chapter 3—where I demonstrate how Bacon's method is unable to sustain the barriers his works erect between "words" and "things"—within criticisms of rhetorical excess there lurks the worry that language is not merely a vehicle to depict ideas, practices, or phenomena but a mechanism of knowledge production that can destabilize any understanding of actuality.

Bacon's Idols in *New Organon* (1620) famously describe illusions that have taken "deep root" in the "human understanding" and are difficult to dislodge, and these Idols capture the acute anxiety that language might exceed, even escape, its allocated bounds of description and presentation.[44] Words are threatening because they have the potential to sever ties with actuality—and thereby propose alternate realities—when they refer to "either names of things which do not exist," or "names which result from fantastic suppositions and to which nothing in reality corresponds," or even when "they are names of things which exist, but yet confused and ill-defined, and hastily and irregularly derived from realities" (1.60). Language, through its generative power, catalyzes a kind of existential crisis. Scholars have long debated Bacon's ambivalent relationship to linguistic excess, marking his concurrent dismissal and embrace of ornate language, figures of speech, and allegorical figuration across his works.[45] Bacon's self-proclaimed followers, however, display no such ambivalence when they explicitly denounce rhetorical excess and applications of language that further such perceived extravagance. Written almost half a century later than Bacon's *New Organon*, Sprat's *History* makes clear that the Royal Society "indeavor'd, to separate the knowledge of *Nature*, from the colours of *Rhetorick*, the devices of *Fancy*, or the delightful deceit of *Fables*."[46] To carve out a distinct way to represent the "knowledge of Nature," he collapses matters of style and content, making no distinctions between the "colours of *Rhetorick*" and the falsehoods and "deceit" he associates with fiction. To avoid these traps, the Royal Society resorts to "the only Remedy," which is "to reject all the amplifications, digressions, and swellings of style: to return back to the primitive purity, and shortness, when men deliver'd so many *things*, almost in an equal number of *words*."[47] Sprat also echoes Bacon's ultimate hope that the labors of natural philosophy can restore the human mind to the "perfect and original condition" that was lost at the Fall.[48] In doing so, Sprat's language conjoins the "return" to an original Adamic language to the "return back to the primitive purity" via the labors of experimental philosophy.[49] These words also represent the culmination of the sentiment William Gilbert expressed at the beginning of the century in *De Magnete*, that he "aimed simply at treating knotty questions about which little is known in such a style and in such terms as are needed to make what is said clearly intelligible."[50] The efficient transmission of facts demands a minimalist style, while unrestrained language is suited to the purportedly futile ends of the fanciful and the fictional.[51]

In Sprat's argument, the attacks on rhetoric organically blur into criticisms of imaginative writing.[52] By linking linguistic excess to poesy's dealings with the fabricated and the nonexistent, such criticisms transform an attack that is purportedly about style into one on content; linguistic fabrication becomes inextricably linked to ontological emptiness. Sprat echoes the sentiment of figures like Galileo, who earlier in the century had opposed his own philosophical works to "fiction . . . like the *Iliad* or *Orlando Furioso*, productions in which the least important thing is whether what is written there is true."[53] Such denunciations link *how* fiction works to the *kinds of entities* it produces. The connections that Sidney had used to elevate poesy's potential, Sprat and his contemporaries utilize to label the imaginative as the site of "delightful deceit." This shift in perspective also underscores a key difference in their understanding of the tools of poetic production: where Sidney sees poesy's workings as a cognitive act, Sprat mainly finds linguistic immoderation. By avoiding ornamental language, the Royal Society will "bring *Knowledg* back again to our very senses, from whence it was first deriv'd to our understandings."[54] To expunge the threat of words run amok, one must strip the category itself of intellectual weight. Language must be restricted to the communication of ideas and not accepted as a medium that generates ideas.

In contrast to such unqualified declarations, Bacon's ambivalent relation to language might result from his recognition that *poiesis* names an act whose intellectual processes cannot be so easily dismissed. In the *Advancement*, Bacon defines

> poesy [a]s a part of learning in measure of words for the most part restrained, but in all other points extremely licensed, and doth truly refer to the Imagination; which, being not tied to the laws of matter, may at pleasure join that which nature hath severed, and sever that which nature hath joined, and so make unlawful matches and divorces of things: "Pictoribus atque poetis, &c." It is taken in two senses, in respect of words or matter. In the first sense it is but a character of style, and belongeth to arts of speech, and is not pertinent for the present. In the latter, it is (as hath been said) one of the principal portions of learning, and is nothing else but Feigned History, which may be styled as well in prose as in verse. (Book 2, 186)

Bacon distinguishes poesy's stylistic elements from its "portions of learning" and follows Sidney in claiming that it "may be styled as well in prose as in verse." Even though he acknowledges its capacious reach by equating it to the multifarious image-making faculty of "Imagination,"[55] the eruption of legalistic terms signals his reservations about the "extremely licensed" mode. By stating that poesy "doth *truly* refer to the Imagination" (emphasis mine), Bacon seems to be making a claim of its absolute alignment with the faculty. Poesy's freedom from the "laws of matter," moreover, promotes "unlawful matches and divorces of things."

I am most interested in Bacon's struggles to articulate poesy's relations to truth. Like Sidney he recognizes that literature's power lies in its peculiar relation to the natural world—their definitions are aligned in this respect—but Bacon ascribes the opposite value to poesy. Whereas Sidney locates poesy's liberatory potential in its severance from actuality, Bacon argues that because poesy can "join that which nature hath severed" and "sever that which nature hath joined" its manipulations are unnatural. His application of "unlawful" to rearrangements of matter implies that to *be* poetic is to occupy a state of existence that should be otherwise. In this acknowledgment of poesy's immense jurisdiction, might we detect a hint of respect, even envy? After all, the poet's ability to control and reorganize the "matter" of nature is precisely the dream of the natural philosopher that Bacon will articulate in subsequent works. The disquiet undergirding Bacon's criticisms of poesy—a disquiet that seems to register the expansive potential that Sidney locates in the art—is also the impetus for *Possible Knowledge*, which begins from the premise that early modern poesy had no obligation to the methods of knowledge production and standards of truth so dear to natural philosophers, and that were central to their attempts to legitimize and authorize their own intellectual pursuits.

Thinking with Literature: Making Possible Knowledge

The chapters that follow seek to figure out how early modern literature thinks. I begin by arguing that Spenser's *The Faerie Queene* (1590–96), through engagement with travel literature and cosmological discourse, generates an iterative method of speculation that sparks a desire for ethical action by immersing the reader in the possible worlds of the allegorical epic-romance. I then turn to *Macbeth* (ca. 1606) to explore how Shakespeare's most topical play examines the limits of prophetic certainty as it grapples with the major epistemic shift from *gnosis* to *praxis* at the turn of the seventeenth century. Macbeth treats the witches' prophecies as practical recipes, and his exigency in actualizing their utterances has formal effects, setting in motion the play's march toward its tragic end. My third chapter demonstrates how poetic epistemologies pervade foundational aspects of the "New Science." Baconian induction, as outlined in the *New Organon*, formally resembles a Spenserian romance: Bacon's empirical method is an "endlesse worke," to borrow Spenser's words,[56] that depends on what it claims to erase from natural inquiry—digression, error, interminable delay. The final two chapters explore how writers reject probabilistic and empiricist legacies of Baconianism to reclaim fiction as the domain of metaphysical truth. The fourth chapter traces Cavendish's "physical poetics," in which the material force that she terms "Nature" animates literary production. Creating atomist worlds in *Poems, and Fancies* (1653) and vitalist ones in *The Blazing World* (1666), Cavendish challenges both empiricist technologies and mechanistic theories of the universe and champions literary worldmaking as an alternate mode for explorations of materiality and truth. In the final chapter, I recover Milton's "evental poetics," which tests poesy's capacity to produce any degree of certitude about a fundamentally unknowable world. Epic events in *Paradise Lost* (1667) enact how a singular instance of acute change can produce certain knowledge about states of being that remain beyond the reach of repeatable and probabilistic modes of knowing (such as the Royal Society's experimental methods).

I have organized the chapters in chronological order to outline a literary history of possible knowledge that is concurrent with other narratives in intellectual history (such as the history of scientific probability, or the evolution of experimental philosophy) but that cannot be absorbed into them. This organization also acknowledges the networks of engagement, whether overt or latent, that exist across the texts I study: Cavendish challenges mechanist philosophies and experimental methods in *The Blazing World* by reframing parameters of utopian fiction outlined in Bacon's technocratic *New Atlantis* (1627), and *Paradise Lost* participates in a long history of poetic production—also embraced by Spenser—as it links the poet and the *vates* in the figure of the epic narrator, who claims that only a select few can translate unfallen and divine language for fallen readers and make intelligible a world others can never truly know. But the connections between these chapters are not merely temporal. Chronology is not reducible to teleology, a narrative of progression, say from literary incertitude to scientific probability. The chapters are also connected by a collection of imaginative intellectual modes (including hypothesis, conjecture, and prophecy) that overlap in complex ways. For instance, prophecy serves as a key mediator between past and present, and between certitude and indeterminacy, and I attend to its epistemic affordances in the works of Spenser, Shakespeare, and Milton. These tools of possible knowledge do not map onto specific disciplines (such as dialectic or rhetoric), are not confined to particular topics or methods of inquiry (like calculus or experimentation), are not classifiable under definitive schools of thought (like skepticism or Epicureanism), and cannot be reduced to literary tropes or figures of speech. Together, the chapters of *Possible Knowledge* tell several stories simultaneously: the evolution of a literary epistemology, an alternate version of intellectual history, and how the formal apparatus of poesy served as an engine of producing knowledge.

Tracing the epistemological and ethical ambitions of an art that contemporaries would often dismiss as intellectually vacuous and morally dangerous, the following chapters situate Sidney's claims about poetic production within the broader culture of Renaissance uncertainty. In this introduction's final pages, I elaborate on aspects of Sidney's *Defence* that are crucial to plotting the vectors of possible knowledge and outline briefly how these elements manifest in the rest of the book. When Sidney declares that "Nature never set forth the earth in so rich tapestry as divers poets have done; neither with so pleasant rivers, fruitful trees, sweet-smelling flowers, nor whatsoever else may make the too much loved earth more lovely" (24), he is ascribing to the poet the kind of creative act that Philippe Lacoue-Labarthe, drawing on Aristotle, terms "general mimesis," "which reproduces nothing given (which thus re-produces nothing at all), but which *supplements* a certain deficiency in nature, its incapacity to do everything, organize everything, make everything its work—*produce* everything. It is a productive mimesis, that is, an imitation of *phusis* as a productive force, or as *poiesis*."[57] In this framing, Sidney's poet resolves nature's "incapacity": poesy perfects what nature lacks. The poet's creation of that which does not exist—"forms such as never were in nature"—results in worlds that "may" and "should" be. This is not to suggest that all writers sever the worlds of fiction from "nature." As the penultimate chapter shows, Cavendish offers an alternate model of "general mimesis" that is inextricable from her theories of nature and physics. But Sidney's distancing of poetic realms from the "brazen" world has particular ramifications for the truth-value of poesy's propositions and thereby for the kinds of knowledge to which it lays claim.

Sidney explores this issue in detail as he compares the work of the poet and the historian, who deals with what Sidney terms the "bare *Was*" of history (36), which, "being captived to the truth of a foolish world, is many times a terror from well-doing" (37–38). The poet, by contrast, has a more flexible relation to truth because of the states of being they can claim as their purview. We see this flexibility in another oft-cited passage from the *Defence*, where Sidney transforms criticisms about poesy's lack of truth-value into its singular asset and rejects the idea that "the poet is the least liar." Only they lie who

> take upon them to affirm. Now, for the poet, he nothing affirms, and therefore never lieth. For, as I take it, to lie is to affirm that to be true which is false. So as the other artists, and especially the historian, affirming many things, can, in the cloudy knowledge of mankind, hardly escape from many lies. But the poet (as I said before) never affirmeth. . . . He citeth not authorities of other histories, but even for his entry calleth the sweet Muses to inspire into him a good invention; in truth, not labouring to tell you what is or is not, but what should or should not be. And therefore, though he recount things not true, yet because he telleth them not for true, he lieth not. (52–53)[58]

Bound to questions of "what is or is not," historians limit their work to determinations that must fit within categories of "true" and "false." But products of *poiesis* need not be similarly restrained: they need only point to entities *within* fictional worlds, limited exclusively by the "zodiac of [the poet's] own wit." Poetry's distance from truth is a feature and not a bug. Transcending the physical world, poesy opens the space for a different kind of existence. To unshackle themselves from responsibility to empirical verifiability and proof, the poet requires a different method; they thereby adopt the process of non-affirmation. Sidney intimates that poesy's unique intellectual merit lies in this lack of "affirmation," as it enables poets to propose how things could or should be. Rather than serving as apologies for poesy's fluid relation to truth, the modal claims of "may be" and "should be" boldly announce a distinctive method suited to the expansive scope of the worlds poets create. To not affirm, moreover, is an ethical act oriented toward the future. It is an answer to the Horatian injunction to both "teach and delight" the reader (25). As "the creator of an independent artistic heterocosm, a world of its own,"[59] the poet possesses a unique set of tools to direct the reader to "well-doing" or *praxis*.[60] In modeling what is possible or ideal, they are also catalysts for remaking reality.

Although Sidney does not separate prescriptive and possible worlds, the disjunction between what "should" and "may" be hints at conflicting notions of truth—historical, logical, and metaphysical—operative in the *Defence*. The diverging notions of affirmation embedded in Sidney's modal claims can be traced to systems of thought in which the "possible" is a key constituent. Early modern grammarians classified "may" and "should" under the potential mood. As Margreta De Grazia notes, the "potential mood" was added by "Thomas Linacre in his English Latin Grammar of about 1525" to the "classical five of indicative, imperative, optative, subjunctive, and infinitive." Traditionally, "the wishing mood that covered wishes put to God and wishes dependent on the speaker's resources" were both classified as optative. But in the sixteenth century, the optative came to represent only "possibility resting in God's hands," while the potential mood came to signify "possibility residing in individual power."[61] The potential was known "bi these signes, maie, can, might, would, shoulde, or ought," and it was standardized in grammar books by the mid- to late century.[62] In this framework, Sidney's two modals—"may" and "should"—align as non-verifiable modes of expression, where both kinds of poetic claims differ from the "bare *Was*" of history; both terms distance poets from actuality. These worlds, in other words, cannot be affirmed empirically or historically but still exist because of the poet's creation.

Logical treatises such as Thomas Blundeville's *The Arte of Logick* (1617), however, indicate something different: a term expresses a "modall proposition" when it "affirmeth or denieth something, not absolutely, but in a certaine respect, sort, or mood." Blundeville explains further how these propositions function: "Mood is a word determining and limiting the signification of some other word whereunto it is ioined," and "of moods making modall propositions, there are but these foure, that is, Possible, Contingent, Impossible, and Necessarie."[63] Modal propositions highlight how certain terms that are the "signes" of the potential mood—"may," "might"—can affirm or signify certain kinds of truth that need not be empirical or historical. Although Blundeville does not mention the potential mood, his formulation of "Possible," as in "for a man to be iust, it is possible,"[64] can be easily translated to "a man *may* be just" to align with Sidney's usage. (Terms such as "shoulde," however, are prescriptive and do not strictly fit into this system.) Instead of maintaining the separation of both possibility and idealization from historical fact, modal logic suggests some verbs in the potential mood can be employed to affirm *and* deny, since they signify forms of truth that are not reducible to the empirical or the historical.

We could dismiss this split between Sidney's two keywords as one of the many contradictions that abound in the *Defence*. Or we might read it as indicative of the complex ideas that early modern writers were navigating as they mapped out contours of nonexistence: the state of being of a fictional world is affected by the intellectual systems, philosophical discourses, and disciplinary constraints with which poets engage in their acts of literary making. The example of the modal proposition also highlights that fictional beings can incarnate different notions of truth. By recognizing the shifting terrains of truth embedded in Sidney's modal claims, we are thus able to recognize that distinct modes of fictional existence have different relations to certitude and plausibility. Vectors of possibility enable a range of modes of being that, I propose, can serve as indexes of kinds of poetic worlds. Extending this idea, we might say that generic categories such as utopia, romance, epic, lyric, and tragedy define distinctive contours of fictional existence. As writers pursue different *kinds* of worldmaking, they calibrate a fictional world's distance from actuality, thereby indexing a literary form's proximity to modalities of non-actualized being (such as probable, implausible, hypothetical, counterfactual, and even impossible). The specific kind of worldmaking in turn determines the knowledge a particular act of *poiesis* produces.

This brief excursion into Sidney's *Defence* points to ontological conundrums embedded in the concept of the "possible"; these conundrums link problems of literary form to questions of epistemology. It also touches on four entangled aspects of possible knowledge that I take up in more detail in the chapters that follow:

1. Particular registers of possible knowledge—like conjectural worldmaking or prophetic utterance—can serve as indexes of the flexible but recognizable conventions of literary genres. I explore this issue across the chapters. Chapter 2, for example, traces the performative epistemology underlying vatic utterances in *Macbeth* to reveal that tragic theater, at its core, is prophetic in form: the play's tragic outcome is predicated on the gambit that potential events—here outlined by the "weird sisters" (2.1.20) in the play's opening act—will be actualized within the delimited temporality of theatrical enactment. Chapter 3 offers a meditation on the epistemological scope of one particular genre. By showing that Baconian induction enacts the formal features that characterize Spenser's "endlesse worke," I highlight how romance, a genre ostensibly intended for pleasure and idleness, comes to inform some of the period's most significant knowledge practices about the natural world. My study in Chapter 5 of the "evental poetics" animating *Paradise Lost* offers yet another vantage, exposing how generic conventions could be mobilized to challenge contemporary methods of learning. Scholars typically read Eve's fall as a representation of experimentation, but I contend that, in this critical scene of transformation, Milton appeals to epic events as units of certainty. The poem condenses into an unrepeatable event the certitude of the Fall to expose the failings of repeatable methods and revealing that experiment, a postlapsarian epistemology, cannot take one back to paradise.

2. In orienting readers toward *praxis*, Sidney's poetics is implicitly attentive to the future, and it indicates how discourses of possibility are intimately tied to questions of time. I focus most explicitly on the futuristic orientation of possible knowledge in the earlier chapters. Chapter 1 excavates how *The Faerie Queene*'s iterative method of speculation propels readers into imagined futures: the Proem to Book 2 suggests that readers might assess the existence of Faerie Land by linking its potential state of being to the "real" realms recently accessed by travelers, and Merlin's genealogical prophecies in Book 3 transport readers, with Britomart, toward a narrative future that produces Elizabeth's reign, only to reveal the absolute contingency of the approaching times, in which the vision of the monarchy after the queen's death remains beyond poetic narration. As Spenser's invocation of Elizabeth hints, the temporality of possible knowledge often mediates contemporary political concerns. Chapter 2 brings this into sharper focus, revealing how the ability to envision a future and act on it has severe implications for political authority. Forecasting the rule of King James, *Macbeth* stages a theory of political futurity, where sovereign authority requires the temporality of performance (the capacity to reveal the "future in the instant" [1.5.56]) to project fictions of the times to come and to legitimize itself.

Chapter 3 captures fantasies of authority that exist beyond court and monarchy, exposing how futuristic imaginings can spill out of works of fiction like the *New Atlantis* and reshape actuality; we witness this in the complex mid-seventeenth-century afterlives of Baconianism, from Royalist continuations of *New Atlantis*, to Hartlibian reforms, to the most tangible legacy of Bacon's thought: the Royal Society's experimental philosophy.

3. Behind these appeals to futurity lurks another figure: the reader or audience, who is an essential participant in the production of possible knowledge. In addition to attending to the ethical implications of readerly engagement in the Coda, I recover various models of participation by readers across the chapters. Spenser's iterative method requires readers to transport themselves into situations in the fiction (as I will examine in Chapter 1). Shakespeare implicates audiences in the theatricality of prophecy, as we will see in Chapter 2, when the spectacle of the line of kings leads them from the staged world inhabited by Banquo's character to the historical moment of James's reign, that is, to their current existences beyond the dramatic plot. My study of Cavendish's *Poems, and Fancies* in Chapter 4 illustrates, through an inverse logic, the power of readerly insight to sustain fiction. Her lyric poetry reveals how the reader's obliviousness to the products of *poiesis*—the microscopic worlds of the poetic corpus—may contribute to the destruction of the atomist realms themselves.

4. *Possible Knowledge* invokes the apparatus of early modern natural philosophy to reveal that, as astronomers proposed new theories of the universe, and as experimental philosophers came to privilege probabilistic methods over metaphysical truth, literary works expanded the epistemic scope of the imagination. Each chapter touches on this—as I consider Spenser's attention to cosmological discourse, or as I trace how *Macbeth* engages with *secreta* or occult knowledge—but this focus on natural philosophy is clearest in the final two chapters of the book. In Chapter 5, I explore how *Paradise Lost*'s prophetic certitude stands as a contrast to the Royal Society's probabilistic experimental methods, while in Chapter 4, I examine how Cavendish's vitalist physics fictionalizes the authorial rejection of mechanist theories of the universe proposed by thinkers like Thomas Hobbes and René Descartes. Indeed, as Cavendish creates possible worlds out of her theories of matter, she implicitly offers a materialistic response to Sidney's entanglements of epistemology and ontology: in her early work *Poems, and Fancies*, she builds worlds out of atomist physics. When she renounces her belief in atomism in favor of vitalist materialism, this physics manifests as a blazing world of utopian fiction. Imitating not objects but the *motion* of nature, her physical poetics realizes the radical promise of *mimesis*.

Together, these chapters revise inherited ideas on various aspects of the "Scientific Revolution," such as the relation of experience and experiment, the processes that constitute Baconian induction, and even the tensions between atomism and vitalism in the seventeenth century.

Possible Knowledge recovers a literary culture that celebrates the power of contingent methods of thought and action during a period of remarkable epistemological shifts, rather than one stymied by the uncertainties of religious controversy, political turmoil, scientific innovation, or environmental disaster. By attending to imaginative modes that were incipient to early modern thinking, this book uncovers an intellectual paradigm in which knowledge is understood as that which is conceivable rather than verifiable or certain. Such modalities of knowing demand their own standards of evidence and strategies of evaluation. The subsequent chapters demonstrate why literary studies is key to this evaluation. A study that focuses on the relations of literature and science is especially intriguing in this context. Literary scholars have increasingly embraced both the language of the sciences and many of their methods. This trend is most prominent in approaches in the digital humanities, but leanings toward such modes of criticism are more widespread in literary studies—in methods such as network analysis, for instance, or in the application of terms like "literary data." In the process, we have also accepted scientific standards of evaluation as universal. *Possible Knowledge* proposes that we need to reassess literary knowledge through the parameters being produced by literary texts—which were not striving for facts, generalizable precepts, or objectivity—because they have implications for how we interpret and value modes of discourse that do not fit into binaries of truth and lies, reality and falsehood.

Possible Knowledge is also an invitation to think about how a focus on poesy can help us uncover the intertwined histories of politics, ethics, and philosophy. My argument aligns with claims that our discipline's "evidentiary standards" are premised not on what was "thought" but on what was "thinkable" at a historical moment.[65] I would further argue that modern scholarship's tendency to describe the literary as the domain of the "thinkable" is itself one of the many legacies of possible knowledge. We see the traces of this kind of thinking in emerging literary modes: counterfactual narratives, allohistories, and conjectural histories serve as formal vehicles to address questions of identity, creation, and being—of lives unled, of paths not taken, and even of idealistic modes of existence—in realist fiction.[66] And sometimes the modes of thinking that govern early modern poesy escape the bounds of fiction and impact actuality: imaginative works generate ideas that provoke scientific research, fabricated places engender actual travel, and in what is surely the awful side of possible knowledge, the created worlds of poesy offer models that incite the destructive practices of colonization. By recovering how literary writing was grappling with pressing philosophical problems—on kinds of existence, on limits of knowledge, and on the nature of truth—during a time of remarkable intellectual tumult, *Possible Knowledge* asks us to consider whether what we have termed a history of science might ultimately be a history of the imagination.

CHAPTER 1

Edmund Spenser's Speculative Method

> He was always to love the journey more than its end, the landscape more than the man, and reason more than life, and the tale less than its telling.
>
> —William Butler Yeats, *The Cutting of an Agate* (1912)

In his "Letter to Raleigh," the retrospective prospectus to the 1590 edition of *The Faerie Queene*, Edmund Spenser ruminates why "Xenophon [is] preferred before Plato, for that the one in the exquisite depth of his iudgment, formed a Commune welth such as it should be, but the other in the person of Cyrus and the Persians fashioned a gouernement such as might best be: So much more profitable and gratious is doctrine by ensample, then by rule."[1] Aligning his own poetic project with what "might best be" and the "ensample," rather than the "should be" and the "rule," Spenser embraces the maximal scope of thinking with possibility, that is, the "best" possible world the poet "might" produce without prescribing precepts. Through this distinction, the "Letter" seems to reinstate the gaps between the possible and the ideal, or the "may" and the "should," that Sidney elides in *The Defence of Poesy*. While the actual narrative strategies of *The Faerie Queene* repeatedly refute the neat order that is outlined in the "Letter," this document does seem to offer the best example of what *Spenser*'s defense of poesy—a serious articulation of his poetic theory and method—would look like.[2] (Sidney's *Arcadia* similarly contradicts the theories of poesy he elaborates in the *Defence*.) This chapter begins to delineate the contours of "possible knowledge" by taking seriously the claims about poetic ontology that Spenser outlines in the "Letter," examining how the desire to create worlds that "might best be" demands particular speculative methods—ways of thinking and knowing that perpetually drive the allegorical epic-romance to the brink of, but then withhold, the precept.

I use the term "speculative method" to describe such practices of knowledge production—the iterative processes of conjectural inquiry that spark myriad forms of imaginative projection in *The Faerie Queene*, where poetic experiments with multiple temporalities via engagements with problems of history, cosmology, and the occult give form to the potential states of being that constitute the fictive world. In the "Letter," Spenser himself invokes the term "method" when he discusses "the Methode of a Poet historical,"[3] and I propose that thinking about the poem's operations as a method can help us reconcile its ethical goal, or its "generall end" (that is, "to fashion a gentleman or noble person in vertuous and gentle discipline"),[4] with its narrative propensity toward excess and proliferation. William Butler Yeats famously observed that Spenser was a poet who liked "the journey more than its end . . . and the tale less than its telling."[5] But the expansive acts of "telling" are in constant struggle with the stated "end" of "vertuous and gentle discipline" in a poem whose inaugural quest is derailed by an entry into the "wandring wood" (1.1.13.6), and whose final completed book concludes as the blatant beast "broke his yron chaine, / And got into the world at liberty againe" (6.12.38.8–9). As this chapter will demonstrate, speculative method denotes the processes through which the poem harnesses contingent habits of thought and action in order to ensure that acts of "telling" can be directed to the desired "end." Such a speculative method thus aims to foster a kind of practical ethics, demanding the reader's active participation in holding at bay the existential stasis and narratological closure that would halt the "endlesse worke" (4.12.1.1) of romance worldbuilding.

Britomart's encounter with "the glassy globe that *Merlin* made" (3.2.21.1) offers a vivid enactment of Spenser's speculative method. Merlin constructs the mirror "By his deepe science, and hell-dreaded might," and Britomart, the titular knight of the Book of Chastity, sees her future husband, Artegall, in this "looking glasse" (3.2.18.7–8). But instead of showing "in perfect sight" (3.2.19.1) what she wishes to see, the speculum reveals only "the shadow of a warlike knight" (3.2.45.6). She sees "Nor man," "nor other liuing wight . . . But th'only shade and semblant of a knight" (3.2.38.1–3). This imprecise "shadow" demands interpretation, prompting her to seek out Merlin and thereby activating her quest. The narrator intimates that the mirror can provide an accurate representation of actuality: "It vertue had, to shew in *perfect sight*, / What euer thing was in the world contaynd" (3.2.19.1–2, emphasis mine). But this "wonderous worke" (3.2.20.1) is related to the phenomenal realm by similitude rather than by identity because it is "Like to the world" and "seemd a world of glas" (3.2.19.9). Its obscured visions do not bridge the chasm between different modes of existence, of "world" and "glas." Indeed, the gap between artifact and actuality is constitutive to the mirror's function, which is to provoke viewers to navigate various temporalities and ontological domains: providing virtual projections of an uncertain future, it intimates to its audience what actuality *could* look like, its cryptic images animating spectators to close the "interpretive gap" to which it exposes them.[6] Britomart's future, the poem suggests, depends on her willingness to actively shape it based on the "shade and semblant" she views. As the mirror interacts with the desires of its viewers—tailoring images to their particular aspirations—its predictive worldmaking becomes the engine of plot-making.[7]

By directing its audiences toward a potential future whose actualization depends on their acts of interpretation and participation, the mirror serves as a corollary to Spenser's own poetic project, where one learns by stepping into the wandering narrative-scapes of romance. But this encounter between speculum and spectator—where Britomart's vision activates the subsequent envisioning of her possible future—also offers us a striking example of how Spenser's speculative method works: by translating problems of imperfect perception, and the attendant gaps between seeming and being, into iterative modes of knowing. It is perhaps unsurprising that a "glassy globe" is the most potent symbol of Spenser's poetic method, given that the term "speculative" is etymologically linked with *speculum* as well as—through the Latin *spectāre* (to look)—with notions of "spectacle" and "speculate."[8] Following these connections, I read speculative method as comprising the processes involved in the work of "speculation," a word that Spenser's contemporaries would have associated both with the senses—"The faculty or power of seeing; sight, vision, *esp.* intelligent or comprehending vision"—and with suppositions such as "a conclusion, opinion, view, or series of these, reached by abstract or hypothetical reasoning."[9] Capturing the complex associations of sensory perception and imaginative projection that drive Spenser's "endlesse worke," speculative method thus names the ways in which incomplete knowledge instigates imaginative projections that determine the form and scope of the ensuing poetic world.

Spenser scholarship has repeatedly marked the poem's opposing pulls toward expansion and inconclusion on the one hand and containment and closure on the other.[10] I align my study of speculative method with research that excavates *The Faerie Queene*'s pluralistic and open-ended formal apparatus—work that ranges from Patricia Parker's examination of the poem's inventive practices under the rubric of romance, "a form which simultaneously quests for and postpones a particular end," to Jonathan Goldberg's study of how the "very endless quality [in Spenserian narration] denies hermeneutic closure."[11] I expand on these insights to link questions of literary form to problems of epistemology and ontology. At the same time, my approach differs from scholarly treatments that, in order to explicate the philosophical disposition of *The Faerie Queene*, document its indebtedness to classical schools of thought—such as Platonic forms, Lucretian materialism, Aristotelian intuition, or Galenic medicine.[12] Rather than asking what sources motivate Spenser, or turning to moments where the poem obviously engages with natural history and philosophy (such as *The Mutabilitie Cantos* or the Garden of Adonis),[13] I attend to the techniques of profusion that enact *The Faerie Queene*'s ways of becoming.[14] The premise of this chapter is simple: to recover the intellectual processes that drive the work of a poet "whose main concern," according to Gordon Teskey, "was to think," we must attend to the cognitive and formal techniques through which the poem itself comes into being.[15]

I propose, then, that Spenserian poetics is a theory of knowledge production in action. This chapter attends to how the poem develops its method of speculation, what kinds of knowledge ensue from this method, and how these forms of knowing shape the possible worlds of the fiction. I focus on three types of actors: the narrator, the allegorical figure or character, and the imagined ideal reader. I first examine how the narrator models appropriate forms of speculation. Appealing to potential existences in both terrestrial and cosmological realms, he enacts the iterative procedures of imagining that could lead to the revelation of a best possible world. Then, I explore how figures in the allegorical landscape embody both the scope and the limits of a speculative method, raising the prospect of their disappearance from fictional space-time if they fail to inhabit the poem's ethos of psychosomatic mobility. Finally, I study how the poem coopts readers into its own framework of polychronic existence. It constructs an ideal reader, whose encounters with the past (in history) and the future (in prophecy) can push the ethical work of fashioning from the sphere of fiction to actuality.

"More inquyre": Narrating a Speculative Method

To examine how the created world's ontological instability animates the narrative's methods of iterative inquiry, I focus on the Proem to Book 2, where the narrator mobilizes notions of polychronic time to lay bare processes that indicate to readers that potential realms exist beyond the visible and the tangible. Experiments with time permeate *The Faerie Queene*: narration buckles under the stress of numerous chronologies; time stops; past, present, and future collapse into an instant; and events perpetually strive toward unrealizable ends. But the narrator's words in the Proem illustrate why this attention to time is so pervasive: worldmaking requires traversing multiple temporalities. Spenser makes evident in the "Letter" that his poetic "Methode" is predicated on manipulations of time: "the Methode of a Poet historical is not such, as of an Historiographer. For an Historiographer discourseth of affayres orderly as they were donne, accounting as well the times as the actions, but a Poet thrusteth into the middest, euen where it most concerneth him, and there recoursing to the thinges forepaste, and diuining of thinges to come, maketh a pleasing Analysis of all."[16] Unlike the "Historiographer," who deals with past events "orderly as they were donne," the poet is simultaneously oriented toward past and future occurrences. This relation to time enables the poet to perform a "pleasing Analysis," a phrase that underscores the affective dimensions of poetic labor. To grasp this emphasis on polytemporality, it is useful to compare the "Letter" with Sidney's *Defence*, which also contrasts the poet and the historian. As we saw in the Introduction, Sidney locates the distinction in the truth-value of their enterprises—the poet's freedom lies in their ability to escape the "bare *Was*" that traps the historian. Spenser, however, argues that the poet's "Methode" is formally different because of their relation to time.[17] The "Poet historical" subverts the chronological order of historical narratives. Entering "into the middest, euen where it most concerneth him," the poet emerges as a vatic figure who deals with "thinges forepaste" and "diuin[es]" the future.[18]

Rather than trying to match this theory of poetic production with the content of the poem, I suggest we read the Proem to Book 2 as an attempt to narrativize the worldmaking impulse Spenser lays out in the "Letter." The narrator models the act of imagining into existence a realm that "might best be," named Faerie Land. This "world" is not an inert object but mediates between past and future, offering an alternative to the absolute distinctions between fictional and historical states: a condition of potential existence. Although Samuel Taylor Coleridge characterized Faerie Land as a "mental space," "in the domains neither of history or geography,"[19] the Proem reveals how the existence of this space is tied to several topical discourses on being and knowing—historical events, natural philosophy, cartography, and cosmology—as it presents Faerie Land as one of many "possible, nonactualized worlds" that are not yet but might be known.[20]

Framed between references to the "most mighty Soueraine" (2 Proem, 1.1) as though she anchors "thine owne realmes" and the "lond of Faery" (2 Proem, 4.8), the narrator's words privilege possible existence over phenomenal experience. He worries

> That all this famous antique history,
> Of some th'aboundance of an ydle braine
> Will iudged be, and painted forgery,
> Rather then matter of iust memory,
> Sith none, that breatheth liuing aire, does know,
> Where is that happy land of Faery,
> Which I so much doe vaunt, yet no where show,
> But vouch antiquities, which no body can know.
>
> But let that man with better sence aduize,
> That of the world least part to vs is red.
> (2 Proem, 1.2–2.2)

The narrator uses the trope of travel to demarcate Faerie Land as an entity that remains unmapped but that might be found if readers' lack of certain knowledge motivates them to search for it—that is, readers must resist accepting absence of evidence as an absolute sign of nonexistence. The Proem thus models a simultaneous capacity to be and not be. Arguing that its potentiality should activate rather than foreclose desire to locate it, the narrator underscores that epistemology is inseparable from ontology: the status of Faerie Land depends on readers' ability to know it.

The Proem's arguments about the plausible existence but as-yet-unknowable location of Faerie Land echo contemporary travel narratives that combined factual and fabricated accounts of realms newly "discouered" (2 Proem, 2.4) by Europeans. Travel writing often relied on description and narrative modes that privileged information, cataloging, and experience.[21] Analyzing works such as Humphrey Gilbert's *A Discovrse of a Discouerie for a New Passage to Cataia*, Thomas Harriot's *A Briefe and True Report of the New Found Land of Virginia*, Richard Hakluyt's *Divers Voyages Touching the Discoverie of America*, and Walter Ralegh's *The Discovery of Guiana*, Mary C. Fuller argues that these works aimed "to establish certain realities—the possibility of discovery, the lands discovered, the experiences or intentions of oneself or fellow travelers."[22] These authors intimated "that the trajectory of the intended action would arrive at a place that existed in fact," despite the lack of assurance that their efforts would be successfully rewarded.[23]

Travel writers, often reliant on the conventions of fiction—for instance, Harriot's work "inserts hypothetical goods and actions into its catalogues of American abundance"[24]—were aware that their narratives, too, might seem fabricated. To suppress perceived alliances with fiction, they actively distanced their endeavors from imaginative writing. The titular island of Thomas More's *Utopia* became the signal example of a realm whose hypothetical state of being was contrary to the actuality of explored or navigable lands. Mary Baine Campbell argues that although *Utopia* was "in a sense the first piece of travel literature produced in England's Age of Discovery" and "appears to have worked some changes . . . in the rhetorical situation of the travel writer," by the mid-sixteenth century "the *fictionality* of Utopia (Noplace) had been established as an opposite pole to the geographical reality of the lands they hoped to discover and exploit."[25] In *A Discovrse of a Discouerie*, Humphrey Gilbert foregrounds this distinction: "*SIR, YOV* might iustly haue charged mee with an vnsetled head if I had at any time taken in hand, to discouer *Vtopia*, or any countrey fained by imagination: But *Cataia* is none such, it is a countrey, well knowen to be described and set foorth by all moderne *Geographers*, whose authoritie in this art (contrarie to all other) beareth most credit."[26] While Utopia is "fained by imagination," Cataia is "well knowen"; the "authoritie" of "moderne *Geographers*" is supposed to serve as guarantee of its actuality. Nominally nonexistent, "*Vtopia*" is also in ontological contrast to the lands described by travel writers. This distinction will sharpen in the next century: while in 1576, Gilbert's "or" distinguishes the specific fabricated entity that is Utopia from any other "country fained by imagination," by the time Edward Phillips's *The New World of English Words* is published in 1658, "utopia" is more firmly "taken by Metaphor for any imaginary, or feigned place."[27] These accounts, in different ways, emphasize the fictive island's distance from the physical world. Yet, Campbell demonstrates the unsustainable nature of the absolute oppositions that were proposed by early modern writers: *Utopia*'s "pedagogical moralism" prevents it from becoming a version of an "'other world,'" since the "normative examples of utopia exert a pressure on readers to alter their own worlds in its direction."[28] *Utopia* is ultimately a hypothesis that, as I have argued elsewhere, by "claim[ing] a provisional obligation to the actual" sparks "desires to alter reality."[29]

Although he is not writing utopian fiction, Spenser pushes conflations of imaginary and geographical existence to their logical limit, as he makes the case that the narration of possible worlds—of worlds that "might" be—can activate processes that would lead to locating them. Fictions, in other words, can direct readers toward potentially mappable lands. The trope of discovery, of course, as much masks the purely fabricated status of Spenser's project as it occludes the horrors of colonizing inhabited places. The narrator sidesteps these issues by presenting ontological absence as epistemological lack, a strategy that allows him to focus on the contrast between the authority of narrative and readers' misplaced trust in their senses. To contest claims about Faerie Land's nonexistence—and related claims that it cannot be known—one need not seek evidence of geographical specificity. The revelation of so-called new worlds depends as much on readers who are willing to expand the limits of their imagination as on cartographic mapping. As this narration reorients approaches to knowledge production from acts of seeing to conjecture, from spectating to speculation, it argues that Faerie Land's existence cannot be apprehended through empiricist epistemologies. Although the complete knowledge of the "happy land of Faery" can be obtained by learning "Where" it is, this certitude is secondary to establishing the possibility that it might exist. Rejecting the need to prove Faerie Land's exact location—a key element of humor in texts such as *Utopia*—the narrator hints that potentiality is the ontological condition of fictional existence.[30] This stance dissociates the capacity to "know" Faerie Land fully from establishing its status as a world. By modeling how one evolves from what one has witnessed to imagining things that "might" exist, the Proem's narrator enacts in miniature the conjectural leaps that structure Spenser's poem. The condensed modal claim, what "might best be," then, denotes a limit case. While a process ending in a precept would result in narrative stasis, the ongoing attempt to construct what "might best be" keeps propelling the narrative, directing one to the edge of a perfected existence that remains just beyond reach.

To facilitate such ongoing processes in the absence of proof, the potentiality of Faerie Land must be established by analogical relations between what is known at present and what might be known in the future. After admitting that only the "least part [of the world] to vs is red," the narrator demonstrates how evidentiary lack can become the basis of learning:

> daily how through hardy enterprize,
> Many great Regions are discouered,
> Which to late age were neuer mentioned.
> Who euer heard of th'Indian *Peru*?
> Or who in venturous vessell measured
> The *Amazons* huge riuer now found trew?
> Or fruitfullest *Virginia* who did euer vew?

> Yet all these were when no man did them know,
> Yet haue from wisest ages hidden beene
> And later times thinges more vnknowne shall show.
> Why then should witlesse man so much misweene
> That nothing is but that which he hath seene?
> What if within the Moones fayre shining spheare,
> What if in euery other starre vnseene
> Of other worldes he happily should heare?
> He wonder would much more, yet such to some appeare.
> (2 Proem, 2.3–3.9)

Citing examples of places (from "Indian *Peru*" to "fruitfullest *Virginia*") recently "discouered" by European travelers, the narrator imagines a constantly expandable world in which present absence does not symbolize nonexistence. Recent knowledge of "all these" lands that "haue from wisest ages hidden beene" indicates that one can know beyond what one sees or accesses. Although empirical verifiability is sufficient to prove existence, it is not necessary. The belief that these lands did not exist, or that Faerie Land does not, arises from a limited cognitive capacity that trusts what is "seene." This moment is symptomatic of the poem's distrust of visual epistemologies—Artegall criticizes the Giant for misinterpreting the cause of "things vnseene" as he "misdeem'st so much of things in sight" (5.2.39.1, 3), for instance, and Red Crosse is deceived when Archimago shows him a false sight of Una (1.2.5). To counter the aura of legitimacy that not only the "vew[ed]" but also the "heard" and the "measured" exude, the Proem shows how to move away from what is known to the imagination of what is possible. In doing so, the Proem posits that poetry operates on a different evidentiary standard that can enable us to grasp what cannot be comprehended by other means. Establishing Faerie Land's potentiality is predicated on contesting the sensory, primarily optical epistemologies—from Aristotelian ideas of intuitive experience to the controlled experiments of naturalists[31]—that limit one to verifiable facts. This methodology also dismantles the binaries set up by contemporary travel writers. While Humphrey Gilbert claims that his quest will end when a place called Cataia is located, Spenser suggests that Faerie Land's potentiality facilitates the perpetual possibility of being "discouered."[32] Gilbert opposes travel literature to fiction by focusing on "worlds" as objects. In contrast, Spenser exploits their similarities by turning worlds into instruments for generating knowledge.

Spenser's embrace of potentiality as a catalyst for learning crystallizes into an iterative method when the Proem's speculative process shifts scales, culminating in an analogy between terrestrial and cosmological realms. The imaginative projections about (potentially accessible and, in theory, empirically verifiable) terrestrial entities lead the narrator to inquire about something that must have seemed unprovable to sixteenth-century travelers: the habitability of realms beyond earth. Instead of stopping after establishing the potential existence of Faerie Land by comparing it to recently "discouered" places, he invites readers to continue speculating by scaling up from cartography to cosmology. To extend his analogical argument, the narrator echoes debates in astronomy, making readers test the limits of their imaginative faculties: "What if within the Moones fayre shining spheare / What if in euery other starre vnseene / Of other worldes he happily should heare?" The repeated "What if" is a methodological tool, inviting readers to make new demands of their cognitive faculties: they must make the imaginative leap from historical memory to contingent futurity. Readers can replicate the query with other examples to test the maximal scope of their projective desires. The interrogatory conjecture thus facilitates speculations that propagate, rather than end, inquiry. Moving from the assertion that "later times" will be revelatory, the narrator supplements actual or possible resolution with speculations about an ever-expanding cosmos. He shifts to questions of habitability on the moon to "other worldes" in "euery other starre," that is, to what cannot be visually captured. The "What if" formalizes as speculative methodology the desire and impetus for continual questioning—the idea that we "*may* it fynd" (emphasis mine) if we "more inquyre" (2 Proem, 4.3, 1).

The narrator, ultimately, is not invested in proving (non)existence. Instead, the Proem's imaginative projections transfer focus from objects (locatable lands) to process (speculation as a mode of coming to learn about the unknown), enacting a method that requires shifting scales of inquiry. As the Proem's analogous worldbuilding links the recovery of forgotten history to the "disco[uery]" of the unknown, it also promises to make potentially present what was never historically in existence. In this way, the possibility of a land being "discouered" becomes a rallying cry for imagined projections: worldmaking is replicable as readerly remaking because of one's ability to envision the invisible and the intangible. This is not to suggest that such imaginative projections are isolated intellectual exercises that are ideologically neutral and devoid of political concerns. As David Read has demonstrated, "Spenser's continuing association with individuals involved in English colonial projects in the New World" enabled him to treat the project of colonization itself as a projection: "Colonization in this account becomes an aesthetic endeavor in which artists naturally perform better than laymen."[33] J. K. Barret has complicated the charge of such readings that see in the poem "a glimpse of an impending British imperial future" by arguing that Spenser "writes a history of the future, but rather than defining that future in terms of a projected imperial interest, the poem imagines itself as reinventing the history it offers to the reader."[34] I would argue that these aspects are inseparable: the imaginative projections the narrator performs serve as a spark to colonizing projects that embrace the invitation to transform imagined possibilities into reality as a vital element of their violent will to power.

Reveling in the iterative principles of the "What if" and the associated capacity to "more inquyre," the narrator also traverses ground that we now consider cross-disciplinary. His argument for an infinite cosmos represents tropes common in imaginative fiction, such as the trip to the moon in Ludovico Ariosto's *Orlando Furioso*, and it engages with contemporary philosophical debates on cosmology and cosmogony. In projecting this cosmos, the poem draws not only on theories of the plurality of worlds, as recently represented in the "hyperbolic possibility" of Giordano Bruno's *De l'infinito*, but it also draws from a long philosophical tradition (including Lucretius's *De rerum natura* and Plato's *Timaeus*) that contemplates the nature of the universe.[35] At a time when philosophy, cartography, and theology all claim to tackle questions of physical and metaphysical realms, the Proem dramatizes how fictional worldmaking, too, participates in, indeed thrives on, the production of knowledge about different scales of being.

At stake in the Proem is not only the ability to imagine multiple worlds—a radical act in itself—but also an argument for the critical role of *poiesis* in reshaping one's understanding of reality. The poetic imagination can generate methods to explicate the unstable status of entities that cannot be represented by the conventional calibrations of space and time. This moment also exemplifies why Spenser's poem is a key contribution to his era's worldmaking project—he is one of the many thinkers who reflected, to borrow Ayesha Ramachandran's words, "a new recognition of [their] existence in a radically uncertain world where [they] must create [their] own order": in other words, an inventor.[36] Spenser expands the typical counterfactual stance, which asks what could have been, and explores what conjecture about things that do not (and may never) exist can trigger in the reader's mind. Faerie Land's absence facilitates the Proem's formal proliferations, asking readers to adopt and adapt the narrator's mode of inquiry. Proving actual existence remains subsidiary to this performance of continual imagining. Faerie Land's potentiality thus represents a state of suspension between fabrication and history, and this state of liminal existence is a vital resource, rather than a liability, for further inquiry. The calls to replicate the act of speculating highlight that in an ever-expanding world that is never fully knowable, fictional worlds asymptotically approach, but do not converge with, historical facts, yet their epistemic status may be as significant as the real worlds that lie over the horizon. At their most expansive, then, poetic worlds are projections that provoke one to "more inquyre."

Embodying Speculation: Allegorical Figures and the Mobility of Narrative

In the previous section, I examined how the narrator models the poem's speculative method through a form of worldmaking where projections from mapped spaces to unmapped landscapes to an unmappable cosmos expand modalities of existence. In this section, I turn from the narrator to the allegorical figures who embody the scope, and test the limits, of this method within the formal apparatus of the narrative. Spenser's fiction is populated with unperfected allegorical figures who undergo transformations as they pursue their romance quests. Although Red Crosse, Guyon, or Britomart might represent abstract notions (Holiness, Temperance, and Chastity, respectively), readers come to understand these concepts through events unfolding in narrative time. And the events that constitute their acts of fashioning—the instances of wandering, learning, and alteration that occur within the temporality of the narrative—draw attention to the mechanics of poetic making. By examining how the unpredictable journeys of these virtues *in potentia* toward their purported ends open multiple routes that the poem may or may not complete, and that may even halt its forward movement, we can see speculative method unfolding in narrative time via a particular poetic technique: allegory.

Allegory in *The Faerie Queene* facilitates the conditions of constant mutability demanded by an enterprise that thrives on speculation. In the *Defence*, Sidney argues that "the poets' persons and doings are but pictures what should be, and not stories what have been, they will never give the lie to things not affirmatively but allegorically and figuratively written" (53).[37] Allegory, as the "mode of the unaffirmed,"[38] operates in the fissures that separate historical narratives ("stories what have been") from poetic inventions ("pictures what should be"). In this section, I trace how Spenser's narrative dramatizes the "temporal passage" of this gap through figures whose actions put pressure on the relations between narrative and symbol, as well as between history and fictionality.[39] My attention to Spenser's work thus foregrounds how the "darke conceit"[40] is a technique of formal mobility. As such, I excavate the epistemological potential of what Angus Fletcher identifies as allegory's "incompletable" nature. Attending to the "arbitrary nature of allegorical action," Fletcher marks its "tendency toward infinite extension" and claims that "all analogies are incomplete, and incompletable, and allegory simply records this analogical relation in a dramatic or narrative form."[41] This "incompletable" strain is at the heart of Spenser's worldmaking. As soon as a reader—or a surrogate reader in the fiction—arrives at an idea, the text asks them to move on to a new formulation of this idea. Allegory's "incompletable" nature—inseparable from and at odds with its desire for completion—instigates and perpetuates movements in narrative.

Spenser's "*continued* Allegory"[42] taps into the intellectual possibilities inherent in an "incompletable" mode, which, as early modern writers indicate, already contains within its definition notions of movement, protraction, and deferral. Like Puttenham, who defines allegory as "a long and perpetual *metaphor*,"[43] Spenser draws on Quintilian's definition of allegory as a "continued metaphor."[44] Puttenham's argument that allegory distances thought from speech, or cognition from articulation, also hints at the figure's temporal reach. It is a figure of dissembling, "when we speak one thing and think another, and that our words and our meanings meet not."[45] Such dissemblance occurs through the perpetual displacement of thought in expression: "we do speak in sense translative and wrested from the own signification, nevertheless applied to another not altogether contrary, but having much conveniency with it, as before we said of the *metaphor*."[46] Puttenham had defined the metaphor's "translative" capacity as the "wresting of a single word from his own right signification to another not so natural, but yet of some affinity or conveniency with it."[47] The "sense translative" is produced through signifiers "not altogether contrary," through those that transport signification to words of "some affinity or conveniency." This "conveniency" suggests that allegory's point of stability is perennially shifting: chains of "affinity" carry over a concept from signifier to signifier. By linking what Puttenham terms the "sensable" figure—which "alter[s] and affect[s] the mind by alteration of sense" and "abuses, or rather trespasses, in speech"[48]—to the dynamism of narrative, Spenser's poem uses multiple degrees of transference to dramatize how speculative thinking unfolds in the process of poetic making.

To examine how the poem's formal techniques serve as instruments of interpretation and knowledge production, I explore in what follows how Spenser's poetic project grapples with the struggle between figuration and narrative expansion. Two figures capture with particular clarity how the allegorical narrative wrestles with the possibility of ending: Malbecco vanishes, while Arthur's disappearance from the poem is foreclosed even before he appears in the narrative space-time. As Malbecco's material de-formation leads to narrative stasis, it also shuts down the interpretive processes proposed in the Proem to Book 2. By contrast, Arthur's polychronic presence collapses the boundaries between fact and fiction to propel such interpretative practices toward active forms of readerly participation. In opposite ways, the two figures expose the intimacy of narrative protraction and existence: only by operating in a paradigm that postpones abstraction can one create patterns for readerly *praxis*.

Malbecco's transformation into the abstract concept "Gelosy" dramatizes the progression from multiple significations toward instrumental meaning. The disintegration of the mind and body of this figure, who has been read as both paradigmatic instance of "a larger Ovidian nexus in *The Faerie Queene*"[49] and a limit case of the "larger concern with paralysis" haunting the epic-romance,[50] begins when he loses his wife Hellenore, his wealth, and his property. As he attempts to escape from things ("ran away, ran with him selfe away" [3.10.54.6]), the narrative "transports" him across phenomenal states:

> His nimble feet, as treading still on thorne:
> Griefe, and despight, and gealousy, and scorne
> Did all the way him follow hard behynd,
> And he himselfe himselfe loath'd so forlorne.
>
> (3.10.55.4–7)

His metamorphosis commingles active and passive forces and reveals the paradox of his transformation—he progresses toward abstractions that chase him even while he is attempting to escape them.

Malbecco's self-initiated metamorphosis leads to his increasing self-alienation:

> Still fled he forward, looking backward still,
> Ne stayd his flight, nor fearefull agony,
> Till that he came vnto a rocky hill,
> Ouer the sea, suspended dreadfully,
> That liuing creature it would terrify,
> To looke adowne, or vpward to the hight:
> From thence he threw him selfe dispiteously.
>
> (3.10.56.1–7)

The hill, "suspended dreadfully," corresponds to his current state of mental indecisiveness and foreshadows his upcoming somatic liminality. The series of oppositions ("forward"/"backward," "adowne"/"vpward") promises further movements, but all sense of suspension is destroyed when he physically disappears:

> through long anguish, and selfe-murdring thought
> He was so wasted and forpined quight,
> That all his substance was consum'd to nought,
> And nothing left, but like an aery Spright.
>
> (3.10.57.1–4)

Malbecco activates his own destruction ("he threw him selfe"), ensuring that his "selfe-murdring thought" becomes a reality. As his "substance was consum'd to nought," he physically disappears. To echo Puttenham, Malbecco is "wrested from [his] own signification." Unlike critics who argue that Malbecco becomes "mere elemental matter,"[51] or who locate in his "final state an uncanny corporeality,"[52] I read his reduction to "nothing" as the erasure of being; the poem registers disintegration not into brute matter but into "nought."

The "aery Spright" ceases to exist as Malbecco crosses over from a liminal state to one representable by a permanent abstraction:

> Yet can he neuer dye, but dying liues,
> And doth himselfe with sorrow new sustaine,
> That death and life attonce vnto him giues.
> And painefull pleasure turnes to pleasing paine.
> There dwels he euer, miserable swaine,
> Hatefull both to him selfe, and euery wight;
> Where he through priuy griefe, and horrour vaine,
> Is woxen so deform'd that he has quight
> Forgot he was a man, and *Gelosy* is hight.
>
> (3.10.60)

Malbecco's loss of humanity culminates in his metamorphosis into an abstraction. "*Gelosy*" is neither an emotion nor a figure chasing him: he *is* jealousy now.[53] Through phrases like "Yet can he neuer dye, but dying liues" and "dwels he euer," the passage underscores the constantly renewing contradictions of suspended animation; chiastic inversions like "painefull pleasure turnes to pleasing paine" also contribute to maintaining this sense of liminality. But the stanza finally reveals that one cannot return from a state of abstraction without a memory of what one was before. The language shifts from narrating the past that documents his activities—Malbecco's mobility is captured by verbs such as "ran," "fled," "threw"—to the eternal present where he "liues" and "dwels." Physical stasis mirrors the linguistic shifts: the figure's permanent confinement in a "caue" (3.10.57.9) replaces the multiplicity of circulating signification. This assault on form results in a manifestation of Teskey's "figure of personification," which "directs our attention beyond it to allegory's positive other in absolute meaning, on which a cosmos depends."[54] The term "deform'd" (signaling physical formlessness and, as disability studies scholars have shown, deviation from what was considered an ideal form)[55] unveils a new creature. The last line intimately couples knowing and being: Malbecco "Forgot he was a man, and *Gelosy* is hight." While the text does not clarify if Malbecco is still a man, the correlation of forgotten humanity and emergent abstraction emphasizes an absolute dissociation: carrying no cognitive traces of a human past, Gelosy forecloses further metamorphosis or reconstruction. This moment is thus a peculiar instantiation of the analogy that the Proem outlines as the means to continual knowing. While Faerie Land's potential status is maintained by our ability to keep inquiring about it, Malbecco's human state disintegrates, and he has no possibility of recovering it, when he loses knowledge of his past.

This scene of psychosomatic loss enacts what we might call a physics of forgetting.[56] Malbecco's cognitive erasure is not only a product of Ovidian interplays of matter and form but also a dramatization of matter in motion, a key element of Aristotelian physics. Malbecco's reduction to "nothing" invokes discussions of nothingness and void in Aristotle (who firmly rejects the void) and in Lucretius (who embraces it).[57] But perhaps the clearest reason to term Malbecco's forgotten humanness a physics lies in the usage of the word "substance": it is a descriptor of the stuff of which things are made—Spenser, for instance, uses it to describe raw materials of composition in the creation of the False Florimell (see 3.8.6.1)—and also refers to *ousia* or being, which represents entities as varied as Platonic Forms and Lucretian atoms.[58] To demonstrate how matter's changing forms can push narrative toward an irreversible ending, Spenser adopts the classical term for foundational units of the physical world, reconfiguring concepts from multiple philosophical schools to highlight the limits of narrative dilation.[59]

Malbecco's disappearance is as close to an ending—of personal history, of humanity, of acts of telling—as we get in *The Faerie Queene*. Here, readers encounter the limit case of "incompletable" allegory. This is also the end of the speculative process, since in rendering Malbecco fully fixed, this moment halts further inquiries about the figure. Susanne Wofford argues that the "instant of forgetting marks the moment in which the figure takes control; the character's submission to it is now glossed as having a story behind it," and the "stanza intimates that behind the allegorical text lies the possibility of an infinite expansion in which every personification would receive its due prior narrative."[60] While the notion of "infinite expansion" takes us to the domain of a figure's past history, I am more interested in how the stoppage engendered by Malbecco's forgetting shuts down the future, aborting the *continued* Spenserian allegory. No more a figure of "transport" across space and time, his transformation halts further signification. It is left to the reader to recollect this past and to analyze whether Malbecco's previous history as a physically decrepit and jealous man prefigured his fate.[61] Here, Spenser's "continued Allegory" encounters an intractable problem in the form of an abstract figuration. What happens, then, to the poetic desire to provoke readers to "more inquyre" when narrative expansion halts?

I propose we read the conclusion of Malbecco's transformation as the limit case of the method of further inquiry that the narrator advocates in Book 2's Proem. In contrast to a figure like Duessa, who outstays her explicit purpose in Book 1 and lingers in the poetic landscape, Malbecco vanishes. His forgetting of his own past shuts him out of the future. The resulting state of permanence also severs his connection to readers, who had been pursuing his story so far. His disappearance offers a glimpse of the shutdown of phenomenal multiplicity that drives the poem. Following Colin Burrow's argument that "metamorphosis is often associated with failure," or that it "frequently signifies a reluctance to participate in the processes of striving and moving and living and breeding which represent the central principles of Spenser's poem,"[62] we can read Malbecco's end as the logical (and existential) extreme of this poetic endeavor. But *The Faerie Queene* does offer a prominent alternate model of how fictional existence can continue despite a halted life. We may contrast Malbecco's arrested being to Arthur's polychronic existence, which emphasizes the narrative strategies that actively admit the reader into the fictional landscape.

If Malbecco resides at the margins of the text, Arthur is at its center. If Malbecco's transformation halts further interpretation, Arthur's entry forcefully marks how readerly knowledge is integral to narrative dilation. The poem defers Arthur's perfection—both his abstract figuration and his quest—to dramatize moments that "by resisting absorption into a homogenous present, . . . brin[g] with [them] the difference that produces the possibility of a new future even as

[they] evok[e] the past."[63] Traversing past, present, and future, the figure of Arthur invites readers to experience his polychronous state as inextricable from theirs and to locate themselves in relation to the poem's plural times. It is through this figure—who might show up at any moment in the narrative—that the poem continually registers skepticism about the desirability of perfectly constructed figures who embody abstract meanings. Spenser does claim in the "Letter" to deliver in Arthur a perfected assemblage[64]: "I labour to pourtraict in Arthure, before he was king, the image of a braue knight, perfected in the twelue priuate morall vertues, as Aristotle hath deuised, the which is the purpose of these first twelue bookes."[65] In the narrative, however, Arthur does not represent abstract precepts and is at best an incompletable amalgamation of multiple virtues in the "image of a braue knight." As an "ensample" that is "clowdily enwrapped in Allegoricall deuises," he embodies the processes of perfection coming into being.[66]

Arthur's abstraction depends on the poem's ability to circumvent the advancing time of narrative, and the tussle between process and abstraction that this figure embodies is evident in the "Letter." Arthur's status as a figure of perfection is complicated, first, by the ways in which Spenser relates "magnificence" to other virtues: "in the person of Prince Arthure I sette forth magnificence in particular, which vertue for that (according to Aristotle and the rest) it is the perfection of all the rest, and conteineth in it them all, therefore in the whole course I mention the deedes of Arthure applyable to that vertue, which I write of in that booke."[67] This statement posits that Arthur should "conteineth" within him all virtues, with each book revealing a particular facet. At the practical level, such "perfection" is deferred till Spenser finishes his project of "twelue bookes." But even in the composed books, the complete picture of magnificence remains unattainable. Arthur gestures to his own transitory state when he tells Una that his identity and lineage are "hidden yitt" from him (1.9.3.4). At a more fundamental level, the impossibility of perfection undergirds the concept of "magnificence" itself: early moderns understood the term not only as "the perfection of all the rest," as Spenser declares, but also as an individual cardinal virtue.[68] It is attractive for an "endlesse worke" primarily because it is oriented to activities that could lead to a virtuous product. Arthur is in a state of suspended performativity, "being brought into action" as the poem unfolds,[69] serving as a perpetual reminder that his transformation into precept would halt the mechanisms of narrative.

Anticipating (or recalling) the Proem to Book 2's speculative disposition, the "Letter" refuses to stop on arriving at a supposed conclusion. Instead, Spenser projects a future work: "[these first twelue bookes] if I finde to be well accepted,

I may be perhaps encoraged, to frame the other part of pollliticke vertues in his person, after that hee came to be king."[70] Even if Arthur comes to represent a fully realized constellation of "priuate morall vertues," his manifestation as an abstract ideal is marked as an impossibility. The "Letter" associates the Arthur of the current text with possible projections of his "polliticke vertues," the name functioning as a "rigid designator" that connects the figure's existence across worlds and works.[71] Intimating that this Arthur is inseparable from the figure's overdetermined intertextuality—a figure of history, myth, and legend, a "priuate" as well as a "polliticke" figure—Spenser suggests that only the exposition of "polliticke vertues" can provide any semblance of closure. But such resolution, too, seems impossible, as Spenser declares his task to be a conditional enterprise. His projected work "may be perhaps" produced "if" the first twelve books find satisfactory reception. The poet's act of writing is contingent on the collective reception of readers. He declares his dependency on a reader similar to the one in Book 2's Proem, who "may it fynd" desirable to venture beyond what is immediately accessible. As Spenser theorizes readerly desire as an essential component of poetic production, the specter of Arthur as the exemplar of private-public perfection recedes from view.

The "Letter" thus primes us to expect an end different from Malbecco's: Arthur's story is ongoing, and the reader's role is vital to his continued presence. Indeed, Arthur serves as a locus of readerly interpretation and action even before he appears in the poem. The narrator's first mention of the "young Prince" creates a temporal disjunction in which the future arrives before the past to disrupt the narrative present:

> It *Merlin* was, which whylome did excell
> All liuing wightes in might of magicke spell:
> Both shield, and sword, and armour all he wrought
> For this young Prince, when first to armes he fell,
> But when he dyde, the Faery Queene it brought
> To Faerie lond, where yet it may be seene, if sought.
> (1.7.36.4–9)

In this zone of simultaneous anticipation and retrospection, the text prefaces Arthur's exploits in the plots of *The Faerie Queene* by inserting the ending of his individual story. The future, in which Arthur dies, is not yet but also already is, and any upcoming plot carries traces of this proleptic past within it. By introducing Arthur through his impending disappearance in this scene, the poem also presents him as the crux of predictive knowledge, forcing readers—who are aware of his presence prior to his entry in the poetic landscape and cognizant of his death before it occurs—to ask: What might one learn from the unchronological descriptions of fictive lives? Or, to echo the Proem, how might they "more inquyre" about such forms of being? This mere acknowledgment of Arthur's existence renders the narrative polychronous, creating an epistemic schism between the world imagined by the poet and the one occupied by the reader, wrenching the "young Prince" out of the former and displacing him across various narrative and historical times.[72]

This moment does not end the narrative even as it points to Arthur's death. The "But" creates a linguistic opposition between life and death, interrupting linear progression by projecting the reader into a future beyond death. While the phrase "he dyde" denotes the stoppage of life, the "But" ensures plots continue, demanding a "when" that propels readers beyond the actual event of Arthur's demise to future endeavors that, oddly, have already happened. In other words, the "But" offers a textual counterpoint to the ontological termination indicated by "he dyde." The Faerie Queene's role in bringing Arthur's arms back to Faerie Land might seem like an acknowledgment of the completion of his quest of finding Gloriana, but this polychronous instant of his introduction actually transfers readers beyond it. The future arrives ahead of the historical past and the fictional present, not to mark Arthur's completed figuration but to provide an entry point for the reader.

As Maureen Quilligan has argued, "the real 'action' of any allegory is the reader's learning to read the text properly."[73] The temporal rupture that is created by Arthur's retrospective/anticipated death exemplifies how the apparatus of narrative can efficaciously shape such learning. Arthur's death does not plainly lead readers to a mapped world where his arms are kept. Instead, the reader is thrust into the digressive, non-chronological Faerie Land of romance, where viewing Arthur's "armes" remains contingent on one's capacity to discover them. The reader's success of locating the arms is dependent on the kinds of action—moving from the visible and tangible to what remains beyond verifiability—that will soon be called for in the Proem to Book 2. This stanza thus brings willing readers to the threshold of a task predicated on the impossibility of fixing Arthur's status, as the poem urges them to reconceptualize their actual states of existence as possible ones. The discovery of arms in Faerie Land is conditional: the yet-unknown space can be found "if sought" by readers who embrace their roles as participants of romance quests, their roles analogous to the reader of the Proem who "may" "fynd" Faerie Land.

This moment, moreover, takes the speculative exercise further than the Proem. It offers a potential enactment of the task that the narrator invites in the Proem, since the ability to locate Arthur's arms demands more than imaginative projection. It also presumes a reader who actively embraces the logic of unpredictable quests, where things might be discovered "if sought." Arthur's polychronous presence occasions the potential experience of the ideal reader who can enter fictional worlds in search of the arms—the moment foregrounds the act of seeking, rather than only imagining, a possible world. Perhaps the most audacious claim here is that learning requires stepping into fictionality, where the reader's hypothetical action is the instrument of knowledge production. While Malbecco's disappearance exposes the absolute severance of abstract figuration and interpretive progression, Arthur's past/projected death oversteps the certain knowledge (of his demise) and points to a potential future (the display of arms). Arthur's present quest gives way to the reader's potential quest, incorporating the latter into fiction. By equating an anticipated past with a potential future, the narrator reveals that events lodged in fictional memory might be "seene" if readers reorient their cognitive frames toward the simultaneity of multiple temporalities and establish interactive relationships with poetic quests. Fracturing chronology and dislocating any firm grounding in actuality, the speculative method transforms the act of interpretation into an experiential immersion in fictive ontology.

To propel imagined readers into possible past/futures, and to suggest to them that they might have a hand in shaping the trajectory of the time to come, the poem makes Arthur himself briefly occupy the role of the reader who experiences the collision of his own history with poetic temporality. Together, the discontinuous historical narratives read by Arthur, prophesized by Merlin, and outlined by Paridell testify to an active, mutable history that, to borrow Barret's words, Spenser "plays a significant role in creating."[74] When he and Guyon read "Their countreys auncestry" (2.9.60.7) in memory's chamber, "th'hindmost rowme" (2.9.54.9) in Alma's castle, Arthur becomes, in Elizabeth Bellamy's words, "both implicit participant in and reader of the chronicles of his ancestry."[75] In the process, he dramatizes how readers confront their own existences as mere potentiality. He stops abruptly at the moment when the historical past intersects with the poem's present:

> After him Vther, which *Pendragon* hight,
> Succeeding There abruptly it did end,
> Without full point, or other Cesure right,

> As if the rest some wicked hand did rend,
> Or th'Author selfe could not at least attend
> To finish it: that so vntimely breach
> The Prince him selfe halfe seemed to offend,
> Yet secret pleasure did offence empeach,
> And wonder of antiquity long stopt his speach.
> (2.10.68)

The abrupt "Without full point, or other Cesure right" highlights the ontological crossings that occur at moments when incongruent temporalities overlap: this disruption provides neither ending nor expected break.[76] The violent language—"abruptly," "wicked hand did rend"—points to an untimely breach in the narrative chronology, as Arthur is, in the words of Jonathan Gil Harris, both "inhabiting a moment but also alien to and out of step with it."[77] The multiple possible explanations—represented by the "Or" that signifies alternate truth-values—function as placeholders for an absent historical narrative that could bridge the past to the reader's present.

As Arthur unknowingly confronts his place in history, his character too approaches a kind of fixity—his own death.[78] But this scene in Alma's castle ultimately refuses to narrate Arthur's encounter with his own annihilation. In the scene when Arthur is introduced, the poem produces a polychronicity that leads beyond his past/future death. And in Eumnestes's chamber, Arthur's reading of chronicle history takes us precariously close to the same event. But its chronological orientation forecloses, or at least fails to accommodate, the possibility of narrative to provide any recuperation from death. The surrogate reader, Arthur, exposes a narratological limit: as the narrator refuses to forgo the linear movement of history, he becomes like the historiographer in the "Letter," who tells of "affayres orderly as they were donne, accounting as well the times as the actions." The question then arises: If this framework of chronological telling cannot take them further than Arthur's life, what options are left to the reader being asked to speculate and participate? Readers can know Arthur's future, since they have access to other chronicles and romances that recount his life; they can thereby exceed the narrator's "end" by abandoning the Arthur of *The Faerie Queene* to his ignorance in this moment.

Or, readers might embrace other interpretive modes. After all, the poem has been modeling various forms of speculation, or processes of knowledge-making, that can instigate their own acts of worldmaking. The awareness of Arthur's retrospective/anticipated death had converted a seemingly known past into an event that might occur, drawing readers to experience the full fictionality of the created world. Arthur himself experiences this complex mode of knowing here: the historical past (and poetic future) vanishes as he encounters the fictional present, the aborted reading leaving him no option but to move on from this time and place, that is, to continue his quest. Asking readers to conceptualize time from the vantage of Arthur's existence, such polychronous moments also promise to transfer them from their cognitive frame outside Faerie Land to potential futures within it. The reader's actual world becomes a possible future projected *from* the narrative—one that "may" come to exist.

Arthur's multitemporal presence underscores how Spenserian allegory demands readers follow the "translative" procedures that shift references in narrative. This ethos is embodied by the poem's many "incomplete" figures that strive to, but do not, become abstractions. By contrast, when thought collapses into a perfectly readable figure, such as Malbecco, the narrative reaches an irreversible ending. Such embodiment of singular meaning also halts acts of fashioning and modes of interpretation. Allegory thus functions in the epic-romance not only as the enabler of deferral and endlessness but also as the vehicle that incorporates readers into the narrative process. It is a key mechanism of speculative method: ongoing attempts to collapse the gap between what is said and what is meant, and between what is and what might be, provoke both allegorical figure and reader to continue.

Harnessing Speculation: Prophecy and Readerly Action

We have seen how the narrator explicates a method of speculation, and how allegorical figures test its scope and limits within the narrative landscape. In this final section of the chapter, I turn to the figure of the ideal reader whose practical engagement with speculative methods activates her own role in shaping the future. Returning to the scene with which I opened this chapter, I read Britomart as the exemplary Spenserian reader, whose act of looking through the predictive lens prompts her to know more; her imperfect knowledge motivates her quest, which shapes the events that follow—and, by extension, furthers the ongoing worldbuilding of an "endlesse worke." The scenes explored in the earlier sections transfer the burden of interpretation to readers who must locate themselves in the complex time-scape of the poem. They also suggest that unknowability can motivate inquiry about a contingent future and, even more ambitiously, that one can claim prospective authority through guarantees to shape the times to come.

To demonstrate that a particular future can be actualized based on incomplete knowledge, in Britomart's encounter with Merlin Spenser poeticizes one particular intellectual mode that promises to make the future completely knowable: prophecy.

Prophecy was a hermeneutic mode that promised certainty despite the contingent nature of its object of scrutiny, that is, the future. As I will discuss in Chapter 2, it was also a kind of practical knowledge, connected to occult practices and esoteric pronouncements. I suspect that for Spenser prophecy is a useful tool because it instrumentalizes a nexus of ideas—of polytemporality, of interpretive possibility, and of predictive authority—that are vital to the unfolding of his poetic endeavor. By presenting the historical past as a possible future and by modeling the process of creating a reader aware of her place in upcoming history, the poet deploys prophecy to address a series of questions animating *The Faerie Queene*: In a realm of precarious existence, how can prophecy direct the reader to certainty? Can the *vates*'s ability to look into multiple temporalities enable them to actively fashion their audience? Can prophetic knowledge rein in the incertitude associated with the speculative processes we have witnessed so far?

Spenser's Merlin embodies both the prophetic and operational traditions of poetic production—what Sidney defines as "*vates*" (22) and "maker" (23), respectively. Merlin is an artisan in the tradition of glassmakers and painters, who came to know the world by doing. We catch glimpses of his proficiency as a maker in the artifacts that are dispersed across Faerie Land: he creates the "mirrhour playne" (3.2.17.4), and, as we have seen, he "wrought" Arthur's "shield, and sword, and armour." His ability to enact material changes through words operates in conjunction with this artisanal knowledge:

> he by wordes could call out of the sky
> Both Sunne and Moone, and make them him obay:
> The Land to sea, and sea to maineland dry,
> . . . .
> And hostes of men of meanest thinges could frame,
> When so him list his enimies to fray.
>
> (3.3.12.1–7)

In his character, then, Spenser conjoins literary making and practical knowledge. Merlin also exemplifies the conventional association of poet as prophet. Spenser's own "new Poete" makes bold claims to the exalted Virgilian role of the *vates* in *The Shepheardes Calender*,[79] Sidney identifies the poet as a "diviner, foreseer, or prophet" (21), and Puttenham declares that poets operated "by some divine instinct—the Platonics call it *furor*" and were "the first prophets or seers *(videntes)*."[80] By presenting Merlin as one who not only sees the future but also fashions it, Spenser converts the passive role of the *vates*—a figure who receives divine inspiration or is coerced by external forces[81]—into the active maker who gives form to unformed entities.

Spenser tethers a notorious character of history, legend, and literature (probably familiar to his readers through the works of Sir Thomas Malory and Geoffrey of Monmouth) to the particular fictive world of *The Faerie Queene* as a divining artisan. As a prophet-maker, Merlin foretells the times to come, and his constructed instruments direct intended users, like Arthur and Britomart, to action.[82] His prophecy to Britomart is a continuation of the history that Arthur reads in Alma's castle—it begins at the moment Arthur's reading ends but shifts the focus from him to Artegall. Moreover, by placing Merlin at the crux of Britomart's quest, the poem recuperates into a "providential" role a figure often associated with dangerous aspects of demonic magic and occult knowledge.[83] We see his divining skills at work when he foresees Britomart's (and Glauce's) visit: "He nought was moued at their entraunce bold: / For of their comming well he wist afore" (3.3.15.1–2). This moment serves as a vital instance of the poet's complex work with time that Spenser had outlined in the "Letter"—Merlin deals with "thinges forepaste, and diuin[es] of thinges to come."[84] As a figure oriented to both past and future (or, from the reader's perspective, to the past *as* future), he "conteineth" within him all the capacities necessary to instigate audiences to move from speculation to action.

Merlin's mirror sparks Britomart's quest, and as its vision provokes her conjecture, driving her to "more inquyre," she comes to impersonate the active reader that the narrator imagines in Book 2's Proem. Desiring interpretation of what the "shade and semblant of a knight" might mean, she visits Merlin, who prophesies to her:

> It was not, *Britomart*, thy wandring eye,
> Glauncing vnwares in charmed looking glas,
> But the streight course of heuenly destiny,
> Led with eternall prouidence, that has
> Guyded thy glaunce, to bring his will to pas:
> Ne is thy fate, ne is thy fortune ill,
> To loue the prowest knight, that euer was.

> Therefore submit thy wayes vnto his will,
> And doe by all dew meanes thy destiny fulfill.
> (3.3.24)

Merlin contests Britomart's belief that her vision was an accident. When her nurse, Glauce, puts pressure on his claim that "dew meanes" are necessary to "fulfill" this destiny ("what needes her to toyle, sith fates can make / Way for themselues, their purpose to pertake" [3.3.25.4–5]), Merlin espouses a philosophy of predestination that is dependent on action. This position admits some contingency within the eventual success of the prognosis:

> Indeede the fates are firme,
> And may not shrinck, though all the world do shake:
> Yet ought mens good endeuours them confirme,
> And guyde the heauenly causes to their constant terme.
> (3.3.25.6–9)

As Michael Witmore has argued, accidental events could have providential underpinnings, even though one might not be able to offer metaphysical explanations for them.[85] Merlin's words espouse this kind of claim. What Britomart sees as an accident of history—her "wandring eye" glancing into a mirror—is actually the result of "eternall prouidence." Her actions "confirme" and "guyde" the fates instead of the other way around. To persuade Britomart that her actions will determine the future, Merlin moves her from the "wandring" of romance into the "streight course" of "destiny." Launching into a genealogical prophecy, he makes her "loue" essential, and not one of the poem's many "Accidents," a stance that contradicts Spenser's own claims in the "Letter."[86]

This prophetic narration begins as a performative event, in which Merlin seems to be learning the future as he divines it. His reactions to events seem spontaneous and unmediated, as if the visions are discoveries for him too. His passionate declaration of the "woe, and woe, and euerlasting woe" (3.3.42.1) that awaits the Britons invokes similar passions in Britomart: "The Damzell was full deepe empassioned, / Both for his griefe, and for her peoples sake, / Whose future woes so plaine he fashioned" (3.3.43.1–3). Merlin's performance elicits her empathy, as his "griefe" demands she respond by reacting to this vision of her progeny. As the narrator's words ventriloquize Merlin's emotions, we might ask: Why does this figure who can see into the future accentuate, rather than downplay, uncertainty? We can explain this performance, in which he "fashioned" Britomart's "future woes," by reading it as an example of what Sidney identifies as the key aspiration of the poet: to move the reader.[87] Britomart's emphatic responses reveal how Merlin's words deeply affect his audience. As the poem layers the present with, and as Britomart witnesses herself being projected into, a possible future, she becomes the kind of reader desired by the Proem of Book 2—one who immerses herself in the story she hears, willing to "fynd" what "may" be, or to "more inquyre." Merlin's displays of surprise and grief during his predictions highlight that prophecy, as an ongoing performance that emerges in acts of narrative revelation, can successfully stage a speculative method, since it is open to surprise, suspension of judgment, and prediction of the unknown. Performed prophecy, like poetry, moves its intended audience, in the sense of emotional effect and being provoked into action. Taking this argument to its limit, we could say that prophecy requires the cloak of uncertainty—in this instance, a series of speculations that stretch beyond the visible—to be efficacious. It both produces new knowledge and offers a method for the dissemination and application of this knowledge.

By contrast, when prophecy drives at a certain future, its narrative structure collapses, leaving the *vates* with a failed predictive art. Prophecy, in its orientation toward eventual outcomes, demands that one future come true. Its fulfillment occurs when a proclamation converges with the historical present. But at the crux of any prediction becoming a prophecy is an ontological fissure—till a future is confirmed in history, the prophetic utterance remains a fiction, in the Sidnean sense that it "nothing affirms." Merlin's unfolding revelation tests these gaps between fiction and history. His divination suddenly halts with the proclamation:

> Then shall a royall Virgin raine, which shall
> Stretch her white rod ouer the *Belgicke* shore,
> And the great Castle smite so sore with all,
> That it shall make him shake, and shortly learn to fall.
>
> But yet the end is not. There *Merlin* stayd,
> As ouercomen of the spirites powre,
> Or other ghastly spectacle dismayd,
> That secretly he saw, yet note discoure:
> Which suddein fitt, and halfe extatick stoure
> When the two fearfull wemen saw, they grew
> Greatly confused in behaueoure[.]
>
> (3.3.49.6–50.7)

As Spenser masks chronicle history as prophecy, Merlin's words take Britomart beyond the fictional world of the poem to the actual world, not as a mirror to or a shadow of how Elizabeth "should be" but as a marker of the impossibilities of her reign.[88] The caesura, which denotes the interruption of speech, is the linguistic marker of how this scene dramatizes endings differently from the moment of Arthur's reading: the poem formalizes an absolute break when this instance of fictional future collides with the current historical moment. Merlin's ending, "But yet the end is not," is tinged with interpretive possibilities, indicative of both knowing what the future looks like and acknowledging one's inability to express it. But the collision of the historical present (in the actual world of Elizabeth) with the predicted future (in the fictional one of Britomart) creates a problem. Dilation becomes fixity, and the boundary crossings between fact and fiction that drove narrative speculation suddenly cease. Merlin's performance ends with a caveat that might be the paradox of prophetic utterance itself: by driving toward certainty, vatic language marks its own limit, that is, the point beyond which it cannot continue. Earlier, Merlin was engaged in poetic making—affective, active, futuristic—but in the moment his pronouncement collides with the actual present, his narrative project screeches to a halt. Prophetic discourse, here, offers a stark contrast to ongoing narrative speculation.

Of course, at this moment Spenser acknowledges the limits of his own poetic art that cannot predict the actual future. However, he does so by turning prophecy into uncertain prediction, an unresolvable promise that cannot be completed outside of history. Although Merlin sets out to situate Britomart in *one* future, he ends up dissociating her from all narrations. He does not merely create a gap between thought and articulation, or between the spoken and the known; he halts the possibility of expressing knowledge. The unknown actual future serves as the perfect vehicle for dramatizing the collapse of a prophetic one that was supposed to be representable in narrative. Although Merlin might be witnessing a "ghastly spectacle," he cannot "discoure" of it, that is, he cannot disclose it. The figure who has infinite power with "wordes" is now devoid of language. Spenser does not provide an explanation of what Merlin sees, establishing yet another rift between prophetic and poetic endings. Merlin's behavior is indirectly mirrored in the "confused" reactions of his audience. As his knowledge totters at the edge of the unnarratable, the poem's larger experiment with prophecy, paradoxically, exposes the limitations of the vatic mode—its authority is no more a guarantee of a certain future than the Proem's imaginative projections.

We have arrived, then, at another moment of potential ending. When prophecy encounters the singularity of the present, the resulting ontological rupture dissociates Merlin and his audience from the poem's generative strategies of speculation. The prophecy itself becomes unnarratable. While Arthur's polychronous existence invites readers to experience fictional worlds, Merlin's engagement with a moment of apocalyptic future does not admit any such point of entry.[89] Instead of establishing prophecy as an affirmative mode of prediction, his "suddein fitt, and halfe extatick stoure" take us to the limit of what is knowable. In this moment, prophecy makes knowledge impossible.

Yet, Britomart does move on, undeterred by the aborted ending. Unlike Arthur, she can access her future, but its truth-value hinges on whether Merlin will be able to activate her desires, motivating her to transform what she now sees as one potential future into her singular "destiny." The performative aspect of prophecy, full of affective proclamations, is sufficient to drive her to embark on her quest. Merlin's curtailed revelation presents to her how the future could be, but only if she participates in it. The uncertainty of the future could be resolved, the poem suggests, if she accepts her role as a reader being fashioned into action. Merlin reflects the position of the poet in his implicit advocacy for "good endeuours"—"well-doing" in Sidney's terminology. Just as the artificer's created objects animate audiences (that is, figures in the poem), the poet's worldmaking functions with the expectancy of an engaged reader who acts to secure a particular future by embracing possible knowledge. Merlin's prophecy is analogous to Spenser's poetic project to "more inquyre" when it allows for suspension of truth and gestures to genuinely open acts of interpretation, narration, and action. Prophecy can provide possible knowledge when it, like poetry, "nothing affirms."

The limits of prophecy, we might say, also highlight the scope and kind of knowledge that poetry can produce. As Spenser examines when vatic utterance can be analogous to poetic making, he reveals that instruments of speculation—from divining mirror to "continued Allegory"—and not the apocalyptic vision serve as the apparatus of possible knowledge that can move audiences. As such, they are corollaries to the Spenserian *poiesis* that thrives in zones between ignorance and certainty. They show that even the most "divining" poet can offer only conditional knowledge. What readers—and fictional actors like Britomart—require is speculative narration and not conclusive precepts. In an era that was increasingly valuing proof as one key basis of authority, Spenser asks what it means to reside in paradigms of belief and knowing not predicated on verification, and what it means to embrace the "not yet" instead of the "now." The ideal prophet-maker need only dramatize the "Methode" that will allow observers to fashion themselves through the shadowy images or future possibilities they almost perceive. Since possible knowledge in *The Faerie Queene* serves an ethical function, the logical end of narration is the reader's "good endeuours." While many early modern writers grappled with the ethical purpose of their writings, through moments such as this *The Faerie Queene* flaunts how the apparatus of poetic making can actualize an ethics in the formal encounters of its polytemporal romance-scapes. Merlin's work is a hypothesis for the future, and he provides the foundation for how the upcoming time "might" be.[90] This epistemology requires action by readers: they must imagine themselves as entities with indeterminate futures who will realize their anticipated lives from the multiple potentialities created by the poet.

The Faerie Queene here models a generative relation between vatic pronouncements and action, but it also provokes several questions about this association: If prophecy can only be fulfilled by history, do its predictions hold within them the capacity to shape history? What are the limits to action when a prediction invites one to embrace—and even predicate one's desires on—the inevitability of a particular outcome? These questions would become of increasing importance as early modern thinkers grappled with prophecy not only as a mode of esoteric knowledge accessible to a few but as a kind of knowledge that could shape social, cultural, and political conditions more broadly. In this shifting paradigm, prophecy might not only put forth desired political genealogies, as Spenser imagines in Britomart's case. It could be the source of much destruction. From Merlin's direction toward romance proliferation, I turn in the next chapter to early modern theatrical action. As prophetic insight collides with the protagonist's political ambitions in *Macbeth*, forms of prediction whose ends remain contingent on readerly participation become catalysts for tragedy.

CHAPTER 2

William Shakespeare's Prophetic Recipes

> And do they not know that a tragedy is tied to the laws of poesy, and not of history; not bound to follow the story, but having liberty either to feign a quite new matter or to frame the history to the most tragical conveniency?
>
> —Philip Sidney, *The Defence of Poesy* (ca. 1581)

As Spenser's Merlin dramatizes the power of vatic utterance—its ability to drive audiences to action based on incomplete revelations about an uncertain future—he also exposes its limits: prophecy's guarantee of certainty is futile. At the same time, his enigmatic words establish the prophet as an agent who knows the future and can manipulate this knowledge toward desired ends, in this case, the continuation of Britomart's lineage. In this chapter, I put pressure on the animating force behind Merlin's inscrutable language: the fact that vatic discourse aims to construct, and not merely predict, the future. I argue that by examining prophecy as a contingent mode of knowledge production that engages with contemporary ideas of action and prescription, and by focusing on the predictive strategies and esoteric utterances that comprise the vatic mode, we are better able to understand why it was such a powerful mode for shaping political futurity. This chapter, then, expands on what I identified in Chapter 1 as the performative aspect of prophetic narration, the moments of ongoing surprise and affective response that highlight the contingent nature of vatic revelation. To examine how prophecy performs its knowledge production—in the early modern sense of "perform," namely, "'to carry through to completion; to complete, finish, perfect'"[1]—I turn from the "endlesse worke" of romance worldbuilding to tragedy's drive toward an ending. The temporality of tragic drama, I propose, calibrates the dilated processes through which unconfirmed prediction is translated into confirmed prophecy. By recovering how theatrical worldmaking is predicated on wedding the promise of vatic revelation to the efficacies of action, this chapter locates possible knowledge in the prophetic orientation of tragic form, whose action privileges contingency and ambiguity even as it is governed by pre-scripted ends.

To work through this hypothesis, I consider *Macbeth*, a tragedy in which the form of action is shaped by the ambiguous utterances of the "weird sisters" (2.1.20), the play's "instruments" (1.3.126) of occult knowledge. Often perceived as the most topical of Shakespeare's tragedies, *Macbeth* repeatedly turns to questions of time as it imports to the stage sections of Holinshed's *Chronicles* that were particularly relevant to the new English monarch, James, who traced his lineage to Banquo.[2] From the witches' opening query, "When shall we three meet again?" (1.1.1), to Lady Macbeth's claim that she "feel[s] now / The future in the instant" (1.5.55–56), to Macbeth's musings on time's passage ("Tomorrow and tomorrow and tomorrow / Creeps in this petty pace from day to day" [5.5.19–20]), *Macbeth* enacts what David Scott Kastan has identified as the "play's very insistence upon its future" and its refusal to "stop."[3] As moments such as Duncan's naming Malcolm "The Prince of Cumberland" (1.4.39) and Macbeth's worrying about his "barren scepter" (3.1.62) become inseparable from questions about political lineage, the titular character feels the necessity of knowing the future and to this end inquires how potential, and therefore unconfirmable, events might be understood and manipulated to authorize his own ambitions of kingship. The play's obsession with the future is haunted by the question of the relation between knowledge and action that also underlies Merlin's implicit charge to Britomart: Can prophecy instrumentalize the unknowability of the future to desired ends? *Macbeth* enacts a potent answer to this question and reveals that the theatrical medium is ideal for addressing this conundrum because it formally mimics the mode of contingent knowledge that promises certainty about approaching times. In Shakespeare's tragedy, prophecy becomes the thematic and formal vehicle that reveals the "future in the instant."

The mode of prophecy is a key instrument of possible knowledge, one where the processes involved in the actualization of a prognostication determine contours of existence. *Macbeth* dramatizes this entanglement of epistemology and ontology as its exercise of worldbuilding becomes an ongoing response to the questions the play poses at the outset: Can the enigmatic predictions that inaugurate the play come true? And how can these predictions be actualized within a dramatic form? As the witches' predictions spark both interpretations and "acts" (4.1.148), the impetus to fulfill their opening prognostications drives the theatrical action: the titular character interprets the predictions as prescriptions for action, activating the tragic worldbuilding that culminates in the transformation of equivocal words into prophetic truth. Approaching the enigmatic prophecies as prescriptive recipes, Macbeth not only engages with the realms of the supernatural and the marvelous, with the "weird sisters" serving as embodied manifestations of these realms,[4] but also enacts a perverted form of "maker's knowledge," where one knows by doing.[5] The chasm between the witches' enigmatic utterances and their ultimate resolution, then, serves an analogous function to the gap between the "might" and the "should" that gives poetic form to Spenser's speculative method. Possible knowledge in *Macbeth* resides in the tensions between the certainty assured by prophetic utterances and the suspension of this surety within the temporality of theatrical action. The negotiations between the impossibility and promise of prophetic certainty form the tragic world of *Macbeth*.

The topicality and transgressiveness of the vatic mode, its popularity in contemporary writing, and its particular associations with political genealogy make prophecy an ideal instrument for examining concerns about sovereignty (especially the future of kingship) that lie at the heart of *Macbeth*. Unsurprisingly, scholars have noticed the role that prophecy plays in Shakespeare's tragedy, but it is more typical to focus on its political import and contemporary resonances than to inquire how the processes or methods of knowing that underlie the visionary mode are formal instruments of dramatic worldbuilding.[6] *Macbeth*'s prophecies are not merely instances of "hindsight masquerading as foresight," which Marjorie Garber identifies as the condition of dramatic prophecies, where "the audience knows that these 'impossible' things will prove true, and it can do nothing with that knowledge but wait for the fulfillment of the future anterior—the future that is already inscribed."[7] Such a reading cannot accommodate how, in *Macbeth*, the interpretation of prophecy triggers theatrical action and incorporates spectators in the play's temporality. To adopt but also adapt Garber's terms, *Macbeth*, I argue, is about "foresight" rather than "hindsight": its open-ended predictions expose how the resolution of dramatic prophecy is contingent on its audience's interpretations. The play's predictions demand that audiences, both the characters hearing the prognostications and the external spectators, react to them. Prophecy in *Macbeth* produces neither fact nor a predestined future before the fact but augurs forms of analysis that shape our understanding of how characters earn places in history and on the stage, that is, in the extra-theatrical future and in the theatrical present, respectively.

In adopting prophetic knowledge-making as a mechanism of theatrical worldmaking, *Macbeth* grapples with a major epistemic shift occurring in the early seventeenth century, which scholars such as William Eamon and Elizabeth Spiller mark as the movement from knowing to doing, or from *gnosis* to *praxis*. A crucial aspect of this shift is that various forms of occult knowledge become more firmly associated with practices and developments in natural history and philosophy. This association in turn transforms the utilitarian value attached to even occult and esoteric knowledge.[8] As esoteric knowledge came within the purview of popular discourse, and as increased value was placed on *praxis*, one could approach futuristic ways of knowing, including prophecies, as recipes for action—or, as Eamon describes, "a prescription for an experiment, a 'trying out.'"[9] Such changes in the parameters of occult and practical knowledge are contemporaneous with ideas generated by works on poetics such as Sidney's *Defence*, which claims that even poetry, which "nothing affirms," aims toward *praxis* or "well-doing" over *gnosis* or "well-knowing."[10] *Macbeth*, then, captures a moment in early modern culture when action and uncertainty intersect in a variety of discourses about imaginary and natural worlds. Keeping these contexts in mind, we can see how *Macbeth*'s ambiguous sources of esoteric knowledge and the play's exploration of the complexity of prophecies serve as crucial examples of changing concerns over the access to, use, and scope of the "occult forces" of "nature."[11] By inviting characters to interpret, understand, and act on the witches' equivocal prophecies, the play ultimately transforms prophetic revelations into transgressive recipes for political, and theatrical, action.

Prophecies and Recipes: Prescriptions for Action

Prophecy had a complex status in Elizabethan and Jacobean England, where the prophet was defined in two different ways: "either for a shewer, or foreteller of things to come" or "for a preacher, or interpreter of the scripture," in the words of John Harvey, the Elizabethan astrologer.[12] Prophecy did not signify a single concept or action but a range of practices from diverse arenas, including astrology, mathematics, and religion.[13] It was an art of prediction, but the contingency of outcomes typically associated with predictive arts was countered by claims that prophets had divine inspiration and could thus foretell the truth. Since prophets claimed unmediated or immediate access to a privileged, esoteric source of knowledge, their practices were always seen as potentially subversive: "whether oracular or interpretive in nature, [prophecy] was dangerous because it implied that the Holy Spirit still spoke immediately to individuals, even though most Protestants maintained that the age of miracles and inspiration had ceased and that the Holy Spirit only spoke to men *mediately* through the holy scriptures."[14] At its most threatening, prophecy could destabilize established hierarchies and conventions. It disseminated elite knowledge to commoners, and its practices were associated with utilitarian ends that carried with them the potential for tremendous social and political upheaval.[15] Although often employed as a tool for maintaining political authority, it could just as well undermine current power structures by pointing to a reformed future, or one that marked a disruption from structures of present-day life.[16] Worries about the accuracies as well as the effects of prophecies had led Elizabeth I to issue various proclamations against prophesying throughout her reign. To a large extent, prophecy was seen as a powerful and threatening instrument—as a means either to subvert or to consolidate positions and ideas—in both politics and religion because of its implicit promise of certainty, that is, that it will be fulfilled in history. In the essay titled "Of Prophecies," even a skeptical Bacon acknowledges that people give prophecies "grace, and some credit" because some have proved true: "men mark when they hit, and never mark when they miss."[17]

Prophetic discourse was an amalgamation of occult, practical, and poetical epistemologies, and realizing a prognostication in history entailed reconciling theories of *praxis* or "doing"—where "action contains its end in itself"—with the productive aspects of *poiesis* or "making," where "making finds its ends in its object."[18] Since a prediction becomes a prophecy only when it is fulfilled in history, the end-oriented activity of *poiesis* was inextricable from the act of vatic utterance: prophecy, then, refers to both action and product. Prophetic expressions were also poetic in the sense that poetry purports to move readers toward action, invites audiences to interpret figurative modes of representation, and yet "nothing affirms," as Sidney put it. Spenser's Merlin exemplifies these connections of prophet, poet, and maker, as we witnessed in Chapter 1. In a different arena of early modern knowledge-making, naturalists often classified prophecies under the category of *secreta*. Unlike *scientia*, where only regularities or unchanging elements of nature were deemed worthy of philosophical inquiry, "*secreta*" encapsulated diverse phenomena, including the "manifestations of occult qualities" and "events that occur unexpectedly or idiosyncratically"; they "could be experienced" but were "not demonstrable."[19] *Secreta* manifest themselves in the form of wonders, marvels, and accidents, and they find expression in diverse practices and media, including natural magic, practical arts, alchemical experiments, recipe books, and prophetic utterances.

Prophecy's promises of certain knowledge inevitably led to the scrutiny of its epistemic claims. Works such as Henry Howard's *A defensatiue against the poyson of supposed Prophesies* (1583) and John Harvey's *A Discoursiue Probleme concerning Prophesies* (1588) grapple with the incertitude that undergirds the mode. Harvey compares it to an "vncertaine collection of mans inuention, without any further diuine instinct, angelicall illumination, or propheticall gift of foreknowledge, either mediate, or immediate, either sensible, or intellectuall."[20] Howard, too, stresses the ambiguous underpinnings of the vatic voice: "Wee knowe that men are images of God: but no gods indeede, that our wittes may deeme: but not diuine, forcast vppon occasion: but not prefigure without certainty."[21] Both writers highlight the limitations of the human understanding and argue that divine intervention might not be accessible to all. There might be some "true" prophets, but the majority of prophecies are false. Claims of access to, and the desire for, certain knowledge could lead to folly, since men and women seek "deeper knowledge, of the future causes" and are stung by "curiositie" when "it pleaseth God to discouer and reueale by ordinarie meanes."[22] Exceeding the limits imposed by God, human beings try to appropriate God's unique knowledge "of declaring future thinges before they come to passe: [which] belongeth onely to the deepth of diuine wisedome."[23] What false prophets claim as divine calling is actually learning gained by "the light of long experience" and "signes of observation."[24] Howard and Harvey underscore how grandiose claims of prophetic knowledge lead to multiplying predictions, which run counter to the promise that history would fulfill one true prophecy. As prophecies multiply—or in Keith Thomas's terms, with the increase in "counter-prophecy"[25]—possible *true* futures also multiply.

The proliferation of prophecies and diverse interpretations of individual predictions undermine the promise of singular, certain knowledge, but they also suggest that one potential future could be made to come true. Revelations, then, held within them the capacity to shape the actions and experiences of those who were privy to them. The possibility that predictions could be realized through active intervention would seem especially appealing in a moment when esoteric knowledge that used to be accessible only to a few was being popularized through the dissemination of practical "how to" books,[26] which were comprised of recipes that served as "a prescription for an experiment, a 'trying out.'" The term "recipe" refers most specifically to a technical category—a "list of ingredients" and a "set of instructions describing how they were to be employed"—in which the "accumulated experience of practitioners [was] boiled down to a rule."[27] As Eamon notes, "*recipe* is the Latin imperative 'take'"; the term more broadly "prescribes an action" where "the recipe's 'completion' is the trial itself."[28] Julia Reinhard Lupton argues, "the recipe is a kind of script, in both the dramatic and the software sense: the forms of thinking it 'directs' are pragmatic,"[29] capturing both the prescriptive and active aspects that are incorporated into the mode. Until it is actualized through a trial, a recipe—or what early modern practitioners typically termed "receipt"[30]—is incomplete and imperfect. As such, recipes instantiate the active fulfillment of futuristic knowledge only when they are performed. Reading prophecies and recipes—two modes of knowledge production whose outcome was contingent on external factors such as interpretation and trial—as intertwined intellectual forms, we can better perceive how the colliding forces of prescription and action were reconfiguring the relations between occult and practical knowledge in early modern England.

Technical recipes offer a particularly complex insight into the wide scope of early modern practical knowledge. They could be classified as medicinal, culinary, cosmetic, or occult; they engaged with topics as diverse as natural magic, home remedies, and cosmetics; and they were practiced by cooks, housewives, artisans, and alchemists. For scholars, they offer a rich archive for a variety of entryways into thinking about the past. Some were "depersonalized" and therefore had a "more general, universal quality,"[31] as Eamon notes, while others offer glimpses into individual lives and even serve as repositories of family histories.[32] Technical recipes are thus key to representing a diversity of practices (and practitioners), exposing how those at the margins of academic institutions or societies of learning were essential to the phenomena we have come to term the Scientific Revolution. Moreover, as the recovery, preservation, and reproduction of household recipes have revealed, they also give us access to the intricacies of home ecologies, to the inventive practices of women, and to female knowledge communities in much more detail and variety than perhaps most other documents from the period.[33]

While the basic elements of the technical recipe—ingredients and instructions—remain the same, they vary widely in their details. Susanna Packe's cookbook contains a recipe for "A cake" that offers detailed and specific measurements of ingredients: "3 quarts of flower 6 yolks of eggs i pin=t of Barme, halfe a pint of good milke i = pound & halfe of Butter" (Figure 1).[34] Hugh Plat's instructions on how "To boyle a chine of veale, or a chicken in sharpe broth with hearbes," on the other hand, rely on the reader's ability to interpret more ambiguous calibrations: "TAke a *little* muttō broth, white wine and veriuyce, and a *little* whole mace, thē take lettuce, Spinage, and Parsley, and bruise it, & put it into your broth, seasoning it with veriuice, pepper and a *little* sugar, and so serue

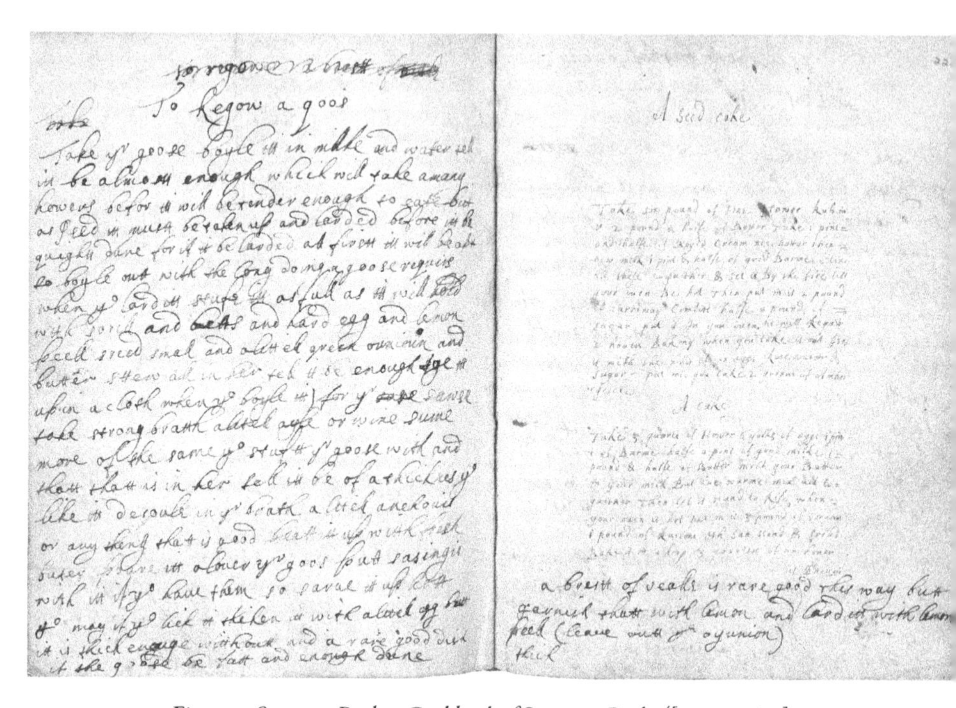
To Regow a goos

A Seed cake

A cake

Figure 1. Susanna Packe, *Cookbook of Susanna Packe* ([manuscript], 1674), 221. Call #: MS V.a.215. Used by permission of the Folger Shakespeare Library.

it" (emphasis mine).[35] The recipe "To preserue Walnutts" in the receipt book attributed to Mary Granville and Anne Granville D'Ewes is similarly ambiguous, but it also seems to assume knowledge on the reader's part of external conditions as well as their familiarity with the task—or at least seems to assume the reader's discretion about how and when to perform the task: it instructs them to "Take some walnuts about the latter end of June or the beginning of July when the shells are like a Jelly or before they bee tuffe then pare the vpper thin skinne off like an apple."[36] Yet other recipes are fascinating not only for their directives about the activity itself but also because of their ability to capture reactions of users: the recipe for "damson wine" in a manuscript of "Choyce receits collected out of the book of receits" contains a note stating that it is "the worst in the world" (Figure 2).[37] As Wendy Wall's research on the intellectual aspects of recipes reveals, these documents expand our understanding of "what it meant to be a maker, knower, creator, artist, artificer, worker, and preserver."[38] Products of

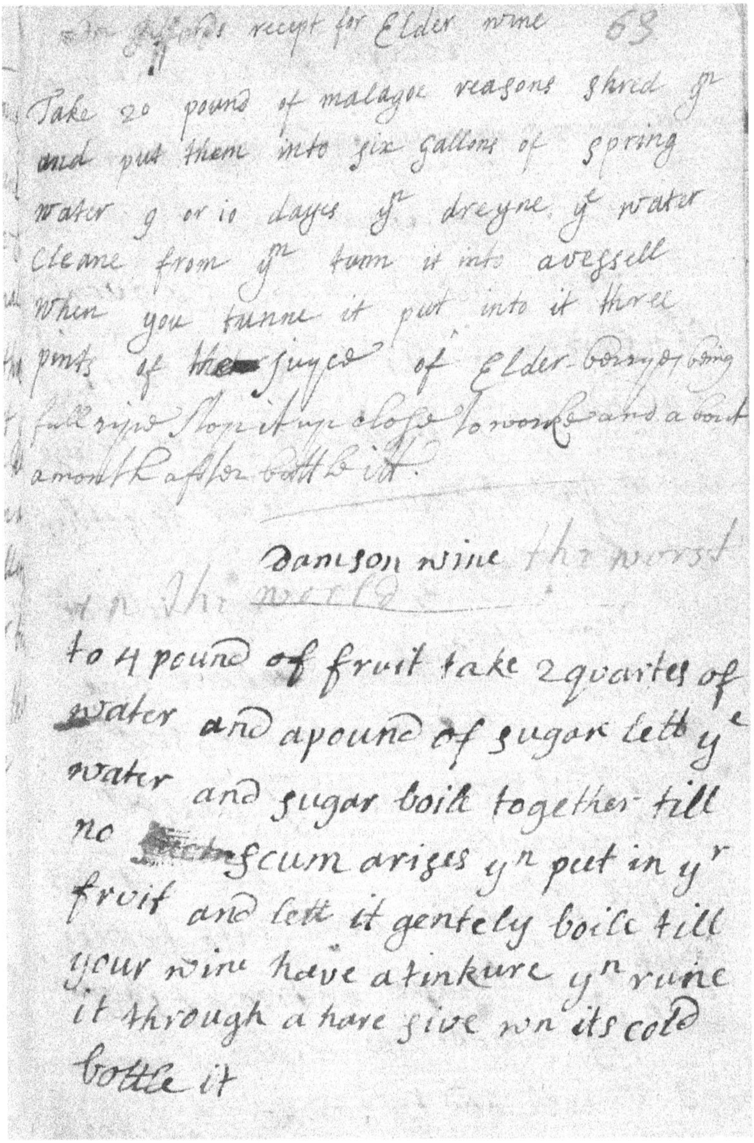

[illegible] recept for Elder wine 63

Take 20 pound of malagoe reasons shred ym
and put them into six gallons of spring
water 9 or 10 dayes yn dreyne ye water
cleane from ym turn it into a vessell
when you tunne it put into it three
pints of the juyce of Elder-berrys being
full ripe stop it up close to worke and about
a month after bottle it.

Damson wine the worst
in the world

to 4 pound of fruit take 2 quartes of
water and a pound of sugar lett ye
water and sugar boile together till
no scum arises yn put in yr
fruit and lett it gentely boile till
your wine have a tinkure yn rune
it through a hare sive wn its cold
bottle it

Figure 2. Anne Carr, "Choyce receits collected out of the book of receits, of the Lady Vere Wilkinson ([manuscript]/begun to be written by the Right Honble the Lady Anne Carr," 1673/4), 63. Call #: MS V.a.612. Used by permission of the Folger Shakespeare Library.

individual and collective hands, recipes, by their very existence, insist that practical knowledge intertwines prescription and action.

In *Macbeth*, the witches' trials with "ingredients" (4.1.34) provide the clearest example of the promise, and the ultimate limitations, of the technical recipe.[39] Staging how this instrument of practical knowledge provides directives, they call for

> Scale of dragon, tooth of wolf,
> Witches' mummy, maw and gulf
> Of the ravined salt-sea shark,
> Root of hemlock digged i'th' dark,
> Liver of blaspheming Jew,
> Gall of goat, and slips of yew
> Slivered in the moon's eclipse,
> Nose of Turk and Tartar's lips,
> Finger of birth-strangled babe
> Ditch-delivered by a drab,
> Make the gruel thick and slab.
> Add thereto a tiger's chawdron,
> For th'ingredients of our cauldron.
> (4.1.22–34)

As they list their ingredients, the witches dehumanize non-Christians, reducing their bodies to objects to be used.[40] At the same time, they echo the contents of actual recipes. Recipes from Alessio Piemontese's extremely popular *De' secreti del reuerendo donno Alessio Piemontese* (initially published in 1555), for instance, often contain unorthodox ingredients, such as animal parts: "wild boar's teeth, skin of a dog, and 'dung of a blacke Asse, if you can get it; if not, let it be of a white Asse.'"[41] *Macbeth*'s catalog mimics this form and lists similar content, but this technical recipe does not outline the goals of the actions undertaken by the witches; their obscure aims and unexplained practices transgress the boundaries between practical and occult arts. Emphasizing the actions one must undertake in the moment through words such as "Make" and "Add," the witches' trial stages a kind of suspended performance, rather than modeling a prescription that works toward a definite end. The witches act not to realize goals but to enact a process without a direction, and the recipe becomes a performance for performance's sake. As this scene offers a momentary respite from the questions of futurity that are omnipresent in the play, suspending audiences in a spectacle of pure presence, it serves as an extreme example of how the recipe is first and foremost a performative event, a "trial" that owes nothing to the future it is supposed to bring into being.

In a play fixated on actualizing the future and intent on dramatizing diverse modes of secret knowledge, the ambiguities surrounding the efficacy and the purpose of technical recipes effectively signal the limits of this mode of prescription. Encountering the witches during this trial, Macbeth asks, "How now, you secret, black, and midnight hags! / What is't you do?" The witches only reveal they perform "A deed without a name" (4.1.47–48). Instead of divulging the meanings or purpose of what they "do," they mark the impossibility of explicating their labor, of boiling it down to a rule, of even giving it a "name." The technical recipe, in *Macbeth*, seems insufficient to harness occult forces to the particular ends that its titular character desires. When a recipe can only take us so far, the play suggests, one needs to approach other kinds of *secreta* as prescriptions. I propose that, in the play, the witches' prophecies emerge as recipes—implicitly "prescrib[ing] an action"—that invite characters to fulfill their utterances. They intimate that the potential political futures they reveal will become true only when one acts to ensure a certain outcome. As Macbeth's action becomes the vehicle for this transformation, *Macbeth* tests what constitutes proper *praxis* in the play's created world.

How to Interpret a Prophecy

From its very first scene, the play yokes questions of political futurity and the temporality of drama:

> FIRST WITCH. When shall we three meet again?
> In thunder, lightning, or in rain?
> SECOND WITCH. When the hurly-burly's done,
> When the battle's lost and won.
> THIRD WITCH. That will be ere the set of sun.
> (1.1.1–5)

While their repeated "When"s demand a fixed meeting time, the witches suggest that this future cannot be defined merely in terms of calendrical units or seasonal cycles. They de-privilege chronology and instead link knowledge about the future to fluctuations in political power. Even as they use environmental markers ("thunder, lightning, or in rain") and note daily changes ("ere the set of sun"), their anticipated time of meeting is primarily calibrated through a current event, the battle. They delimit the timing of their next gathering to the resolution of the war, such that the phrase "when the battle's lost and won" becomes the only specific marker of the future in an exchange otherwise characterized by ambiguity. The chiastic inversions in their departing statement, "Fair is foul, and foul is fair" (1.1.11), are also indicative of how their propensity to speak "not affirmatively," to borrow Sidney's words, is ultimately a strategy of marking the temporality of the dramatic plot. This pronouncement, as they step offstage, linguistically prefigures an approaching scene. Macbeth makes his entrance on stage by echoing their language: "So foul and fair a day I have not seen" (1.3.39).[42] His juxtaposition of "foul and fair" draws attention to the unsettled state of the environment, where the foulness of the physical world is countered by the fairness of victory in battle. Repeating the language of the witches, Macbeth's words also echo the connections they make between the natural and political spheres. His appearance and utterance resolve the ambiguity of the opening exchange, revealing how the future "When the battle's lost and won" has become the theatrical now. These words, then, offer a succinct fulfillment of prophecy through theatrical action: through Macbeth's entrance, the play reminds audiences that they have arrived at the moment "when" the witches were supposed to "meet again."

These opening scenes indicate how, in *Macbeth*, supernatural forces, the discourse of prophecy, and environmental signs are intimately tied to the temporality of dramatic plot. These connections are evident at other pivotal moments in the play. For instance, Lennox underscores the natural environment's prophetic force when he recalls the weather at the moment that, as the play's external audience knows, was the time of Duncan's murder—Lennox himself is unaware of the murder and will learn of it immediately after this speech:

> The night has been unruly. Where we lay,
> Our chimneys were blown down and, as they say,
> Lamentings heard i'th' air, strange screams of death,
> And prophesying with accents terrible
> Of dire combustion and confused events.
>
> (2.3.48–52)

When he describes actions of the natural world, Lennox extends the witches' association of environment and event to his own sphere of human interactions: nature's foreboding signs index the horrors at Inverness. The environment "prophes[ies] with accents terrible" about human events through natural anomalies, portending the "combustion" and "confused event" that is the impending (for Lennox and his companions) tragedy of the King's murder. Nature dissembles through "Lamenting" and "strange screams," but its explanation can be most easily found within Macbeth's castle. Lennox's recollection enacts the palimpsestic relation between the physical, political, and dramatic worlds that is crucial to the unfolding of the plot of Shakespeare's play, and his confusion takes on a vatic force as it uses nature's "prophesying" to point to an event, yet undiscovered, that has already been actualized within the playworld: Duncan's death. The ambiguous semiotics of prophecy, then, reveal the future through the trajectory of tragic plot.

The entanglements of political futurity, prophecy, and the dramatic present are most apparent in the scene in which Macbeth and Banquo encounter the witches and hear their dissembling verbal predictions. As they pronounce that Macbeth, as well as Banquo's descendants, will be kings, the witches reveal potential futures that will only retroactively become "prophecies" if they are fulfilled. They "hail" Macbeth as "Thane of Glamis!" (1.3.49) and "Thane of Cawdor!" (1.3.50), and declare that he "shalt be king hereafter!" (1.3.51), revealing his present status and predicting his future rule. Only Banquo's request that they reveal his future makes them prognosticate again. Their initial predictions for Banquo, "Lesser than Macbeth, and greater" (1.3.66), "Not so happy, yet much happier" (1.3.67), are, like a lot of their utterances, enigmas. The "*Enigma*," according to Puttenham, is the figure through which "we dissemble again under covert and dark speeches when we speak by way of riddle (*enigma*), of which the sense can hardly be picked out but by the party's own assoil."[43] The enigma obscures the evidentiary quality of language. Puttenham notes that the "riddle is pretty but that it holds too much of the *cacemphaton*, or foul speech, and may be drawn to a reprobate sense."[44] The enigma might approach "foul speech" but by dissembling makes it "pretty"—or "fair," to borrow a term from the witches. This ability of the riddle to make the "foul" seem "fair" is a particularly threatening instance of the power of language to draw one into "a reprobate sense." This ambiguity demands the constant work of interpretation, since "the sense can hardly be picked out but by the party's own assoil." The witches' final promise, "Thou shalt get kings, though thou be none" (1.3.68), augurs possibilities and transports Banquo into a distant future through the revelation of his descendants' fates, even as the phrase "though thou be none" shuts him out.

The different temporal and genealogical concentrations of the two prophecies are indicative of the play's struggles between multiple possibilities and a singular certainty. The prophecies directed to Banquo—predicated on the erasure of his character—are futuristic, while the witches' predictions for Macbeth are limited to and realized within the play's temporal limits. These differences provoke the two men to pursue opposing ways of interpreting prophecies, and their acts of interpretation direct attention to different structures of dramatic action and character. While Macbeth repeatedly turns to denotative language and reads the prophecies conclusively—and, subsequently, as prescriptions—Banquo's implicative reading, in which he marks ambiguities and gaps among seeming, being, and knowing, positions him as the character most clearly able to decipher dissembling language.[45]

Macbeth's initial question, "Speak, if you can. What are you?" (1.3.48), attempts to designate unsuccessfully what the witches *are*, an effort he variously repeats in this scene. He seeks literal answers to the question "what are you" and posits zero-sum conclusions. He oscillates between declaring the prognostications completely true or false, beyond "prospect of belief" (1.3.75). In his desire to establish the truthfulness of their words, he does not consider whether what they *are* might serve as evidence of their credibility. While early modern audiences might consider them either supernatural beings or even examples of "marvels" and "wonders,"[46] whose anomalous states of being were of increasing interest to contemporary naturalists, Macbeth accepts them primarily as repositories of esoteric knowledge. His growing curiosity manifests in his wish for more information after the witches have made their pronouncements about Banquo and him:

> Stay, you imperfect speakers, tell me more.
> By Finel's death I know I am Thane of Glamis,
> But how of Cawdor? The Thane of Cawdor lives,
> A prosperous gentleman, and to be king
> Stands not within the prospect of belief,
> No more than to be Cawdor. Say from whence
> You owe this strange intelligence, or why
> Upon this blasted heath you stop our way
> With such prophetic greeting.
>
> (1.3.71–79)

Macbeth tries to fathom the witches' ontological state and uncover the meaning of their riddles, but he ultimately ends up separating the two issues. He demands specific answers—"from whence," "tell me more," "why"—in order to literalize their riddling language. It is through these specifics that they can perfect their "imperfect" predictions. Their imperfection, suggests Macbeth, lies in their inability to reveal all, but in using the adjective, he also implicitly responds to his own question of what they "are"—the word thus serves as a warning. Yet, he subverts the inkling that he should know more about these peculiar beings before he accepts their word and instead piles up interrogatives and imperatives to fully divine their "prophetic greeting," which he understands as knowledge that may be extricable from the speaker. Macbeth's separation of knowledge and being—his distinction of what the witches know from what they are—is also a harbinger of his inability to fathom the conditions of his own existence. Later in the same scene, he will recognize how the revelations produce imperfection in him and split his unified self: "Shakes so my single state of man / That function is smothered in surmise / And nothing is but what is not" (1.3.142–44).

The witches' revelatory greetings, in which they demonstrate their awareness of Macbeth's present and future titles, contrast with his own incomplete knowledge that "the Thane of Cawdor lives" and his yet-unshaken conviction that "to be king / Stands not within the prospect of belief." His unawareness of the multiple roles he presently occupies, which the play's audience knows of but he does not, will convert ignorance into the ontological doubt that "Shakes so [his] single state of man." His knowledge of his newly acquired title thus remains dissociated from the recent events of the battle that have already altered his roles. Macbeth's conflicting reactions also underscore that he perceives markers of identity as relatively stable in this moment. Although he knows that the titles of Thane of Cawdor and King are transferable, he reads both, and the former in particular, as attached to a specific individual. This perceived fixity initially leads him to dismiss the prophecies as unequivocally false, but the revelation nevertheless "Shakes" his sense of self. Banquo comments on Macbeth's affective and cognitive condition immediately after the witches tell him that he "shalt be king": "why do you start and seem to fear / Things that do sound so fair" (1.3.52–53). Marking the disjunction between tidings that "sound so fair" and their unexpected effect on Macbeth, "fear," Banquo underscores the troubling effect of these revelations on Macbeth. These predictions direct attention to past occurrences—of which these characters are still ignorant—as well as to his possible future roles. But in this moment in the play, their ambiguous truth-value has the capacity to propel Macbeth to wonder as well as doubt. Splitting his subjectivity and leading to self-construction and self-doubt, his interpretation of the uncertain prognostications—as prescriptions to shape his future—alters his present state of existence.

Macbeth's first question, "Speak, if you can. What are you?" interrupts Banquo, who, by contrast, was elaborating on the ambiguous ontology of the "weird sisters" as he assessed the veracity of their predictions:

> What are these,
> So withered and so wild in their attire,
> That look not like th'inhabitants o'th' earth,
> And yet are on't?—Live you? or are you aught
> That man may question? You seem to understand me
> By each at once her choppy finger laying
> Upon her skinny lips. You should be women,
> And yet your beards forbid me to interpret
> That you are so.
>
> (1.3.40–48)

Although Banquo's initial question ("What are these") is designative, he immediately recognizes the inadequacy of this query. The "yet"s and "or" capture the multiplying possibilities that arise from attempts to name them or to identify their meaning; these terms underscore the impossibility of fixed signification. Interlocutors cannot conclusively "interpret" what they are, because of the absence of any distinctive markers of classification: Banquo wonders whether they are terrestrial or living, men or women? The multiple, often contradictory signifiers of identity establish their ambiguous ontology. Their complex forms, like their riddled language, "forbid" us from "understand[ing]" them or concluding "what" they "are" and remain irreconcilable with the multiple things they "seem." At this juncture, in his drive for conclusions, Macbeth interrupts Banquo, leaving no room for a response to these questions.

Banquo associates the witches' capacity to produce veiled knowledge—or, in Puttenham's words, their ability to "dissemble again under covert and dark speeches"—with their states of being. In his refusal to separate epistemology from ontology, he stands in contrast to Macbeth. Banquo couples their unique cognitive capacity to prognosticate from multiple options—"If you can look into the seeds of time / And say which grain will grow and which will not, / Speak then to me" (1.3.59–61)—to speculations about the potential significations of their marvelous states of being: "Are ye fantastical or that indeed / Which outwardly ye show" (1.3.54–55). Banquo's conditional appeal to the witches that they "Speak" to him is predicated on the belief that their peculiar states of being signal their unique ability to interact differently with time, to look into its "seeds" and divulge one secret from various possibilities. Banquo notes how their revealed secrets to Macbeth of "present grace and great prediction / Of noble having and of royal hope" (1.3.56–57) mark potentialities that can divulge the future from myriad options ("which grain will grow and which will not"). Banquo's use of "or" and conditionals such as "If" signal his refusal, yet again, to draw conclusions from the riddling evidence before him. As fitting response to his recognition of their ontological and linguistic multiplicity, and to his refusal to draw conclusions from the riddling evidence produced by their esoteric pronouncements, the witches present Banquo not with clear information, as they do to Macbeth, but with enigmas.

Even after the prophecies are partially fulfilled, Banquo keeps reiterating that one should not dissociate what the witches know from what they are. He cautions Macbeth that since "oftentimes to win us to our harm, / The instruments of darkness tell us truths" (1.3.125–26), one must not accept the predictions as already fulfilled knowledge; they are not yet actualized events. He sees what Macbeth cannot, that even true prophecies might have "reprobate sense" hidden in "pretty" language. Banquo's words explain what remains implicit in Macbeth's speech: the witches are "imperfect," to transport Macbeth's word, because their being cannot serve as evidence of the truth; indeed, it belies their credibility. Their being, as well as their predictions, invites suspicion and skepticism rather than an impulse to act out the predicted futures. Banquo's words thus also function as a warning: although the "instruments of darkness" use words to incite exigent means, one must not succumb to their calls. He sees these prophecies as verbal promises that "Might yet" (1.3.123) come to fruition. Yet, unlike Macbeth, Banquo remains wary of venturing beyond the act of interpretation. As I explore later, *Macbeth* uses these contrasting responses to dramatize the conflicting possibilities of action and inaction available to characters.

Macbeth, on the other hand, accepts the affirmed status of the revelations as soon as he learns of their partial fulfillment. Immediately after his encounter with the witches, he receives reports from Angus and Ross that entail a shift of attitude toward the witches, leading him to apply an analogous logic and credit the rest of the witches' unaffirmed prognostications. Angus informs Macbeth that the Thane of Cawdor has been "overthrown" (1.3.118). In a moment that highlights the play's exploration of how knowledge is intimately tied to the temporality of drama, Ross's words then transform Macbeth's acquisition of titles from an established fact into an act of *becoming* the Thane of Cawdor when Ross draws on the King's power to bestow titles through speech: "for an earnest of a greater honor, / [Duncan] bade me, from him, call thee Thane of Cawdor"

(1.3.105–6).[47] In the temporal logic of the play's created world, Macbeth already was Cawdor before being "call[ed]" so by Ross. Indeed, the witches might only have been reporting something that was already established. However, from Macbeth's perspective, he becomes Thane only when he is so declared by the words of the King, by proxy, *in this scene*. Macbeth's illusion that he sees a prophecy being fulfilled in Ross's words depends on the play's pronounced separation of plot, "the order in which events are related," and story, "the actual order in which events occur."[48] Ross's speech act, however, produces an instance where plot and story converge, and this peculiar meeting of Macbeth's actual status and his knowledge of these conditions situates the dramatic character in a polychronous moment in which he experiences an inaccessible and unknown past—the Thane of Cawdor's displacement—actualizing itself in his own experience of being elevated to the position. The performative nature of bestowing a title retroactively, after it has already been transferred, escapes Macbeth's notice, and he embraces this specific moment as proof of a prophecy being fulfilled.

This polychronous moment "o'erlea[ps]" (1.4.49) Macbeth's epistemic conundrum by transporting a past event into the theatrical present. His resulting new knowledge leads Macbeth to adopt an analogical strategy in which he extrapolates from what he knows to what cannot yet be known.[49] His acceptance that the "prophetic greeting" has been partially fulfilled entails a change in attitude toward the witches and a shift in belief. Since "Two truths are told, / As happy prologues to the swelling act / Of th'imperial theme," therefore "This supernatural soliciting" has "given [him] earnest of success" (1.3.129–34). He can now see a pattern in which a prophetic utterance is being translated into fact. As he constructs an associative relationship between what has happened ("Two truths") and what should happen ("earnest of success"), he reimposes his own linear chronology upon the play's polychronous temporality—what are reports of past events in preceding scenes he perceives as the witches' predictions of a contingent future. It is important to note that Macbeth *requires* this reinscription of linear chronology so he can read his present titles of Glamis and Cawdor as precursors to his future rule, "The greatest is behind" (1.3.119). When he sees an unknown past—the displacement of the Thane of Cawdor—actualizing itself in the current moment in Ross's speech act, Macbeth also feels the need to extricate himself from the polychronous nature of this scene so that he can map the contours of a sure future; he embraces a linear logic of prophetic causation over the complex temporalities being invoked by his interlocutors. From this moment, Macbeth's agency is increasingly predicated on the truth-value of prophecy, which fuels his obsession with the question of *how* he may actualize the predictions. His ultimate solution, to act out the unknowable "future in the instant," produces the terrible violence of the play that seals its tragic status.

Theatrical Action and the Actualization of Possibility

After Macbeth and Banquo's encounter with the witches, the play explores how characters transform the unaffirmed occult speeches into theatrical events. As Macbeth accepts that the "prophetic greeting" has been partially fulfilled, he asks Banquo a question that will haunt him for the rest of the play: "Do you not hope your children shall be kings, / When those that gave the Thane of Cawdor to me / Promised no less to them?" (1.3.120–22). His analogical reasoning here betrays his perception of a relation between prophetic utterance and action: since they "gave" the "Thane of Cawdor" to him, they will also keep their promise to Banquo. The speculation on positions bestowed, moreover, leads Macbeth to consider whether he might *take* them by acting on the predictions. In Macbeth's unequivocal acceptance of the predicted future lies an undercurrent of the incertitude of ever knowing it, except as a performed act. We see this preoccupation in the time leading up to Duncan's murder, when his mind is filled with images of potential action, "suggestion" (1.3.136) of "horrid image" (1.3.137), and thoughts of "murder" (1.3.141). He oscillates between the position of not acting ("If chance will have me king, why, chance may crown me / Without my stir" [1.3.146–47]) and the necessity of doing so, especially when Duncan chooses his son Malcolm as his heir (Malcolm's title of "Prince of Cumberland" becomes an obstacle on which Macbeth must "fall down or else o'erleap" [1.4.48–49]). He does consider the ways in which he might become king, but his reasoning always resolves into a binary logic—to act or not to act—and with Lady Macbeth's encouragement, he increasingly rationalizes the necessity of "deed" (1.7.24). Embracing the logic that one can only guarantee the success of prophecy through direct intervention, the Macbeths treat the witches' words as recipes; the partial fulfillment of one prophecy makes them approach vatic discourse as "a prescription for an experiment, a 'trying out.'" The political future thus becomes a product of the compulsion to fix it with complete surety.

Although he continues to equivocate about "trying" to actualize the predictions, by the time Duncan and his party visit Inverness, Macbeth is no longer focused on "interpretation" but on the goal he must realize. Through Lady Macbeth's insistence on "deed," and her urging that Macbeth privilege his "own act and valor" (1.7.40) over mere "desire" (1.7.41), he comes to see acting as the only way to eliminate alternate future outcomes:

> If it were done when 'tis done, then 'twere well
> It were done quickly. If th'assassination
> Could trammel up the consequence and catch
> With his surcease success—that but this blow
> Might be the be-all and the end-all!
>
> (1.7.1–5)

In spite of the conditional "If" with which he begins this soliloquy, Macbeth primarily concerns himself with questions of action and performance: the "assassination" must be "done" as a "blow" to his desired end (the "end-all"). And instead of waiting for the prophecies to resolve themselves on their own in due time, he believes their immediate fulfillment, their being "done quickly," will provide an ideal resolution ("'twere well"). He also comes to believe that this form of acting on predictive knowledge is necessary to reintegrate his split self and ensure his rule. Rejecting his faith in "chance" and focusing solely on the promised ending, Macbeth seeks means, or a course of action, in a prophecy that does not prescribe one.

It is Lady Macbeth who underscores the importance of exigent means. Her insistence on the absolute necessity of swift action is essential to the conversion of one potential outcome into actuality. We see her role as a catalyst for action early on, when she first enters the stage with Macbeth's letter. After reading about his meeting with the witches, Lady Macbeth echoes her husband's analogical logic—"Glamis thou art, and Cawdor, and shalt be / What thou art promised" (1.5.13–14). But she is more emphatic about the desired result, as she replaces what he perceives as mere possibility into future-certainty: you "shalt be" what you are "promised." This is a logical extension, she suggests, of his present state: what "thou art." She claims that his words have activated her own ability not only to envision the future but to experience it coming into being: "Thy letters have transported me beyond / This ignorant present, and I feel now / The future in the instant" (1.5.54–56). The knowledge she gains has "transported" her from the realm of the potential to the certain, enabling her to locate herself beyond the "ignorant present" in the "future." Of course, what Lady Macbeth presents as a certain "instant" is ultimately a projection that ignores the contingency of the future—in order to close the gap, she will go on to argue, one needs to move from the sphere of language to that of action. Lady Macbeth's "beyond," then, serves as a forward-looking corollary to Macbeth's imagined future that lies "behind" him.

As she imagines herself "transported" from her current act of reading to a time when Macbeth is king, Lady Macbeth's extension of her husband's analogical reasoning exposes what makes their desires so transgressive: only by acting in this "ignorant present" can the "future" be changed into an experienceable moment. When she stresses that some action must be undertaken in order for the prophecies to be realized, she implicitly argues that approaching a prediction as a prescription is always antithetical to maintaining multiple possibilities. The gaps between prescription and immediate action must be dissolved, and this collapse is vital to bringing the political future into the immediacy of the present. Yet, Lady Macbeth also worries that Macbeth's "nature" (1.5.14) lacks that which will make him act. It is "too full o'th' milk of human kindness / To catch the nearest way" (1.5.15–16). Of course, Macbeth has also identified the "nearest way" to the throne: the "murder" of the King. While this realization leads him to envision self-division, Lady Macbeth identifies the "nearest" way as the *only* way to remake the future and convert what he might be into what he "shalt" be.

To find the most expedient means to the throne, she thus revises the signification of prophecy in an absolute way: instead of asking what a prophecy means, as Banquo does, one must strategize *how* it can be actualized. Seeking the "nearest way" becomes the sole means to resolve the contradictions she perceives in Macbeth's state: "Thou wouldst be great, / Art not without ambition, but without / The illness should attend it" (1.5.16–18). She, like him, recognizes the opposition in his nature that "Shakes" his "single state of man" and prevents him from exercising the necessary means. The double negative in her statement, however, introduces an element of flexibility in the state of things that will enable Macbeth to overcome his initial doubt. Containing the wandering thoughts that simultaneously direct him to and prevent him from regicide, she will guide him toward sure results.

As Lady Macbeth meditates on the contrasts in their propensity to act, she suggests that her active role and "spirit" will catalyze Macbeth from passivity into action:

> What thou wouldst highly,
> That wouldst thou holily; wouldst not play false,
> And yet wouldst wrongly win. Thou'dst have, great Glamis,
> That which cries, "Thus thou must do" if thou have it,
> And that which rather thou dost fear to do

> Than wishest should be undone. Hie thee hither,
> That I may pour my spirits in thine ear
> And chastise with the valor of my tongue
> All that impedes thee from the golden round,
> Which fate and metaphysical aid doth seem
> To have thee crowned withal.
>
> (1.5.18–28)

Lady Macbeth imagines actions that will motivate Macbeth to renounce his circumspection and move from what he "wouldst" do to what he "must." By shifting the emphasis from the realm of the possible to that of necessity, she performs a logical shift and simplifies the prophecies into a series of steps that can be performed without seeking to understand their causes, or their sources. While the witches offer no guidance on their enigmas, Lady Macbeth reinterprets their words as a kind of recipe. Her language of doing and undoing leads her to conclude that she must remove all impediments toward which "fate and metaphysical aid" both "seem" to lead him; seeming becomes a *reason* for acting, and not a deterrent as it was for Banquo. She applies the active verbs—"pour," "chastise"—to define her actions, but they also model for Macbeth the practical ethos that will propel him toward the "golden round" and actualize the one outcome he so desires. Macbeth is the conduit that will act on these directives. Drawing on the language Eamon uses to describe early modern technical recipes, we might say that she prescribes deeds as she "[boils] down to a rule" what Macbeth must do and triggers the actions necessary to ensure that the predictive becomes the prescriptive—and then the actual. In this way, Lady Macbeth's translation of prophecies into recipes invites Macbeth to do the same, leading him to murder, to the throne, and then to further speculations.

Showing History, Making Tragedy

After Macbeth follows the "nearest way" and becomes king, he returns to the predicament of the original prophecies and his own place in the future that will become history. Having accepted and ensured that the prophecies are partially true, he cannot discount the witches' revelations to Banquo. As a result, he begins to grapple with his temporary role in history. Macbeth contrasts his limited reign with the "happier" Banquo's projected futurity and meditates on the latter's transgressive role:

under him,
My genius is rebuked, as it is said,
Mark Antony's was by Caesar. He chid the sisters
When first they put the name of king upon me,
And bade them speak to him; then, prophet-like,
They hailed him father to a line of kings.
Upon my head they placed a fruitless crown
And put a barren scepter in my grip,
Thence to be wrenched with an unlineal hand,
No son of mine succeeding. If't be so,
For Banquo's issue have I filed my mind,
For them the gracious Duncan have I murdered,
Put rancors in the vessel of my peace
Only for them.

(3.1.55–68)

Macbeth identifies Banquo as an oppositional figure ("under him / My genius is rebuked") and realizes that his deeds, including regicide, serve Banquo's descendants. Unlike Macbeth's "fruitless crown" and "barren scepter" that mark the temporal limit of his reign, Banquo's "issue" continues. Banquo himself emerges as a "prophet-like" actualizing force. Because he "chid" the sisters and "bade them speak," they produced the prophecies about his future, concludes Macbeth. By this logic, Banquo generates, indeed creates, his own prophecies. Macbeth's knowledge thus "filed [his] mind" and led him to "[murder]" only to secure another's political legacy. While the play initially separates Macbeth and Banquo through their opposing modes of interpretation, Macbeth's rumination demonstrates that these interpretive differences have formal effects because they orient the characters differently to time. Macbeth's conclusive stance leads him to act to secure a crown that permanently delimits him in the present and shuts out his descendants from kingship, while Banquo's interpretations facilitate his emergence as a figure of history.

This distinction between Macbeth as a figure with an end and the projected endlessness of Banquo's line is formalized in the spectacular revelations produced in Macbeth's second meeting with the witches. Speculating about his own place in future histories, Macbeth returns to the original sources of esoteric knowledge. At the end of the ritualistic scene in which the witches try out technical recipes, they reiterate their status as dissimulators of the occult by refusing to affirm what they know. When Macbeth fails to extract answers about their current actions—his demand to know "What is't you do?" remains unanswered—he returns to the earlier unaffirmed predictions and appeals to the process and source of their knowledge: "I conjure you by that which you profess, / Howe'er you come to know it, answer me" (4.1.49–50). Macbeth's declaration of "conjur[ing]" offers another instantiation of how he aims to collapse utterance and action, as his speech act sparks the fulfillment of his desires: his words influence the witches to bring forth answers he desires to know.[50]

This command results in a series of visual and aural cues from three apparitions: "*an armed head*" (4.1.67, sd), "*a bloody child*" (4.1.75, sd), and "*a child crowned, with a tree in his hand*" (4.1.84, sd). The first warns directly: "Beware Macduff, / Beware the Thane of Fife" (4.1.70–71). The others speak in riddles. The second prescribes limited inaction but also advises him to occupy various habits of thought and action—"Be bloody, bold, and resolute; laugh to scorn / The power of man, for none of woman born / Shall harm Macbeth" (4.1.78–80). The third warns him to refrain for a certain time from active intervention:

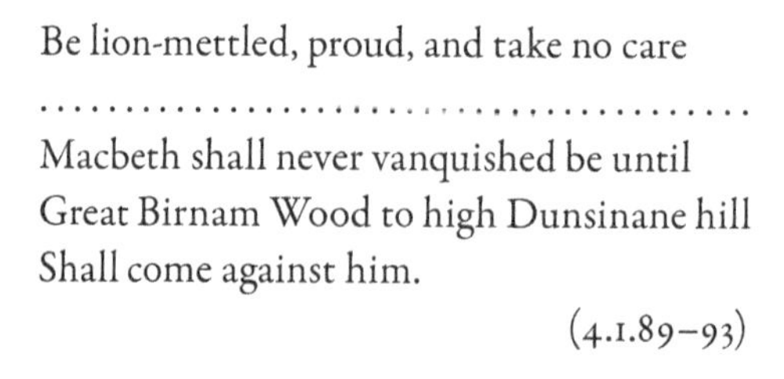

> Be lion-mettled, proud, and take no care
> . . . .
> Macbeth shall never vanquished be until
> Great Birnam Wood to high Dunsinane hill
> Shall come against him.
>
> (4.1.89–93)

In response to Macbeth, the apparitions *prescribe* attitudes, virtues, and mental states of being. Macbeth heeds the first warning but responds ambiguously to the second: "live, Macduff. What need I fear of thee? / But yet I'll make assurance double sure / And take a bond of fate. Thou shalt not live" (4.1.81–83). His diminishing worry coincides with his failure yet again to mark the enigmatic nature of the prophecies. Focusing on "paraphrases,"[51] he ultimately will discount both equivocal and direct warnings and reject the final prediction as an impossibility: "That will never be" (4.1.93). Before, he mistook predictions for prescription. At the end of this scene, he misinterprets these multimodal enigmas as unequivocal truths and immediately proposes action: "The very firstlings of my heart shall be / The firstlings of my hand" (4.1.146–47). Earlier he had equivocated on the relation of prophetic utterance to action, unsure of how, and even if, he should try to realize his ambitions. Now the "hand" works simultaneously with the "heart," erasing gaps between cognition and action. Macbeth will immediately "crown [his] thoughts with acts" (4.1.148) and leave no space for alternate outcomes.

Macbeth takes his conclusive stand after the witches provide him their most transparent prophecy in this scene. As he seeks resolution to the one question that cannot be completely answered within the play's compressed temporality ("shall Banquo's issue ever / Reign in this kingdom?" [4.1.101–2]), they "Show" (4.1.106, 107, 108) him the approaching time. In a scene that Peter Stallybrass identifies as *Macbeth*'s "emblematic centre,"[52] the play enacts the "future in the instant."

> *A show of eight kings, and* BANQUO *last;* [*the eighth king*] *with a glass in his hand.*
>
> MACBETH Thou art too like the spirit of Banquo. Down!
> Thy crown does sear mine eyeballs! And thy heir,
> Thou other gold-bound brow, is like the first.
> A third is like the former.—Filthy hags,
> Why do you show me this?—A fourth! Start, eyes!
> What, will the line stretch out to th' crack of doom?
> Another yet? A seventh! I'll see no more.
> And yet the eighth appears, who bears a glass
> Which shows me many more; and some I see
> That two-fold balls and treble scepters carry.
> Horrible sight!
>
> (4.1.111–21)

Instead of reinserting Macbeth into history as he desires, the "show" shuts him out of an anticipated future by making present the outcome of the ambiguous verbal prophecies. The glass functions as a perspective device that takes Macbeth into a theatrical future in which he witnesses the culmination of Banquo's line in James, alerting him of a time from which his lineage has been erased.[53] Macbeth's earlier imagination that the future is "behind" had produced a visual understanding of time: unknown futures cannot be deciphered because they remain obstructed from our vision. In a peculiar enactment of Lady Macbeth's claim that she can perceive what is "beyond" in the "instant," this scene makes visible such a future by transforming what was "behind" into an immediately observable theatrical scene.

The spectacle, moreover, completes a process of abstraction that fixes Macbeth's roles within the dramatic framework. As a ruler who is oriented toward the present and who focuses on the immediacy of his own rule, he stands as a contrast not only to Duncan but also to the English king, Edward the Confessor, who himself is a source of futuristic knowledge—he "hath a heavenly gift of prophecy" (4.3.157).[54] Macbeth's encounters with the complex temporality of prophecy spur his obsession with his own future but also transform him into an emblem as well as evidence of anomalous misrule. This procession of kings, in particular, visually culminates the verbal process of emblematic erasure that begins when characters such as Macduff and Malcolm refuse to acknowledge Macbeth's particular identity; instead of using his name, they increasingly employ general terms including "tyrant" (4.3.178) and "usurper" (5.7.85) to define him.[55] As the play progresses, his position shifts from the particular "brave Macbeth—well he deserves that name" (1.2.16), to the titular figure of "Thane" and "King," to the general category of "tyrant." Malcolm's refusal to name Macbeth—"This tyrant, whose sole name blisters our tongues" (4.3.12)—suggests how the act of naming indexes an individual. Macbeth's individuality, encapsulated in his "name," disappears with its gradual erasure. The spectacle does not merely "split" Macbeth's "single state." It underscores his emblematic political status as *type*—that of a tyrannical ruler—and prefigures his disappearance from history.

This progression dramatizes the effects of how a character's interactions with vatic discourse constrain them within, or propel them beyond, the theatrical present—an "instant" that encapsulates how the actualization of prophecy can paradoxically define the bounds of the theatrical world while also modeling ways for fictive beings to escape such limits. Macbeth's exigent actions earn him a place on the stage and in the theatrical now, while Banquo's character disappears from the staged present only to emerge as a figure inaugurating a future beyond the temporality of dramatic action. In contrast to Banquo's potential endless futurity, Macbeth's rule is transformed into a singularity where he acts on his conclusive interpretations—his execution of prophetic utterances produces the events in which "thoughts" become "acts"—and facilitates the negation of his own lineage. Through Macbeth's staged immediacy, then, the play offers one model of actualizing prophecy: the theatrical action that makes present different temporalities as well as fabricated states of being. But *Macbeth* also suggests that the act of interpretation might itself actualize prophecies and take characters (and audiences) beyond this staged action. As Macbeth's place in a continuing future gives way to the procession seen through the glass, the audience implicitly gains entry into this polychronous moment when the theatrical future intersects with the historical present. The visual clarity of this emblematic scene, ending in the representation of James, overcompensates for the enigmatic language and urges audiences to recognize their own moment as an actualized future projected from the play. In this instance, the theatrical future morphs into the extra-theatrical present and serves to perpetuate sovereign lineage. By pitting the titular character against Banquo, and then by staging the projection of Banquo's line into the future, *Macbeth* solidifies the distinctions that had begun to emerge when the two characters reacted differently to prognostications about, and about their places in, an uncertain future. The play's promise of continuity transports audiences, with Banquo, to their futures beyond the spectacle of an "instant."

Political Futurity and Tragedy's Prophetic Form

The "Show" of the line of kings is an instrument of sovereign ideology—a vivid manifestation of how political authority legitimizes and sustains itself by envisioning its own continuity.[56] But this "instant" is also a product of the theatrical apparatus that can bring forth an envisioned future on stage, exemplifying how the formal constraints and possibilities of dramatic genres mediate such negotiations of power. Theatrical time, in its constant evolution of events, counters the sense of ending to which the historical past has already led; as Brian Walsh argues, it is "the 'eventness' of theater" that "underlines the unique temporality of drama."[57] The temporally delimited events in a play such as *Macbeth* are not offering a chronological unfolding of history but are simultaneously looking ahead and behind to multiple times to bring forth the spectacle of kings.[58] While the contradictions of polychronic time are most prominent in Shakespeare's history plays, *Macbeth*'s visual prophecy takes us to the logical end point of such performances of history. By drawing on the words from the epigraph to this chapter, we might say that as Shakespeare adapts events from Holinshed's *Chronicles* only "tied," in Sidney's words, "to the laws of poesy," the tussle between contingent theatrical action and prophetic certainty "frame[s] the history to the most tragical conveniency" (*Defence*, 66). From the moment Macbeth acts on the vatic utterances, he accelerates the collapse of multiple possibilities into a singular future. In narrowing the options of how the future can unfold, he also directly contributes to the transformation of his life into fiction. History, in other words, *becomes* a tragedy for Macbeth, and prophecy's epistemological conundrums become recipes for the construction of the tragic form.

Macbeth's inability to escape his tragic end and Banquo's projected endlessness are thus formal effects of the intellectual stance each takes. The titular character's conclusive interpretations concern only him and lead him to act in the confines of theatrical time. Banquo's implicative interrogations extract futuristic revelations, so that his death does not signify the end of the prophecies. Instead, they generate new projections that, in their gestures to a future arising from the plot, lead to the historical present and subsequently to future histories. The procession of kings provides evidence of Macbeth's worst nightmare, crystallizing a visual response to his queries and compressing the earlier prophecies into a singular spectacle. Operating within the confines of dramatic time, Macbeth becomes a spectator to his own eventual erasure; he witnesses in the spectacle, we might say, history erupting into tragedy. At the same time, the play suggests that Banquo as an interpreting character must disappear from the present before his reappearance can symbolize the start of a new line of kings. In his study of history plays, Walsh argues that history "is defined by its inalienable absence" and that early modern plays are operating within dialectics of presence (of characters, living bodies on stage) and absence (of actual historical figures).[59] Adapting this framework, we might say that to be formally delineated a figure who persists in history, Banquo's erasure from the dramatic present is necessary. This spectacle is thereby the culmination of the witches' initial opposition between Banquo and his descendants: "Thou shalt get kings, though thou be none." Just as Macbeth's interpretations and character give way to Banquo's questions and the procession of his descendants, the titular character's position as spectator gives way to an external audience that can glimpse the future beyond the one permitted by the dramatic temporality.

The visual prophecy does offer a prescription, but not the one Macbeth grasps. Instead, it demands that audiences perform a simultaneously reassuring and unnerving prospective act. The spectacle provides them both "ingredients" and "instructions," in the shape of the figures and in the form of the projected line. Instead of ending on conclusive evidence, *Macbeth* incites spectators to project from what they witness on stage—Macbeth's head on a stand, the spectacle of Banquo's lineage, the image of James—to the implicit promise of their own futures extending beyond the boundaries of the dramatic form. They must adopt the play's implicative interpretive strategies, which will lead them to the certainty of the historical present and subsequently toward the imagination of *their* possible futures. Such a projection will enable them to extend this vision toward the extra-theatrical future that might come into being after James's reign itself expires. *Macbeth*, then, prophesies to its audiences that their imaginative acts of projection, when "tried" out, will exceed the exigencies of the fictional present and go even beyond their singular, historical moment of interpretation; it teaches them *how* one imagines a future "in the instant."

This scene thus raises the prospect that the story of Banquo's line need not stop when the plot of *Macbeth* reaches its ending. The spectacle of an elaborately staged fictional future invites the external spectators to extend this "instant," as the performers in the line of kings concretize the difference between the historical past of Macbeth's rule and the present time of James's reign. Indeed, *Macbeth*'s repeated emphasis on the future provokes audiences to anticipate, and to contemplate, the continuation of James's line. As this imagination exceeds the temporal limits of the play, it converts the audience's actual existence into a fictional one, in the sense that it makes them occupy the same position as Macbeth does and experience a future unfolding in front of their eyes: their existence, current and future, becomes dependent on their interpretations and on their acceptance of the play's construction of political lineage. This spectacle propels audiences into possible past/futures that need not end with the play, and the continuation of the Stuart line can realize the historical possibility implied in the plot.[60] I am not arguing that a Jacobean audience *would* imagine such a continuation, but the play's interpretive tactics, its juxtapositions of reality and fiction, and its scrambling of past and future strategically set the stage for such a projection in a manner that Spenser could not achieve by blurring the lines between history and poetry. When in 1590 Spenser's Merlin suddenly halts his prophecy, the figure punctuates the fraught question of succession during Elizabeth's reign to mark the absolute impossibility of knowing a sovereign lineage beyond the poet's current moment. But the political prophecy that Banquo "shalt get kings" need not stop with James, because of the prospect that his children will extend this dramatized projection. The play capitalizes on an opening created by its current historical moment and exploits James's own propensity to equate the king with the father as well as his identification of Banquo as an ancestor. Reimagining a historical past recorded in the *Chronicles*—Macbeth's present—as the originary moment of the audience's current state, *Macbeth* invites audiences to understand fictionalized narratives as part of their own history, where the moment of their existence emerges from events unfolding within the constraints of the theatrical form.

The play's provocation to audiences does offer an odd echo of *The Faerie Queene*'s method of inviting readers into fictionality that I explored in Chapter 1. But by implicating its audiences in an act of projection that is predicated on the erasure of its titular character, *Macbeth* forces us to ask whether the conjuration of possible futures that constitute Spenser's romance worldbuilding—a future in which Faerie Land "might" be found, or where Arthur's sword is locatable "if sought"—also encodes tragic potentialities in the vexed relations between interpretation and action. It is crucial to remember this twinned logic of creation and destruction that undergirds imaginative invitations to readers and audiences, especially when we turn in Chapter 3 to Bacon's ambitious great instauration of learning, an endeavor that treats projection from written text to actual world as a disinterested intellectual act. Bacon's natural philosophical works urge readers to participate and complete the incomplete work of natural history and philosophy that the author can never accomplish alone. Instead of treating projection as a value-neutral component of knowledge production, as Bacon will do, or even foregrounding the ethical impetus of incorporating readers in poetic production, as Spenser does, *Macbeth* flaunts the political aspirations that underlie the injunction to readers and spectators to embrace the task of projection: the play's audience can only imagine its future through an extension of the line of kings, and specifically the lineage of the current monarch.

As *Macbeth*'s insistence on the future invites audiences to continue beyond this staged moment, it also brings into focus a series of questions about knowledge and existence that more broadly drive the play's explorations of futurity, certitude, and action. The scene of Macbeth witnessing the visual prophecy serves as a stark reminder that one cannot *know* the future with certainty, since it is contingent and dependent on interpretations, decisions, and actions in the present. At best, one can access the future *as* an epistemic concept; it cannot come into existence independent of the modes of knowledge that allow one to tend toward it. Since it is impossible to fully grasp the future, and appeals to a historical past remain inadequate to produce an experiential understanding of it, one must necessarily draw on predictive epistemologies and alternate ways of conceiving temporality. Such issues are not unique to drama, of course. In the next chapter, we will see how the hope of future certainty drives Bacon's program of natural philosophical reform; scientific certitude, at a fundamental level, is a futuristic desire, rather than something achieved in the present. *Macbeth* reminds us that any question of futurity—and especially political futurity—cannot be separated from the modes one uses to know it. The unknowability of the future can itself serve as an authorizing principle, since characters (and, by extension, Baconian natural philosophers) can assert access to knowledge about the approaching time, and even promise to control and shape it. Because it promises precisely such knowledge, prophecy becomes the perfect vehicle to resolve and counter the threats of contingency and unknowability.

One facet of possible knowledge, then, operates in the temporal and conceptual gaps between an uncertain prediction and its prophetic resolution, gaps that actively incorporate readers and audiences into the project of knowledge-making. If the mode of prophecy is a vital mechanism of this dimension of possible knowledge—one that furthers political genealogies within the constraints of fictional forms—what role is left for, and what power imbued by, the prophet divorced from considerations of genealogy and political futurity? Bacon might argue that the figure comes to serve "merely as a guide to point out the road" (*New Organon*, Preface, 35), one who leads practitioners toward a reformed natural philosophy. Bacon professes that he is one such "guide," his humility obscuring a kind of hubris: the guide is a prophetic figure who will direct other naturalists to the end he foresees, toward "certain and demonstrable knowledge" (Preface, 36). This coexistence of pride and frailty is symptomatic of the struggles between the ideals and realities of Bacon's scientific corpus. Baconian method continues to function as a shorthand for one of the most ambitious forms of *praxis* in the seventeenth century; as a rich body of scholarship also demonstrates, Baconian science co-opts, legitimizes, and reconfigures vernacular, occult, and practical knowledges into the project of the "great instauration." In the next chapter, I examine how the method of induction that Bacon outlines in the *New Organon* is also predicated on elevating—and claiming—for natural philosophy a set of practices that were constitutive elements of early modern romance. The epistemological conundrums of prophecy, as well as the mobilization of *secreta*, that we encountered in this chapter serve as potent reminders that there lurks within every act of knowledge production specific investments: for Bacon, one of those investments is to absorb into his "New Science" methods of early modern poesy.

CHAPTER 3

Francis Bacon's "Endlesse Worke"

> From that day forth I cast in carefull mynd,
> To seeke her out with labor, and long tyne,
> And neuer vowd to rest, till her I fynd.
> —Edmund Spenser, *The Faerie Queene* (1590)

The Faerie Queene and *Macbeth* offer two contrasting examples of the relation of action to certitude. While Britomart's pursuit of Artegall, the shadowy figure of her vision, extends her genealogy and dilates the romance narrative, Macbeth's interventions to realize the witches' utterances abort his lineage, converting his life into a tragedy. By testing the limits of attempts to fulfill prophecy through the teleologies of different literary forms, both Spenser and Shakespeare emphasize that certainty remains contingent on readerly action. This chapter turns to another thinker whose promise to "establish progressive stages of certainty" (*New Organon*, Preface, 33) galvanized a generation of active practitioners and transformed him into a prophet-like figure: "Bacon," writes Abraham Cowley, "like *Moses*, led us forth" to "the very Border" of "the blest promis'd Land, / And from the Mountains Top of his Exalted Wit, / Saw it himself, and shew'd us it."[1] For Francis Bacon, the vehicle of certitude is the method he terms "true induction" (1.40). This method "for the invention, of knowledge" (Preface, 36), outlined in the *New Organon*—published as *Novum Organum*, the second of the six-part and incomplete *Instauratio Magna* (*Great Instauration*)—engages with "particulars" with the aim of "searching into and discovering truth" (1.19). Bacon invites those who "seek, not pretty and probable conjectures, but certain and demonstrable knowledge" (Preface, 36) to abandon the induction "now in fashion" and embrace his "yet untried" (1.19) method. This method will not only correct extant ways of learning but also contribute to repairing humanity's loss of knowledge and "dominion over creation" that resulted from the Fall (2.52).

Bacon's method was not "yet untried," scholars repeatedly indicate, as they situate his scientific works within myriad philosophical traditions or juxtapose them with other disciplines, from dialectic and law to the mechanical arts and the maker's knowledge tradition.[2] This chapter proposes that we also know the method's workings under different names: narrative, poetry, and—as I elaborate below—romance. This argument pursues the varied forms of literary thinking that are constitutive of the method of "true induction" and that undermine its stated goal of "certain and demonstrable knowledge." I examine several features of "true induction" and its related strategies of collecting natural history to demonstrate that Bacon's project is expectant and expansive, not enclosed and contained: an "endlesse worke"—to return to a phrase from *The Faerie Queene*—that defers fulfillment and in which uncertainty, digression, dilation, endlessness, and error are integral to the method. Bacon's "true induction" relies on techniques we more often associate with one of the most capacious modes of early modern *poiesis*. And herein lies its paradox; induction revels in elements of a "romancical" poetics that it is designed to eradicate.[3]

This chapter thus elaborates on my brief discussion in the Introduction of the conflicted status of poesy and its relation to "nature" in Bacon's works. Bacon's natural philosophical writing oddly echoes concerns about the roles of poesy and the poet that we encountered in Sidney's *Defence*. For instance, Bacon's natural philosopher, who pursues a "middle course" between "the experimental and the rational" (1.95), seems to have a similar orientation to his work as the poet, who, in Sidney's words, is a "moderator" between the philosopher (who dwells in abstraction) and the historian (who studies particulars). Moreover, to describe the natural philosopher's moderating role between theory and practice, Bacon embraces a standard trope of poetic production: the metaphor of "the bee," which "gathers its material from the flowers of the garden and of the field, but transforms and digests it by a power of its own" (1.95).[4] The labor of the bee is similar to "the true business of philosophy" because its work differs both from that of the "men of experiment," who, like the ant, "only collect and use," and the "reasoners," who "resemble spiders" as they "make cobwebs out of their own substance" (1.95). Bacon adapts the standard image for synthetic poetic making into an emblem for the method of "true induction." Although the poet molds a general idea into a specific instance and the natural philosopher produces general notions from particular examples, their comparable investments of mediating *gnosis* and *praxis*, and their corresponding roles as moderators, raise the prospect that poetic endeavors might have methodological commitments comparable to "true induction." Such alliances also raise fundamental questions about Baconian method: If the natural philosopher, like the poet, operates by coupling the general and the particular, can they really produce "certainty"? Or do they, like Sidney's poet, deal with knowledge that "nothing affirms"?

To underscore the connections of Baconian philosophy to literary writing, scholars typically focus on particular linguistic or representational elements, discussing how figurative language, rhetorical tropes, and allegorical images pervade his writing.[5] In this chapter, I foreground the larger issue of which these elements are specific manifestations: the centrality of imagination, both as concept and procedural instrument, in Bacon's writing. Induction is paradoxically composed of, even shaped by, the very imaginative techniques that Bacon purports to eliminate from natural inquiry. By focusing on how imaginative processes constitute the method, we are also able to better understand the tensions underlying key discussions in Bacon scholarship, such as the debates on the rhetoric of violence against "Nature" that recurs in Bacon's work—including his claim that one must "dissect [nature] into parts" (1.51), or the urge to "find a way at length into [nature's] inner chambers" (Preface, 36)—that have dominated feminist critiques of Baconian natural philosophy. I align my work with this historiography of early modern science that critiques the misogynist implications of Bacon's call to master a feminized Nature.[6] To be sure—as critics of this feminist historiography have argued—Bacon's own words often invite readers to ignore his fantasy of absolute mastery: he identifies "Man" as "the servant and interpreter of Nature" (1.1) and even suggests that directives for action emerge from "Nature, [which,] to be commanded must be obeyed" (1.3).[7] Yet, the presence of such statements does not divorce the method from the *New Organon*'s repeated charge to exploit the natural world; to study the discourse of obedience and submission at the expense of the violent rhetoric is to foreground only one aspect—and perhaps the ideals—of the method. I propose that the oscillations in Bacon's language between obedience and mastery—the language at the heart of these scholarly debates—are particularly vivid examples of the frictions between the actions and ends of "true induction." As this chapter shows, the imaginative techniques that Bacon employs forestall his ultimate ambitions of "dominion," not only enhancing the rift between processes and goals but also destabilizing the method itself.

The gap between the purported ends of the Baconian project ("certainty") and its actual procedures (which, to borrow Patricia Parker's phrase for Spenserian romance, are "all middle")[8] is a product of Bacon's desire to offer an exhaustive study of "Nature." Bacon makes the maximalist claim that "whatever deserves to exist deserves also to be known, for knowledge is the image of existence; and things mean and splendid exist alike" (1.120). But he weds the method's efficacy to an unpredictable object of inquiry: "Nature," in Bacon's understanding, is a pluralistic and vibrant entity that shapes its own intelligibility.[9] This vision of the natural world makes Bacon's natural philosophy (despite his desire for certitude) distinct from the category of *scientia*, which was concerned with necessary and immutable things and provided demonstrative knowledge and metaphysical explanations. Bacon's comprehensive "history and experiments" deal not only with objects already deemed worthy of study but also with "things which are trivial and commonly known; many which are mean and low; many, lastly, which are too subtle and merely speculative, and that seem to be of no use" (1.119). Yet, even as the naturalist undertakes this totalizing task, "Nature's" particulars function as actors and mediators, to use Bruno Latour's terminology, that have the capacity to divert knowledge producers from their ambitions of absolute order, control, and knowledge. They also derail the method's ideal, "gradual and unbroken" (1.19) progression toward certainty and the naturalist's ambitions for order and control. Baconian method—in its engagement with a pluralist natural world—unwittingly generates its own unpredictable interminability. Each encounter with "Nature" produces new particulars, operations, and what Bacon terms "forms."[10] The surety promised by the method must be adapted each time the naturalist encounters the material world. Induction purportedly outlines a way to realize human triumph over the physical world, but it ultimately enacts a surrender to the vibrancy of the natural realm. While the distance between the "might best be" and "should be" animated Spenser's poetic imaginary, and the gaps between vatic utterance and their prophetic ending shaped the tragic world in *Macbeth*, in Bacon's writing it is the chasm between the method and its goals that animates possible knowledge: the endless processes that are supposed to guide the quest for certainty constantly make Bacon's readers encounter the poetic impulses shaping the method of "true induction."

Baconian *Poiesis*, or, the Method of Romance

The dilatory tendencies of Baconian method gesture to the formal resemblances between "true induction" and the early modern genre of romance, which deviates from established genres of epic, tragedy, and comedy and remains distinct from its successor narrative, the realist, character-centric novel.[11] Because of its propensity to narrative proliferation, romance has often been understood as a source of readerly pleasure and personal affect, interpreted as a failed novel, and described as lacking neat structures. It is classified as genre, mode, or even as a set of "memes."[12] Perhaps because the form is characterized by variety rather than unity—it is episodic in nature, lacks a single hero and unified action, and contains multiple plots and digressions as well as unending quests—we have increasingly come to discuss what a romance does rather than what it is. Readers, claims Barbara Fuchs, "know it when they see it."[13] Parker's influential argument that romance is "a form which simultaneously quests for and postpones a particular end" gestures to the ways in which the romance narrative self-reflexively theorizes its own practices: variety, wandering, error, and endlessness. Revisiting key studies of romance in twentieth-century literary criticism, Fuchs describes romance as a "strategy" that represents "a concatenation of both narratological elements and literary topoi."[14] Fuchs's study makes explicit the practical epistemology undergirding the form: romance represents distinct operational procedures, enacting a plan that will shape the directions of the narrative.[15]

English romances such as *The Faerie Queene* were deeply influenced by Continental models. The sixteenth-century epic-romance debates in Italy capture how writers, grappling with the tensions between convention and innovation inherent in the genre, tended to define romance through its operational parts and processes. Centering on Ariosto's *Orlando Furioso* (1516) and Tasso's *Gerusalemme Liberata* (1581), these debates highlight how conventions of the mode challenged traditional premises of poetic writing.[16] For instance, romance was initially defined as a genre in opposition to epic: while epic inevitably followed rules and precepts, romance relied on multiple models and exemplars. In addition to their distinct political, ethical, and intellectual aims, the tension between epic unity and romance multiplicity played out at the level of form. Critics lamented romance's deviation from Aristotle's definitions of poetry in particular and its refusal to adhere to rules in general. In 1554, Battista Pigna raised an objection that would continue to haunt the genre's claims to seriousness: "*Digressions* are too numerous."[17] While "Aristotle had demanded a single action of one man, Ariosto has treated various actions of a multitude of knights."[18] In these ways, romances undermined the main features of epic: its focus on the single action of a single hero, as well as its unidirectional and comparatively rigid form, enabled the poet to observe the decorum of the noble style.

Criticisms such as Pigna's highlight how romance undercut the fundamental premise of epic, in the process formalizing the necessity of deviating from a unified plot. Defenders valorized romance by relying on the same characteristics that troubled its critics. They argued that the criteria of one form could not be applied to the other, celebrating the emergence of romance as a new national genre—"[*Orlando Furioso*] as a narrative poem, is the Italian equivalent of the ancient epic," stated Giraldi Cintio (1554).[19] These writers thus reframed the hierarchy of genres to make broader statements about poetic value and scope. Gioseppe Malatesta (1589) defended Ariosto, and by extension romance, through his theory of "new poetry": that "forms of poetry . . . change with the times, that authoritarian principles tending to render them immutable and eternal are not acceptable, that rules should follow upon forms rather than forms upon rules."[20] Ariosto was successful because he "disdained the precedent of the great epic poets" and "espoused a genre better suited to his own language and his own times."[21] Defenders also celebrated romance's ability to generate pleasure—which critics considered one of its primary problems—by rejecting hierarchies of genre. Fransesco Caburacci in 1580 suggested that, since Ariosto was not writing an epic, "it is clear that he wished to produce a different effect from that of the epic" and that "he was aiming at variety in the feelings of the audience."[22] Since poetic writing should, as per the Horatian dictum, both teach and delight, the genre with the most variety and appeal to the senses could provide the greatest pleasure. Romance's novelty enabled it to overcome perceived limitations in epic. It was suited to its "own times," dispensing with ancient rubrics and using its formal features to generate its own rules. By the end of the sixteenth century, the divisions between the two genres were more fixed, and formal elements of Ariosto's romance—multiple plots involving multiple characters, digressions, nonteleological trajectories, and incompletable quests—were typical of many English romances such as *The Faerie Queene* and the *Arcadia*.

As evidenced by the Italian debates, early modern writers were acutely aware that romance was a literary genre with distinct methodologies and aims. I propose that to fully recover romance's intellectual force we consider its "strategies" and "memes" as methods: as repeatable, iterative, portable procedures organizing the form. Such an approach helps us see that the capacious concept of "the possible" structures *how* romance works, and that this structuring logic also marks romance's formal divisions from other genres. For instance, consider David Quint's influential claim that "to the victors belongs epic, with its linear teleology; to the losers belongs romance, with its random or circular wandering"; the "loser's epic," whose "narrative structures approximate and may explicitly be identified with romance," tends to "valorize the very contingency and open-endedness that the victor's epic disparages."[23] We might read the contingency driving the trajectories of epic's non-conquering heroes as embedded pieces within the "victor's epic" that present readers with ways that things *might* have gone. The "loser's epic" is a recurring reminder of the narratological possibilities latent in, even suppressed by, other literary genres.

This focus on methods can also help us recover how instruments of romance infiltrate other intellectual practices. Bacon, of course, is not writing a romance in the *New Organon*. Yet the digression, error, and deferral that pervade his method evoke the untheorized habits of thought and action that structure the fictional mode. By exploring particular techniques of Baconian induction—including the middle axiom, aphorism, error, and prerogative instances—we can observe how it was formally and procedurally as romancical as *The Faerie Queene*. Romances generate their own instruments of inquiry through narrative strategies of diversion and delay; "true induction" utilizes the same strategies but suppresses its tendencies to affective extravagance and pleasure. Baconian method taps into the capaciousness of a literary narrative that privileges *praxis* over precepts and contingent ways of knowing over sure ends.

My approach of uncovering homologies between Baconian method and literary mode also relates to larger issues of language, being, and knowledge that I raised in the Introduction: when natural philosophers worried about the substantiality of language, they were often commenting on the ontology of poesy. In Bacon's writings, this problem manifests as twinned concerns about the wide reach of words and the uncontrollable generativity of the human mind. Bacon worries that words can distort and change matter, and that privileging *verba* over *res* can have serious consequences for learning. We see early documentation of this anxiety in Book 1 of *The Advancement of Learning*'s "distempers of learning" (138) when he notes how excess focus on Ciceronian style over content leads men to "study words and not matter" (139), or even worse, when scholastic disputations privilege endless debate over new knowledge: "as substance of matter is better than beauty of words, so contrariwise vain matter is worse than vain words" (140). The *New Organon* gives the struggle between words and matter a distinct focus: unchecked language is especially dangerous if it impacts the methods and organization of *natural* knowledge. Even though words are merely "symbols of notions" (notions are "the root of the matter" [1.14]), they have potency dissociated from both ideas (or notions) and the physical world (or matter). This critique is most prominent in Bacon's concept of Idols (as we saw in the Introduction), which captures the dangerous ability of words to function completely severed from actuality.

Language becomes the vehicle that exacerbates fallacies of the "human understanding," which, as Bacon notes in the *New Organon*, is "prone to abstractions and gives a substance and reality to things which are fleeting" (1.51). Because of such tendencies, the "understanding must not therefore be supplied with wings, but rather hung with weights, to keep it from leaping and flying" (1.104). Words, however, work against this security, as they bolster the understanding's propensity to give in to "influence of the imagination" (1.65). After all, when it "gives a substance and reality" to "fleeting" things, the understanding effectively performs a poetic act—an act that approaches poesy's capacity to, in Sidney's words, bring forth "forms such as never were in nature." For the Baconian naturalist "searching into and discovering truth," the collaboration of immaterial words and a flighty mind is problematic, especially when "dealing with natural and material things": the "definitions" of these things must "consist of words," but "those words beget others" (1.59). In its capacity to proliferate—"words beget others"—language threatens to overwhelm, even shift, the content of knowledge about the physical world. The realm of things (that is, natural philosophy) is always in danger of tipping over into that of words—the fabricated, the imagined, the poetic.

By recovering the ways in which induction is constituted by imaginative techniques typically associated with the linguistic and narrative extravagances of romance, I align my work with scholarship by Mary B. Hesse, Ronald Levao, and others who have documented how certainty in Baconian method functions as an impossible ideal.[24] I expand on these discussions to argue that the impossibility of closure and the deferral of resolution are formal features of the method. Given that a variety of disciplines, schools of thought, and practical methods shaped Bacon's writings, it is easy to dismiss Baconian science as derivative and administrative. Deborah Harkness concludes, for instance, that "Bacon cannot be credited as an original thinker when he argued for the utility of science to the state, and the need for governmental support for science."[25] While I take seriously the critique that marks the bureaucratic aspects of Bacon's project, I suggest that focusing on them at the expense of following its actual practices privileges the ordered, hierarchical, statist *ambition* of Bacon's "royal work."[26] By doing so, as Michael Witmore observes, "we overlook a more subtle dependence on contingency in his philosophical program" and take for granted the success of "the confident rhetoric of advancement which pervades Bacon's work."[27] I thus take this abundance of connections as a signal of the "exquisite mixture," to borrow Wolfram Schmidgen's phrase, that characterizes the inventiveness of Baconian science.[28] Even at the level of appropriating disciplinary expertise, Baconian science was not only derivative, it collected together and lent coherence to varied experimental, practical modes of knowing that were distinct from older, classical, mathematical sciences.[29]

Indeed, the activity of collecting—ideas, practices, disciplinary knowledges—is key to understanding Bacon's methods as a kind of *poiesis*. Approaching collecting as creativity conjures the image of an "original thinker" different from the Sidneian poet, who "freely rang[es] only within the zodiac of his own wit." Like many other early modern "worldmakers," to return to Ramachandran's capacious term, Bacon is a compiler of concepts and practices, one who synthesizes them into a new mode of naturalistic knowledge. This creative practice of compilation enacts what we might consider an alternate tradition of *poiesis*. As Jeffrey Todd Knight argues, "'Compiling' . . . *was* production, . . . in the semantics of Renaissance literary activities"; the gathering of information signified "'to compose,' to produce an 'original work,'" that is, to create.[30] If we approach Bacon's practices through such rubrics of originality, we can not only place his work alongside reading and writing practices such as commonplacing[31] but also find formal analogies in popular genres such as romance. To this end, instead of comparing naturalist methods to state, legal, or institutional networks, and rather than taking Bacon's aspirations of certainty as self-evident, I begin from the components internal to the method—its unpredictable, contingent, messy techniques of inquiry—that both constitute and expose the poetics of true induction.

"Romancical" Induction

In this section, I study several elements that constitute Baconian method—the axiom, the aphorism, the initiatory or probative style, "error," and the prerogative instance. These methodological and representational items formalize how Bacon's method, to transport Parker's description of romance, "simultaneously quests for and postpones a particular end." Ideally, true induction "derives axioms from the senses and particulars, rising by a gradual and unbroken ascent, so that it arrives at the most general axioms last of all" (1.19). But what Bacon terms an "unbroken" process is disrupted by the "middle" or "intermediate" axiom, which exposes that the method is driven not only by ascent but also by descent and lateral movements. Induction dynamically strives toward new "works" and axioms:

> after this store of particulars has been set out duly and in order before our eyes, we are not to pass at once to the investigation and discovery of new particulars or works; or at any rate if we do so we must not stop there. . . . [B]ut from the new light of axioms, which having been educed from those particulars by a certain method and rule, shall in their turn point out the way again to new particulars, greater things may be looked for. For our road does not lie on a level, but ascends and descends; first ascending to axioms, then descending to works. (1.103)

While the discovery of "general" axioms occurs by "ascending," the "descen[t]" into "works" at every stage disrupts such movement and directs the naturalist toward new particulars and applications. The recurring inductive-deductive process of rise and fall negates the promise of "unbroken" ascent,[32] placing on the middle axiom the burden of controlling the variable pulls of the method.

In its intermediary role between particulars and generals, the middle axiom is supposed to function as a necessary corrective to the imaginative tendencies of the human mind:

> The understanding must not, however, be allowed to jump and fly from particulars to axioms remote and of almost the highest generality . . . and taking stand upon them as truths that cannot be shaken, proceed to prove and frame the middle axioms by reference to them. . . . But then, and then only, may we hope well of the sciences when in a just scale of ascent, and by successive steps not interrupted or broken, we rise from particulars to lesser axioms; and then to middle axioms, one above the other; and last of all to the most general. For the lowest axioms differ but slightly from bare experience, while the highest and most general (which we now have) are notional and abstract and without solidity. But the middle are the true and solid and living axioms, on which depend the affairs and fortunes of men; and above them again, last of all, those which are indeed the most general; such, I mean, as are not abstract, but of which those intermediate axioms are really limitations. (1.104)

The "not interrupted or broken" rise supposedly negates the overreaching ("jump and fly") disposition of the understanding. While ideally this process should move from the "lowest" to the "middle" to the "highest and most general" axioms, Bacon undermines the possibility of gradual movements by emphasizing that middle axioms "are the true and solid and living axioms." Their vibrant ("living") quality leads the naturalist to new works, but pointing them toward new particulars also distances them from the original road map of "successive steps." One of the primary aims of middle axioms is to "render sciences active" by revealing the great "subtlety of nature" (1.24). Although Bacon believes that "axioms duly and orderly formed from particulars easily discover the way to new particulars" (1.24), middle axioms, by propelling the naturalist to track the unfamiliar and the "new," transgress the bounds of the "orderly formed."

For the naturalist studying the natural world, each "living" middle axiom generates a new hunt, to follow Eamon's analogy of the seventeenth-century natural philosopher as hunter.[33] Or, we could say, it generates a new quest. In entering Bacon's "Nature," we might as well be stepping into Spenser's "wandring wood" (1.1.13.6), where its profusion of trees introduces "So many pathes, so many turnings seene, / That which of them to take, in diuerse doubt [the characters had] been" (1.1.10.8–9). The middle axiom activates a strategy of wandering that directs naturalists away from their original quests—the unruliness implicit in the description will be encountered by those who pursue the tenets of these "active," "living" axioms in a disorderly natural world. The process stages an epistemology predicated on digression, where the "highest and most general" axiom keeps getting indeterminately deferred, part of an original quest that must wait for the "intermediate" task to be completed. But if one follows this logic, each new finding will produce its own quests, leading the naturalist to further new works and particulars, distancing them from yet another "living" axiom. In practice, middle axioms indefinitely defer goals, generating an endless process.

The middle axiom exposes how the paradoxes of the method unravel as the naturalist steps into the natural world, but the incongruences between certain knowledge and the uncertain means that are supposed to lead to it are already embedded in Bacon's aphoristic style. Writing is a crucial component of "true induction," and as Stephen Clucas argues, Bacon aims to "initiate a *vocabulary of the real* which functions rhetorically as a spur to engage in a natural philosophy which is more properly engaged with the material world."[34] Aphorisms are crucial to the enterprise. These "short and scattered sentences, not linked together by an artificial method," generate incomplete knowledge, enabling the knowledge producer to "not pretend or profess to embrace the entire art" (1.86). The aphorism thus makes legible an imperfected, and potentially endless, natural philosophy. Scholars have discussed it as an instrument of "expectant inquiry" that serves as "a building block in the structure of man's knowledge"[35] and have shown how Bacon's "scientific" aphorism differs from understandings of the term in the humanist tradition.[36] I suggest that the flexibility of the scientific aphorism mirrors the capaciousness of the natural world: to achieve the ambitious task of exhaustively representing a pluralist nature, it has to remain open to the multiplicities inherent in the world it documents.

Even before embracing it in the *New Organon* as particularly suited to studying "Nature," Bacon had identified aphorism as the ideal mode to depict the production of knowledge more broadly. In 1605, in Book 1 of *The Advancement of Learning*, he marks "errors" that arise from "over-early and peremptory reduction of knowledge into arts and methods, from which time commonly sciences receive small or no augmentation" (145). He then declares, "knowledge, while it is in aphorisms and observations, . . . is in growth; but when it once is comprehended in exact methods, it may perchance be further polished and illustrate, and accommodated for use and practice; but it increaseth no more in bulk and substance" (146). Opposing "aphorisms" to "method," Book 2 of *Advancement* suggests that the former ensure intellectual "growth" because they disavow totality and are directed exclusively neither to application ("for use and practice") based on "exact methods" nor to "over-early and peremptory reduction." They textualize open-ended procedures: "Aphorisms, representing a knowledge broken, do invite men to *enquire farther*; whereas Methods, carrying the shew of a total, do secure men, as if they were at furthest" (235, emphasis mine).

The divisions in *Advancement* between method and aphorism—furthest/farther, total/broken, exact/growth, polished and accommodated/bulk and substance—capture the temporality of learning that Bacon privileges at several points in his career, choosing to "invite men to enquire farther" and avoiding "over-early" generalization. The aphorism's expectant quality aligns with his advocacy of deferred theorization in the *New Organon*, which we see in several moments, for instance in his preference for "*Interpretation of Nature*" over "*Anticipations of Nature* (as a thing rash or premature)" (1.26).[37] While "Anticipations" foreclose investigation prematurely, "Interpretation" is a "just and methodical process" (1.26) because it refuses closure and thereby creates conditions for further inquiry.[38] This binary also manifests in his preference for "experiments of light" over "experiments of fruit": the former are "useless indeed for the present, but promis[e] infinite utility hereafter" (1.121). Reading the aphorism alongside these elements of Baconian natural philosophy outlined in the *New Organon*, we see it is not only an instrument of presentation but one that also affects the substance of inquiry; it breaks down distinctions of style and content. Linking the aphorism's forward-looking orientation to the larger project to "enquire farther," we see how, in his desire to map the "unbroken" ascent to certainty, Bacon relies on a technique of "knowledge broken." In rejecting the "shew of a total," the inductive method enacts an aphoristic way of knowing.

The rift between the promise of an unbroken induction and an aphoristic style that represents "knowledge broken" ultimately impacts *New Organon*'s aphorisms themselves. The aphorism's formal capacity to eradicate extraneous details, Bacon notes in Book 2 of *Advancement*, is supposed to deliver knowledge efficiently: "Aphorisms, except they should be ridiculous, cannot be made but of the pith and heart of sciences; for discourse of illustration is cut off; recitals of examples are cut off; discourse of connexion and order is cut off; descriptions of practice are cut off; so there remaineth nothing to fill the Aphorisms but some good quantity of observation: and therefore no man can suffice, nor in reason will attempt to write Aphorisms, but he that is sound and grounded" (234). Theoretically, it is a form that rejects unnecessary information, thereby overcoming problems associated with extravagant language; if one cannot learn without words, one can at best control unpredictable significations through brevity, focusing instead on the "pith and heart of sciences." Aphorisms supposedly make this possible by stripping representation to a bare minimum. However, as scholars have noted, "very few of the *Novum Organum*'s aphorisms are either self-contained or pithy: they tend to grow in length as the book develops, and take their place within interlocking groups."[39] We witness this in the aphorisms about the middle axiom (e.g., 1.104) that I describe above. They are descriptive rather than pithy, expansive rather than contracting. Like other aspects of induction in general and the middle axiom in particular, the aphorism in *New Organon* functions as a "living" entity, extending beyond mere summary, thriving on "discourse of illustration" and "recitals of examples." The expanding aphorism represents, to borrow James Stephens's words, the "fusion of style and content in a method which approximates the true induction."[40] The aphorisms of *New Organon* are composed of examples, experiments, and multiple routes of inquiry. Ultimately, through its form, the aphorism enacts both the deferral of completed inquiry and the impossibility of curtailing extravagances that are inherent to the inductive process.

The Baconian aphorism exposes how the links between words and things cannot be dismantled: its protraction results from attempts to untangle the variety of the natural world in writing. Cross-references and continuations across aphorisms dramatize how natural knowledge is not contained but always "in growth," how it "increaseth" in "bulk and substance"; the "things" he describes control the content and form of his aphorisms (see, for instance, the extensive "investigation" into the "form of heat" that extends from 2.11 to 2.20). Like the inductive method, which defers certainty because of the proliferation of matter, his aphoristic style expands to accommodate nature's multiplying particulars. By breaking the bounds of its definitionally condensed form, the aphorism facilitates inquiry and reflects Bacon's larger investments: "my course and method, . . . [is] this—not to extract works from works or experiments from experiments (as an empiric), but from works and experiments to extract causes and axioms, and *again from those causes and axioms new works and experiments, as a legitimate interpreter of nature*" (1.117, emphasis mine). At the moment the naturalist reaches a stable end point of "causes and axioms," the method asks the "legitimate interpreter" to "enquire farther."

This advocacy to defer theorization emerges as a vital component of method. Lest this be mistaken for "suspension of the judgment," he explains, "better surely it is that we should know all we need to know, and yet think our knowledge imperfect, than that we should think our knowledge perfect, and yet not know anything we need to know" (1.126). This stance extends to his criticism of contemporary philosophers who theorize early believing they have "knowledge perfect"—this is Bacon's complaint with William Gilbert's approach, which Bacon perceives as aiming to encompass prematurely observations under one theory of magnetism (see 1.54). Bacon's claim that an inconclusive style is necessary to comprehensive learning is actually part of a career-long belief that "knowledge imperfect" fuels inquiry. In Book 2 of *Advancement*, he separates the "magistral" method, which is "referred to Use," from the one of "probation," which is "referred to Progression" (233); while the magistral halts inquiry, the probative generates it. In *De Augmentis* (1623), he renames the latter "initiative," arguing that "the magistral method teaches; the initiative intimates. The magistral requires that what is told should be believed; the initiative that it should be examined. The one transmits knowledge to the crowd of learners; the other to the sons, as it were, of science. The end of the one is the use of knowledges, as they now are; of the other the continuation and further progression of them."[41] The initiative method echoes the aims of experiments of light in prolonging inquiry. The "magistral," however, corresponds to immediate gains that are characteristic of experiments of fruit. Brian Vickers's differentiation of these methods—the magistral, whose "dogmatic or doctrinal exposition of information [is] to be believed, and not questioned," and the probation or initiative, which is "to be tested, examined"—in notes in his edition suggests that the difference between the two lie in their aims and intended audiences.[42] The proper method "intimates," as it propels rather than stifles creative study. The "initiative"—a term aligned with suggestion and implication rather than indoctrination—is generative, pushing "sons . . . of science" to "further progression."

Bacon's allegiance to knowledge-making practices oriented toward the future intimates that true and complete knowledge only exists as a possibility rather than as something actualized. He even turns the impossibility of certainty into a virtue of process, admitting in *New Organon* that he has "no entire or universal theory to propound" because "it does not seem that the time is come for such an attempt" (1.116). In a dual tone of self-valorization and humility, he admits amid his "investigation" on heat that it is the method itself that defers conclusion: "I, therefore, well knowing and nowise forgetting how great a work I am about (viz., that of rendering the human understanding a match for things and nature), do not rest satisfied with the precepts I have laid down, but proceed further to devise and supply more powerful aids for the use of the understanding" (2.19). When he admits that he is "frequently forced to use the words 'Let trial be made,' or 'Let it be further inquired'" (2.14), Bacon sounds like the narrator of *The Faerie Queene* (in the Proem to Book 2) telling readers that they "may [Faerie Land] fynd" if they "more inquyre." Such echoes of Spenserian speculation in a centerpiece of Baconian protraction conjure the image of a naturalist performing, like the poet, an "endlesse worke." Thus, instead of focusing on the desire for "order," "progress," and "certainty" as the defining elements of method, I propose we read this call of "'Let it be further inquired'" as the rallying cry of Bacon's "argument of hope" (1.94). True induction is, to transport Bacon's words, an "initiative" form of knowing, one that masks its theory in its inconclusive practices.

Such approaches not only prevent one from knowing immediately what is correct but also thrive on the generative affordances of error. It is perhaps surprising that despite his wariness of an unmanageable understanding, Bacon admits that error is crucial to learning:

> There will be found, no doubt, when my history and tables of discovery are read, some things in the experiments themselves that are not quite certain, or perhaps that are quite false, which may make a man think that the foundations and principles upon which my discoveries rest are false and doubtful. But this is of no consequence, for such things must needs happen at first. It is only like the occurrence in a written or printed page of a letter or two mistaken or misplaced, which does not much hinder the reader, because such errors are easily corrected by the sense. So likewise may there occur in my natural history many

> experiments which are mistaken and falsely set down, and yet they will presently, by the discovery of causes and axioms, be easily expunged and rejected. (1.118)

Induction's forward-looking trajectory will ensure that errors are eradicated when all things "not quite certain" and even "quite false" have been addressed. The errors Bacon discusses are partially products of recording, when experiments are "mistakenly and falsely set down," but method, he claims, is also self-correcting. Hence one can treat mistakes as things that "must needs happen at first" and defer their expulsion; the naturalist need not eradicate them as he perfects a "poor" history (2.14). It is ironical that Bacon defends letting errors in natural history linger by comparing them to scribal or printers' errors and invoking the human tendency to repair "mistaken or misplaced" letters. The reliance on "sense" to ensure amendments brings us dangerously close to aspects of human understanding that he critiques elsewhere: since the "mind longs to spring up to positions of higher generality" (1.20), one is prone to drawing premature analogies.

What is most striking, of course, is any need for error in the *Great Instauration*, which is "the true end and termination of infinite error."[43] Bacon, I argue, begins to construct a theory of error when he acknowledges that it both structures and directs ways of knowing; "infinite error" shapes, rather than distorts, true induction. At one point in the *New Organon*, he even states that "truth will sooner come out from error than from confusion" (2.20). Error serves as a corollary to the digressive routes generated by the middle axiom, its purely negative status receding from view. Bacon ascribes to "error" a value that counters its association with forces such as "confusion" that would place it beyond recuperation. But herein lies the danger of "error" becoming essential, rather than accidental, to method. Despite Bacon's attempts to control multiple strands of inquiry, it is not a given that existing errors will self-correct unless one makes an active effort to trace backward from what one learns at each stage. The concept of error in induction performs its definitional role—the "action of roaming or wandering; hence a devious or winding course, a roving, winding"[44]—that had made "endlesse error" (*Faerie Queene*, 1.3.23.9) a vital conceptual instrument of romance narratives.

Unlike "confusion," "error" is crucial since it supposedly expedites certainty. Bacon, I suggest, gives such expedited procedures of error a particular name: "Prerogative Instances." These twenty-seven instances, trials, and examples in the *New Organon* provide shortcuts from "common instances," with the aim of hastening "gradual" and "unbroken" ascent. They "excel common instances" either "in the informative part or in the operative, or in both"; in the former "they assist either the senses or the understanding" and in the latter "they either point out, or measure, or facilitate practice" (2.52). I argue that the "Prerogative Instances" model "erroneous" ways of knowing. As they provide shortcuts to axioms, they collectively enact how digression is constitutive to the method. Ideally, these instances are supposed to shorten and thereby make more efficient the process of induction. Bacon explains the purview of the prerogative instances in a variety of ways: he notes when they should be performed and in what order; he lists what the subdivisions are; he outlines what examples to use; he marks which particulars have been gathered from experience and which ones still need to be tried; he defines which prerogative instances are essential, which are immediate, and so forth. For example, different instances must be tried at different points (of some "we must make a collection at once, . . . without waiting for the particular investigation of natures. Of this sort are instances comformable, singular, deviating, bordering . . ." [2.52]), while the rest may be delayed until one studies a particular nature. But in their plurality, prerogative instances distribute inquiry into diverse categories and examples, each instance further propelling the naturalist in multiplying directions. They collectively activate dispersion, and by guiding naturalists toward varied shortcuts, they facilitate acts of wandering. We could say that in the "Prerogative Instances" error has become a concept—or "notion," to use Bacon's term—that will ensure a more efficacious method.

Through their presence as a collective and in their varied functions, the prerogative instances ultimately disrupt the stated ends of induction, and by extension, the form of the great instauration. Before he begins to outline these instances, Bacon lays out his plans for the rest of the steps of natural inquiry: "*Supports of Induction*," "*Rectification of Induction*," "*Varying the Investigation according to the nature of the Subject*," "*Prerogative Natures with respect to Investigation*," "*Limits of Investigation*," "*Application to Practice*," "*Preparations for Investigation*," "*Ascending and Descending Scale of Axioms*" (2.21). But he never returns to these anticipated stages. *New Organon* halts at the twenty-seven "Prerogative Instances" even as he suggests in the last aphorism, "now I must proceed to the supports and rectifications of induction, and then to concretes, and Latent Processes, and Latent Configurations," in order to "hand over to men their fortunes" (2.52). This abrupt end produces an unconcluded method. It also formally registers the delay that "Prerogative Instances" injects into the project, postponing the enumeration of the rest of the steps. They deviate by design, and in their proliferation, postpone the enumeration of the whole. The introduction of each instance alters the scope of induction. They professedly outline specific routes that will hasten inquiry but, in practice, sanction deviations and prevent rule-making. Their promotion of wandering exemplifies the contradiction of the method: techniques that most forcefully promise certitude derail movement toward it.

As Bacon acknowledges the productive nature of error, and as he gives it a place of pride in his method, the term comes to refer not only to processes of learning, or even to theoretical concepts: "error" also emerges as a form of being. It defines ontological categories; there lurks everywhere in Bacon's nature a "monster vile" (1.1.13.7) like Spenser's *Errour*, a physical manifestation of the larger forces of digression in *The Faerie Queene*. For instance, Bacon terms "*Deviating Instances*" the "errors, vagaries, and prodigies of nature, wherein nature deviates and turns aside from her ordinary course" (*New Organon*, 2.29), and he uses the term "pretergenerations" to refer to "errors" of nature, or "monsters" that come into being when nature is "forced out of her proper state by the perverseness and insubordination of matter and the violence of impediments" (*Preparative*, 273). Drawing on the work of Daston and Park on how Bacon embraced preternatural philosophy and adapted "marvels" into his reformed natural philosophy, I understand "error" as an exceptional state of being.[45] I also propose that Bacon links his descriptions of anomalous states of being to the digressions of method. He stresses how deviating instances contribute to the process of induction: "he that knows the ways of nature will more easily observe her deviations; and on the other hand he that knows her deviations will more accurately describe her ways" (*New Organon*, 2.29). The relation between "deviations" and "ways" establishes a strict connection between modes of existence and routes of knowing. In their self-propagation, these "errors, vagaries, and prodigies of nature" also multiply knowledge of other deviations ("errors on one side point out and open the way to errors and deflections on all sides" [2.29]), *producing* detours as they "more easily" and "more accurately" describe nature's "ways." In embodying the reciprocity between "ways of nature" and "her deviations," these examples debunk myths about the regularity of the natural world. Instead, they reveal a realm that is constituted by the "perverseness and insubordination of matter." Since errors do not adhere to an orderly concept of "nature," they destabilize ideals about what the "ways of nature" should be. They also reveal that erroneous ways of knowing are necessary to make intelligible "nature's" multiplicity. The capacious method reflects the disorderly and pluralist ontology of the world it aims to represent.

We may consider error a formal principle of induction. It is this specter of "endlesse error" that underlies the author's worry about how words generate notions and things. As naturalists act in and document a vibrant nature, its accidental, provisional, and contingent aspects refashion the method. Yet, in a method predicated on pursuing all of nature's particulars, does not following each deviating instance direct naturalists to yet more directions that are themselves without end? Although the *Great Instauration* imagines the natural philosopher as a second Adam who will restore Nature to its pristine state before the Fall, would it not be as appropriate to see him as a knight-errant? The epigraph from *The Faerie Queene*, in which Arthur describes how he pursues his intended goal "with labor, and long tyne, / And neuer vowd to rest, till her I fynd" (1.9.15.7–8) might as well have been written for the Baconian naturalist, who is on a quest to uncover the secrets of the world, but is required by "Nature's" variety—its deviations, vagaries, or "errors"—wander and err if he is to ever perfect the "endlesse worke" of natural philosophy.

Furthering Inquiry in *New Atlantis*

In spite of his worries about the non-referential capacity of language, and despite his claim in the *New Organon* that by "relying on the evidence and truth of things, I reject all forms of fiction and imposture" (1.122), Bacon writes a utopian fiction. The posthumously published *New Atlantis* (which was appended to the *Sylva Sylvarum* by Bacon's literary executor, William Rawley) has long invited readers to consider the prose fiction as a wish fulfillment of Bacon's enterprise of natural inquiry, where Salomon's House is a precursor to organizations such as the Royal Society and the text "describes the social promise of what will become the 'new science.'"[46] Such readings solidify the text's status as an idealized depiction of the philosophy and method Bacon outlines in earlier writing. His fictional narrative, it seems, just might arrest the interminable techniques of the *New Organon*, as *New Atlantis*'s orderly state—representing a technocratic world that "should be"[47]—begins with the suggestion that one can overcome the contingencies that haunt the actual method of Baconian natural inquiry. The existence of *New Atlantis* also legitimizes Bacon's great anxiety—that the understanding, in its propensity to fabrication, can mobilize words to bring forth an alternate world. In this section I explore this paradox: why is it Bacon's fiction that comes closest to evacuating the error and digression we encounter in the *New Organon*? To this end, I explore how the "vnfinished" (title page, 1627 edition) utopia replaces this continual *praxis* with a philosophy of estrangement and secrecy; *New Atlantis* masks intellectual practices by making "secret" what the philosopher strives to reveal in works such as *New Organon*.[48]

To "[give] a substance and reality to things which are fleeting," Bacon adapts various conventions of utopia, seeking, as Denise Albanese argues, "alignment with More's *Utopia*," which "maintains a difficult balance between revolutionary ideality and practical impossibility, simultaneously affirming and denying its radical agenda."[49] Bacon is less concerned with proving Bensalem's existence than More is of his titular island in *Utopia*, where interlocutors in text and paratexts supposedly attempt to situate the island in their own sphere of existence. *New Atlantis* offers a different model of how, in William Poole's words, "utopianism renders hypothetical those things that cannot be admitted as positive theses."[50] Bacon's work projects an ideal version of his philosophical enterprise onto a distant yet accessible realm. More particularly, it exemplifies how "utopianism is not about being 'no where'; it is about desiring to be elsewhere."[51] *New Atlantis* mediates between the "no where" and "elsewhere" but is ultimately governed by the desire of transforming them into the "here" and the "now." Thus, it is also hypothetical in the sense that, in the words of H. Vaihinger, it "is directed towards reality, i.e. the ideational construct contained in it claims, or hopes, to coincide with some perception in the future."[52] As Albanese argues, *New Atlantis* proposes "the mechanism for change from the present imperfect to the future perfect."[53] The author replaces his current disappointment—the impossibility of realizing his great instauration during his lifetime in England—with a projection of narrative desire toward the unpredictable future, where an ideal technocratic state exists at the edge of his own world, only one expedition and providential shipwreck away.

The narrative begins with a scene of travel, reminding us not only of the voyages of utopian fiction and travel narrative but also of the shipwrecks of romance.[54] The unnamed European narrator enumerates his group's travails at sea, "in the midst of the greatest wilderness of waters in the world" and in the "utterly unknown."[55] In their "beseeching [God] of his mercy," the travelers hope "that as in the beginning he discovered the face of the deep, and brought forth dry land, so he would now discover land to us" (457). In this moment of peril, the narrator's language transfers agency to God. The ocean is a liminal space that curtails human agency, the unmeasurable distance between land and sea symbolizing the limits of experiential knowledge. Yet the mere possibility of land retransfers power to the travelers. From the moment they witness "thick clouds, which did put us in some hope of land," the narrator shifts control back to his group: "we bent our course thither, where we saw the appearance of land, . . . in the dawning of the next day, we might plainly discern that it was a land; flat to our sight" (457). The repeated "we" and active voice intimate how the prospect of "land" reinstates confidence in their power over their surroundings, as the "we" actively "discern," rather than hoping God will "discover land to us." Removed from the "utterly unknown" waters, they begin to detect what might be possible in a strange world. The tensions between the known and the strange in this opening scene, I propose, structure crucial formal elements of the entire narrative.

This struggle is immediately evident as the travelers attempt to understand their standing in Bensalem. Their status as strangers is heightened after they land, and they are governed from the start by rules of containment and prohibition. An inhabitant of Bensalem hands them a scroll of the laws of entry: "'Land ye not, none of you; and provide to be gone from this coast within sixteen days, except you have further time given you'" (458). The command to depart undercuts their supposed deliverance. Their lodging ("Strangers' House" [460]) and sense of being under constant surveillance accentuate their status as outsiders. The narrator announces his concern that "these men that they [the Bensalemites] have given us for attendance may withal have an eye upon us." He contrasts their "deliverance past" with the "danger present and to come." In the initial moments after their arrival, even gestures of hospitality—an abundant feast, excellent lodgings, attendants—garner suspicion, forcing them to wonder, "who knoweth whether it be not to take some taste of our manners and conditions" (461). Their complete estrangement in this land fosters wariness, as structures intensifying uncertainty undermine each instance of reassurance.

The distinction between familiarity and disorientation is reinforced throughout the text. The appearance of the familiar produces momentary relief—a lord's "reverend" (458) look and excellent apparel establish his trustworthiness—but by accepting these markers of status and wealth as signifiers of authority, the narrator cements their own estrangement and liminality: "we are but between death and life; for we are beyond both the old world and the new" (461). The "for" indicates a causality, marking how the inaccessibility of the "old world" has destabilized their status as living. The hosts can grant and withdraw stability. The narrator accepts these hierarchies and subordinates his group's independence by linking their hosts with divinity: "let us so behave ourselves as we may be at peace with God, and may find grace in the eyes of this people" (461–62). The appearance of the governor of the House of Strangers shows "we had before us a picture of our salvation in heaven" (462–63). As they embrace the rules and hierarchies in Bensalem, the travelers accept the imposed estrangement, now identifying their own selves as "strangers" and "servants" (463).

In this place devoid of transparency, the narrator's group embraces any signs of familiarity or references to the "old world" as indicators of stability and authority. The scroll supplies only limited signatory information about laws of entry, "signed with a stamp of cherubins' wings, not spread but hanging downwards, and by them a cross"; its iconography obscures more than it reveals, leaving the newcomers "much perplexed." Yet, signs like the cross provide "a certain presage of good" (458) and, by extension, reassurance. The travelers grasp at these familiar signs in order to understand Bensalem's seemingly absolutist rules of entry and habitation. Even as these reactions enact the mind's tendency to analogize between the familiar and the strange, the travelers' acceptance of such signs and rules cements their estrangement and subordination.

The gap between stable hierarchy and variable practice is clearest when the governor informs the travelers about the strict rules: "The state hath given you license to stay on land for the space of six weeks: and let it not trouble you if your occasions ask further time, for the law in this point is not precise; and I do not doubt but myself shall be able to obtain for you such further time as may be convenient" (462). At first glance, his words are prohibitive. But he immediately erases absolute restrictions, undermining the hierarchies that the narrator's party has come to expect and accept. His declarations echo the competing strategies we have seen in the scroll, which tempers its absolute "Land ye not, none of you" with the moderate "provide to be gone from this coast within sixteen days" and "except you have further time given you," the latter clauses qualifying what seemed like an unconditional declaration. Such phrases introduce contradictions as they both suppress and accommodate strangers in Bensalem. The inability to escape the "further"—what lies beyond prescription—undercuts claims to full estrangement. The logic of "further" and "except" submit inflexible laws to revision; it exceeds the "shew of a total," to borrow a phrase from the *Advancement*. The "further" also gestures to other unknowns: readers, like the travelers, are not aware of the source of authority in Bensalem. Having followed the travelers' analogizing tendencies—which associate things in Bensalem with those in the "old world"—they may fail to perceive how the "except" and "further" surreptitiously instill provisional power in strangers.

The governor's pronouncement acknowledges the necessity of strangers in a state that has not hosted outsiders for thirty-seven years: they are needed to make the unknown familiar. Bensalem's unique position depends on the hierarchy of one-way exchange of information with the outside world. As the governor explains, "by means of our solitary situation, and of the laws of secrecy which we have for our travellers, and our rare admission of strangers, we know well most part of the habitable world, and are ourselves unknown" (463). The travelers find this unidirectional flow of knowledge and power "wonderful strange; for that all nations have inter-knowledge one of another either by voyage into foreign parts, or by strangers that come to them." While "inter-knowledge" facilitates "mutual knowledge" (466) and is the modus operandi in the "old world," Bensalem remains outside such networks of exchange. Here, self-imposed isolation is the norm for natural inquiry, a reverse center-periphery model that attributes more authority to the latter.[56] The governor claims their unknowability enables them to know more, and his exposition propels the narrator to marvel, again, about the merits of absolute difference: "for that it seemed to us a condition and propriety of divine powers and beings, to be hidden and unseen to others, and yet to have others open and as in a light to them" (466). However, even as the narrator accepts the governor's rejection of collaborative inquiry—a form that is central to Bacon's conception of natural inquiry elsewhere—their very interaction undermines the governor's point. Complete isolation is not ideal: Bensalem requires strangers, who create the occasions for the furthering of the Bensalemites' knowledge (of others), even as the strangers themselves ask that the obscure ways of Bensalem be revealed. In this way, their queries further the governor's narrative about the place and its qualities. Storytelling requires estrangement, and the accommodation of strangers is a structurally necessary component for Bensalem's relevance.

To put into practice this theory of knowledge production predicated on lack on the part of the strangers, the governor appeals to a particular grammatical form. He claims that "'because he that knoweth least is fittest to ask questions, it is more reason, for the entertainment of the time, that ye ask me questions, than that I ask you'" (463). We might consider the question a fundamental grammatical unit of possible knowledge. It initiates discussion, supposing the potential existence of an answer and requiring participation from those who "knoweth least." The governor reiterates its importance when he invites additional inquiry from the newcomers, telling them "the questions are on your part" (465). The interrogative structure enacts how knowledge-making is predicated on, but also necessitates, a strategic estrangement from certainty. Framing inquiry as something generated by those with less authority, or in other words from *strangers*, the question both emblematizes how the narrator's party can generate their own avenues of investigation and challenges the stabilized relations that Bensalem's inhabitants seemingly reinforce. Those who "knoweth least" define what is worthy of knowing. The question is the textual form that Bacon's initiative method takes in *New Atlantis*, and the narrator's position is analogous to that of the naturalist of *New Organon* who operates from a position of recognizing that they possess only "knowledge imperfect." The travelers, in other words, embrace the logic of "Let it be further inquired."

The question's capacity to further inquire has formal effects on the laws that structure existence in Bensalem. As the narrator's companions negotiate the fact that "this land had laws of secrecy touching strangers" (466), their questions disrupt the orderly ways in which their hosts control and share knowledge and open up the possibility that this space, too, is available for investigation. After all, it is the travelers' question of how the Bensalemites have escaped detection that shapes the narrative. In a digressive account that takes them away from their original query, the governor provides a history of the "great Atlantis" (467), or America. Only after the narration does he return to tell "why we should sit at home." But he cannot account for the present without digressing again to reveal another history: of Solamona, the king who lived "about nineteen hundred years ago" and was the "lawgiver of our nation" (469). Bensalem's present self-sufficiency can only be understood by delving into its past. There is no unmediated narration of progression but only recursive excursions into different histories. The strangers even learn about Salomon's House, ostensibly the center of Bensalem's uniqueness as well as Bacon's larger project of great instauration, through another digression: "here I shall seem a little to digress, but you will by and by find it pertinent" (471). The questions posed by the narrator's party create the occasions that permit the Bensalemites to paint a complete picture of their land—to elucidate its civic, political, and intellectual institutions and expand what we know of this realm.

These interrogations also lead the governor to admit gaps in what initially seems perfect and fixed. Solamona's laws combine "humanity and policy," where the policy requires detention, "that [strangers] should return and discover their knowledge of this estate." Yet, their "humanity" also prohibits Bensalemites from "detain[ing] strangers here against their wills" (470). This duality extends the law beyond its ideal of containment and prohibition, making it another manifestation of the "except" and the "further" that erupt repeatedly in the text, as it orients the geographical "elsewhere" of Bensalem toward the actual world by accommodating strangers from the "old world." The governor explains their successful concealment by appealing to the qualities of the place itself: only thirteen strangers have returned in all these years, and their accounts "could be taken where they came but for a dream" (470). His words highlight the paradox that is Bensalem: as this place continues to avoid exposure, its state of existence also alters; it becomes a "dream."

The governor's words keep at bay its permanent transformation into a "dream." Even as he controls the flow of information by stating that he may not reveal more, he acknowledges that Bensalemites acquire knowledge only through interactions with the outside world—their scientific enterprise is governed by collecting, travel, and appropriation. This admission necessitates yet another about the flexibility of their laws:

> When the king had forbidden to all his people navigation into any part that was not under his crown, he made nevertheless this ordinance; That every twelve years there should be set forth out of this kingdom two ships, appointed to several voyages; That in either of these ships there should be a mission of three of the Fellows or Brethren of Salomon's House; whose errand was only to give us knowledge of the affairs and state of those countries to which they were designed, and especially of the sciences, arts, manufactures, and inventions of all the world; and withal to bring unto us books, instruments, and patterns in every kind; That the ships, after they had landed the brethren, should return; and that the brethren should stay abroad till the new mission. (471)

The governor's description begins with an acknowledgment of the tussle between the contained and the excess, or the "forbidden" and the "nevertheless." His words also serve as a reminder of the processes of collecting natural history that Bacon had outlined elsewhere as vital to the project of great instauration. As Bensalemites appropriate epistemic practices and entities from the "old world," they reveal that isolation is an unsustainable desire; actual techniques of knowledge-making override restrictive laws to facilitate, to borrow Bacon's language describing the initiative method, "continuation and further progression."

This exposition also underscores the centrality of Salomon's House to the state and the narrative. The scientific society constructs knowledge in Bensalem and structures engagements with the wider world. The governor's earlier, seemingly minor, reference to Salomon's House captures the telos of the narration, the end toward which *New Atlantis* strives. The society called Salomon's House, or the "College of the Six Days Works," is the "lanthorn" (471) and "eye of this kingdom" (464). Its members understand their task to be "to know [God's] works of creation, and the secrets of them; and to discern (as far as appertaineth to the generations of men) between divine miracles, works of nature, works of art, and impostures and illusions of all sorts" (464). While Salomon's House maintains trade for "God's first creature, which was *Light*" (472), the institution itself is cloaked in secrecy: Joabin informs the narrator, "we have seen none of them [the House's members] this dozen years," and although one of its principal members announces that he is coming to the city, "the cause of his coming is secret" (478). Critics often read the institution as a site of bureaucracy, or of the statist culmination of Baconian science, but I propose that its introduction in a digression and its deferred description are also symptomatic of the larger structure of the fiction itself: a struggle between the desire for precision and rule and its necessary succumbing to open-endedness. Salomon's House gives form to what fiction as a "deviating instance" might look like.

The contradictions in the structure of the text orient it toward inconclusivity and expansiveness, even as it seems to be premised on perfected conclusions. Propelled by the continual inquiries of the strangers that break down the isolationist stance, the *New Atlantis* "ends" with a disavowal of secrecy and estrangement. The father of Salomon's House enters into "conference" (479) with the narrator and promises to disclose their activities, resources, and ends: "I will impart unto thee, for the love of God and men, a relation of the true state of Salomon's House." He declares to the narrator that their aim is "the effecting of all things possible" (480). Even though the encyclopedic endeavor exists in a state of contingency, hinging on the hope that what is currently unknown—and which might not even exist—can be investigated, controlled, and utilized, he understands "things possible" as prospective objects of reality. Bolstered by this vision, the narration can "impart" the "true state" of the society and present the full knowledge that has remained inaccessible up to this point. His language now privileges transmission of ideas over secrecy, as had been the norm.

On the one hand, the father's narration in *New Atlantis* stresses knowledge that has been achieved over potential discoveries: his refrain is "We have" and not "we will."[57] On the other, he emphasizes, "I will not hold you long with recounting of our brewhouses, bake-houses, and kitchens" (483), suggesting there is something that inevitably escapes narration. He seems to dissociate presentation from praxis, modes of inquiry from their reporting—the two elements that are entangled in *New Organon*. But his interaction with the narrator, and his one-way dissemination of knowledge, ultimately gives way to the text's most transparent appeal to openness and expansion when he states, "I give thee leave to publish [this relation] for the good of other nations; for we here are in God's bosom, a land unknown" (488). As he brings the "land unknown" into conversation with "other nations," he also opens the possibility of "inter-knowledge." The appeal to "publish"—the term seems to refer to the popularity and increasing influence of print, a kind of "making public," and the legality of announcement and communication—is both an order and a request. In hoping to share their knowledge, the father breaks the codes of secrecy that govern Bensalem. It is striking that he accepts the necessity of the stranger, using the narrator's outsider status as a motivation to convert containment into expansion.

The *New Atlantis* formalizes the pronouncement of publicity and "furthering" through a move that was commonplace in early modern romances like Sidney's *Arcadia*: the narrative stops without ending. The first publication announces the status of the work with the statement, "The rest was not perfected" (488; Figure 3), and this simple declaration seems to propel the fiction into an extratextual realm. The entanglements of Bacon's works of fiction and science are further captured by the list, or the "Magnalia Naturae" that immediately follows the "not perfected" *New Atlantis* (Figure 4). This catalog is a wish list that gestures to the incomplete *New Organon* and, in particular, recalls the wish lists at the end of the *Preparative Toward Natural and Experimental History* (the third part of *Instauratio Magna* that was appended to *New Organon* and served to describe natural and experimental histories). Within the fiction, the father's narration links the labors within Bensalem to the larger Baconian enterprise when he narrates as completed projects several practices and catalogs (see *New Atlantis*, 480–88) that had been laid out at the end of the earlier works: *Preparative*'s concluding catalog includes the "History of Fiery Meteors," the "History of perfect Metals, Gold, Silver; and of the Mines, Veins, Marcasites of the same; also of the Working in the Mines," and the "History of Baking, and the Making of Bread," to mention only a few (285–91). What is potential in *New Organon*, the father makes extant through the fulfillment of utopian desire within the narrative of *New Atlantis*. Yet the appearance of the "Magnalia Naturae" *beyond* the fictional frame also aborts the possibility of closure.

The lists at the end of the "not perfected" utopian work ultimately serve as the seeds of potential topics of further natural inquiry. *New Atlantis*'s lists incite projections into actuality, in the process collapsing some of the gaps between a list of entities to be discovered—or what Bacon in Book 2 of *Advancement* calls "optatives" of operative knowledge (202)—and *desiderata*: wish lists that seek not only desired things, but encompass systems of knowledge.[58] Always intimating

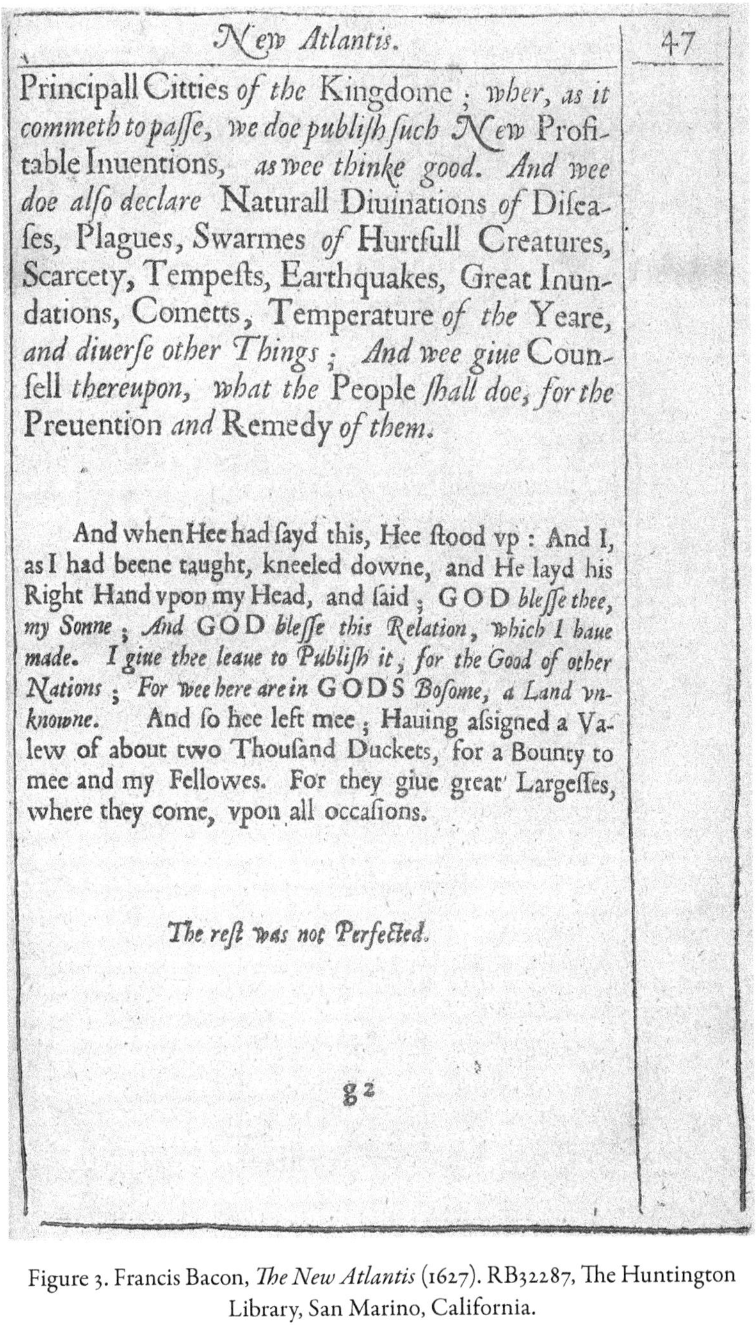

New Atlantis. 47

Principall Citties *of the* Kingdome ; *wher, as it commeth to paſſe, we doe publiſh ſuch New* Profitable Inuentions, *as wee thinke good. And wee doe alſo declare* Naturall Diuinations *of* Diſeaſes, Plagues, Swarmes *of* Hurtfull Creatures, Scarcety, Tempeſts, Earthquakes, Great Inundations, Cometts, Temperature *of the* Yeare, *and diuerſe other Things ; And wee giue* Counſell *thereupon, what the* People *ſhall doe, for the* Preuention *and* Remedy *of them.*

And when Hee had ſayd this, Hee ſtood vp : And I, as I had beene taught, kneeled downe, and He layd his Right Hand vpon my Head, and ſaid ; GOD *bleſſe thee, my Sonne ; And* GOD *bleſſe this Relation, which I haue made. I giue thee leaue to Publiſh it ; for the Good of other Nations ; For wee here are in* GODS *Boſome, a Land vnknowne.* And ſo hee left mee ; Hauing aſsigned a Valew of about two Thouſand Duckets, for a Bounty to mee and my Fellowes. For they giue great Largeſſes, where they come, vpon all occaſions.

The reſt was not Perfected.

g 2

Figure 3. Francis Bacon, *The New Atlantis* (1627). RB32287, The Huntington Library, San Marino, California.

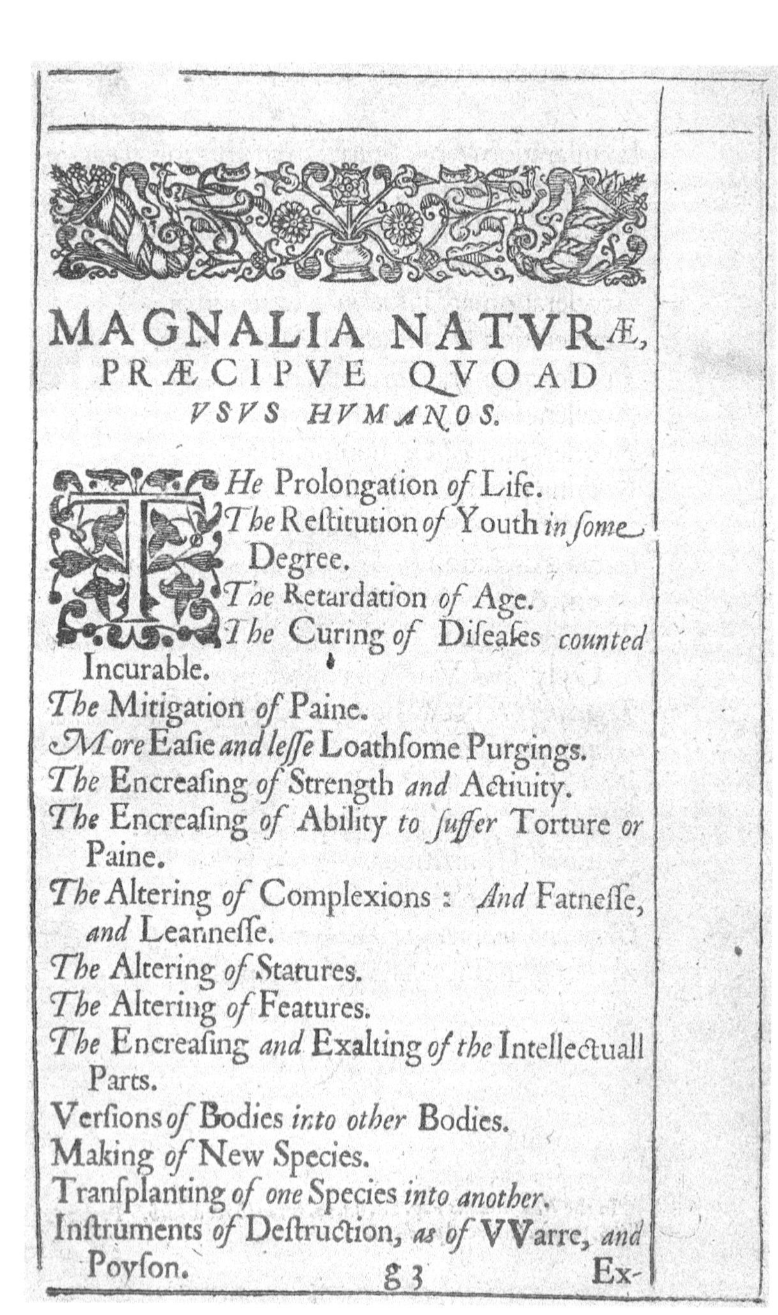

MAGNALIA NATVRÆ,
PRÆCIPVE QVOAD
VSVS HVMANOS.

T*He* Prolongation *of* Life.
The Reſtitution *of* Youth *in ſome* Degree.
The Retardation *of* Age.
The Curing *of* Diſeaſes *counted* Incurable.
The Mitigation *of* Paine.
More Eaſie *and leſſe* Loathſome Purgings.
The Encreaſing *of* Strength *and* Actiuity.
The Encreaſing *of* Ability *to ſuffer* Torture *or* Paine.
The Altering *of* Complexions: *And* Fatneſſe, *and* Leanneſſe.
The Altering *of* Statures.
The Altering *of* Features.
The Encreaſing *and* Exalting *of the* Intellectuall Parts.
Verſions *of* Bodies *into other* Bodies.
Making *of* New Species.
Tranſplanting *of one* Species *into another.*
Inſtruments *of* Deſtruction, *as of* VVarre, *and* Poyſon.

g 3 Ex-

Figure 4. Francis Bacon, "Magnalia Naturae," following *The New Atlantis* (1627). RB32287, The Huntington Library, San Marino, California.

that something exists beyond what is cataloged, Baconian lists expand not only across texts (we see some of them elaborated in the experiments in *Sylva Sylvarum*) but also into the natural world, as they demand the naturalist investigate the topics listed in the catalogs.[59] Moreover, in its intimation of future perfectibility through acts of collecting natural history, the list holds the possibility of composing a "world" from individuals and species that comprise "Nature" itself. As the list abstracts aspects of an uncataloged world, and as it aims to curb the sense of contingency underlying the naturalist's work, it extends their scope of inquiry beyond the page and onto the stage of the world.

Instrumentalizing Imagination and Baconian Continuations

The wish lists that punctuate Bacon's writings on natural inquiry as well as his utopia are inconclusive structures that reinforce how potentiality, rather than actuality, governs his scientific enterprise. As Vera Keller argues, one of Bacon's achievements lies in his success of making the radical literary forms of *desiderata*—which represent the lack of what one currently knows and failures in present learning, rather than extant information—a conventional feature of scientific knowledge. In eliciting the desire for knowledge, such lists "collected and justified incomplete attempts towards shared ambitions" and pushed researchers to realize them through collaboration across generations.[60] Keller argues that "creators of wish lists drew humankind together through an alluring list of shared desires," and in the process naturalists came to envision themselves as part of a collective that serves the public interest across time, rather than only their self-interest in the present.[61] In the decades following Bacon's death, we see enactments and expansions of this form of public interest—as well as, ultimately, its limitations—in the works of Hartlibian reformers, who juxtapose Bacon's operative knowledge with millenarian ideas to formalize improvements in pedagogy, medicine, and agriculture.[62] In their hands, Bacon's "endlesse worke" morphs into a shared, cross-generational enterprise, extending beyond the parameters laid out by the "guide" (the word Bacon chooses to describe his own role). As a result, the Baconian endeavor becomes even more contingent than the Spenserian project: while in the "Letter to Raleigh" Spenser imagines receptive readers who "may" encourage the poet to write more if his first twelve books "be well accepted," the method of "true induction" remains unperfected, its goal postponed until readers transform themselves into practitioners.

While the wish lists are the clearest written forms that transfer the responsibility of perfection onto readers—only if others participate in the collection of histories, they imply, will an item in a catalog graduate from *desiderata* to the real—Bacon embeds invitations to readers across his works. These invitations maintain a delicate balance between pride and frailty, as Bacon both reiterates his role as "guide to point out the road" and acknowledges the impossibility of one individual alone completing his immense project. In *New Organon* he declares, "Neither can I hope to live to complete the sixth part of the Instauration . . . but hold it enough if in the intermediate business I bear myself soberly and profitably, sowing in the meantime for future ages the seeds of a purer truth" (1.116). He reiterates in the *Preparative* that the collection of histories is "so manifold and laborious" that "my own strength (if I should have no one to help me) is hardly equal to such a province" (271). For Bacon, collecting natural history is both necessary task and undesirable burden. Although "one must employ factors and merchants to go everywhere in search of [the materials on which the intellect has to work] and bring them in," such work is "beneath the dignity of an undertaking like mine that I should spend my own time in a matter which is open to almost every man's industry" (271–72). Bacon marks the social hierarchies separating the philosopher and the "factors and merchants," but he cannot fully devalue the latter's work, since "a history of this kind . . . is a thing of very great size and cannot be executed without great labor and expense, requiring as it does many people to help, and being . . . a kind of royal work" (271). But in recognizing the necessity of active agents who will willingly venture out into an unpredictable world, his words hint at an undercurrent of egalitarianism running through the natural philosophical enterprise. The great instauration can only tend toward perfection when released from the hands of its originator.

Bacon's immediate successors revel in this impulse that instigates readers to become knowledge producers, and the mid-seventeenth century is witness to a proliferation of Baconian ideas and methods, as proposed projects escape the secret world of Bensalem into the "not perfected" world of the author; they haunt experiments in the *Sylva Sylvarum*, influence Samuel Hartlib's network of reformers, inform utopian narratives that offer programs of change, and inspire the Royal Society's probabilistic experimental philosophy. I conclude this chapter with a brief survey of the complex afterlife of Baconianism. Instead of offering an exhaustive list of offshoots of Baconian inquiry—a plurality of programs to rival any romance maze that has been richly documented by historians of science[63]—I pause at exemplary instances that retroactively highlight the initiative aspects of Baconian thought. Together, they show how mid-seventeenth-century thinkers found in his writings not only guidelines for specific programs but, more broadly, provocations to "enquire farther."

Bacon's followers expand on the speculative and invitational tendencies of his projects in ways that explicitly tap into the imaginative ethos of his method. For instance, they instrumentalize utopian fiction to address present concerns. Rawley's announcement of *New Atlantis*'s "vnfinished" status seems to have served as sufficient encouragement for others to finish the work. R. H.'s political utopia, *New Atlantis* (1660), claims to describe and perfect the realm that Bacon had introduced. The full title acknowledges its indebtedness to the work "*Begun by the Lord Verulam*" (Figure 5). Its self-presentation as "continued" is also evocative of fictional texts such as Anna Weamys's *Continuation of Sir Philip Sydney's Arcadia* (1651), which completed unfinished narrative threads of an earlier work (Figure 6). R. H.'s text shares very little content with the original, but its author, whose identity is still debated,[64] is at pains to establish the impossibility of the existence of their "platform of monarchical government" (title page) without the worldbuilding undertaken by Bacon. From the opening dedication to Charles II that hopes "You may really become our *Solomona*,"[65] to the identification of Bensalem as an exemplar of monarchical governance ("But he may look on it as calculated for the Meridian of *Bensalem* only; and as but a meer Fiction, aiery speculation, or Golden dream. For such golden things in this Iron age we may rather wish then hope to see wholly effected" [Preface]),[66] R. H. attempts to erase its status as "meer Fiction." Their appeals to Bacon's utopia are key to this framing.

R. H. claims to document something that exists physically, but the limits of "this Iron age" place it beyond our experiential knowledge. R. H. concludes the Preface by presenting their own work as a necessary perfection of Bacon's work:

> This superstructure is only that which he designed and thought to have composed, that is, a frame of laws or of the best state or mould of a common wealth (as Doctor *Rawley* intimates, who knew his mind best) but was never by him perfected. The reason he gives for it was this. His Lordship foreseeing it would be a long work, his desire of collecting the natural history diverted him, which (as he adds) he preferred many degrees before it. Now because he intends not to build a *Solomons* porch before this *Solomons* House: he will summarily discover his Lordships noble design of erecting a Colledg of Light or *Solomons House* (as himself calls it) for the advance of learning. And in case thou canst not find leisure to read his Original . . . he will then open the door, whilst you enter in farther into the Colledg it self.[67]

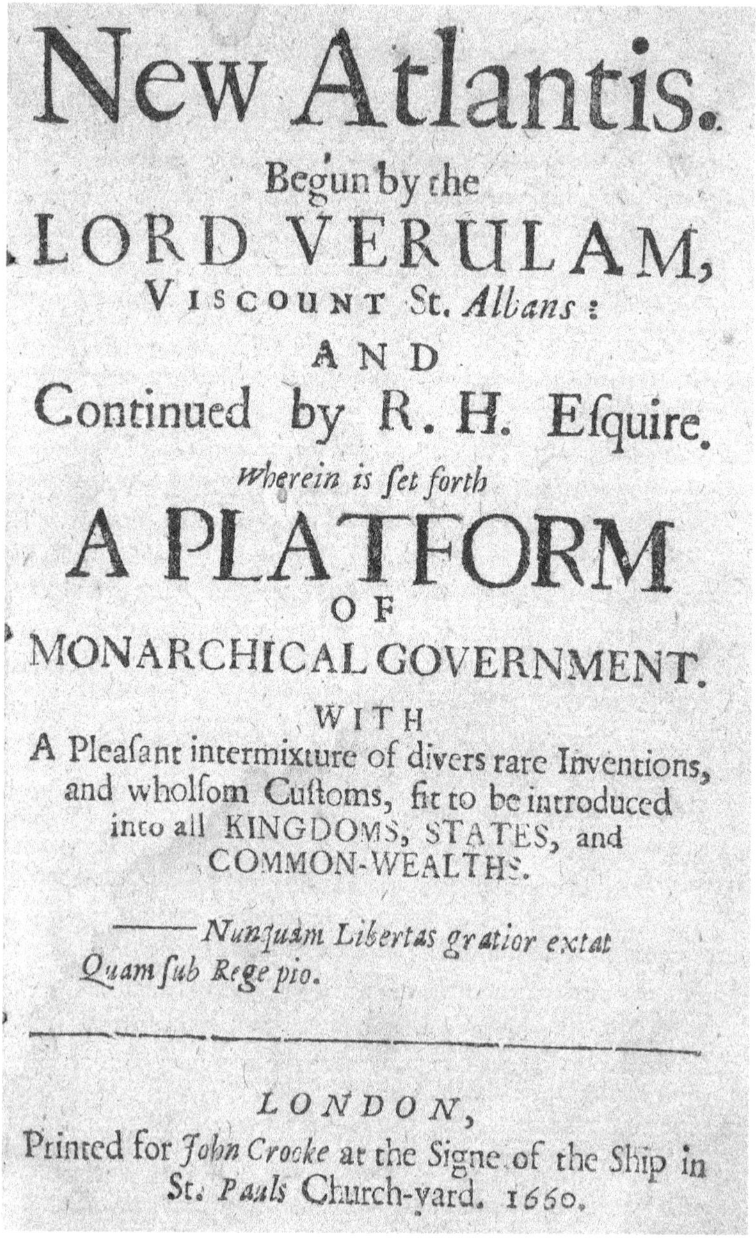

New Atlantis.

Begun by the

LORD VERULAM,

VISCOUNT St. *Albans*:

AND

Continued by R. H. Eſquire.

wherein is ſet forth

A PLATFORM

OF

MONARCHICAL GOVERNMENT.

WITH

A Pleaſant intermixture of divers rare Inventions, and wholſom Cuſtoms, fit to be introduced into all KINGDOMS, STATES, and COMMON-WEALTHS.

——*Nunquam Libertas gratior extat*
Quam ſub Rege pio.

LONDON,
Printed for *John Crooke* at the Signe of the Ship in St. *Pauls* Church-yard. 1660.

Figure 5. R. H., *New Atlantis. Begun by the Lord Verulam, Viscount St. Albans: and continued by R.H. Esquire* (1660). RB373552, The Huntington Library, San Marino, California.

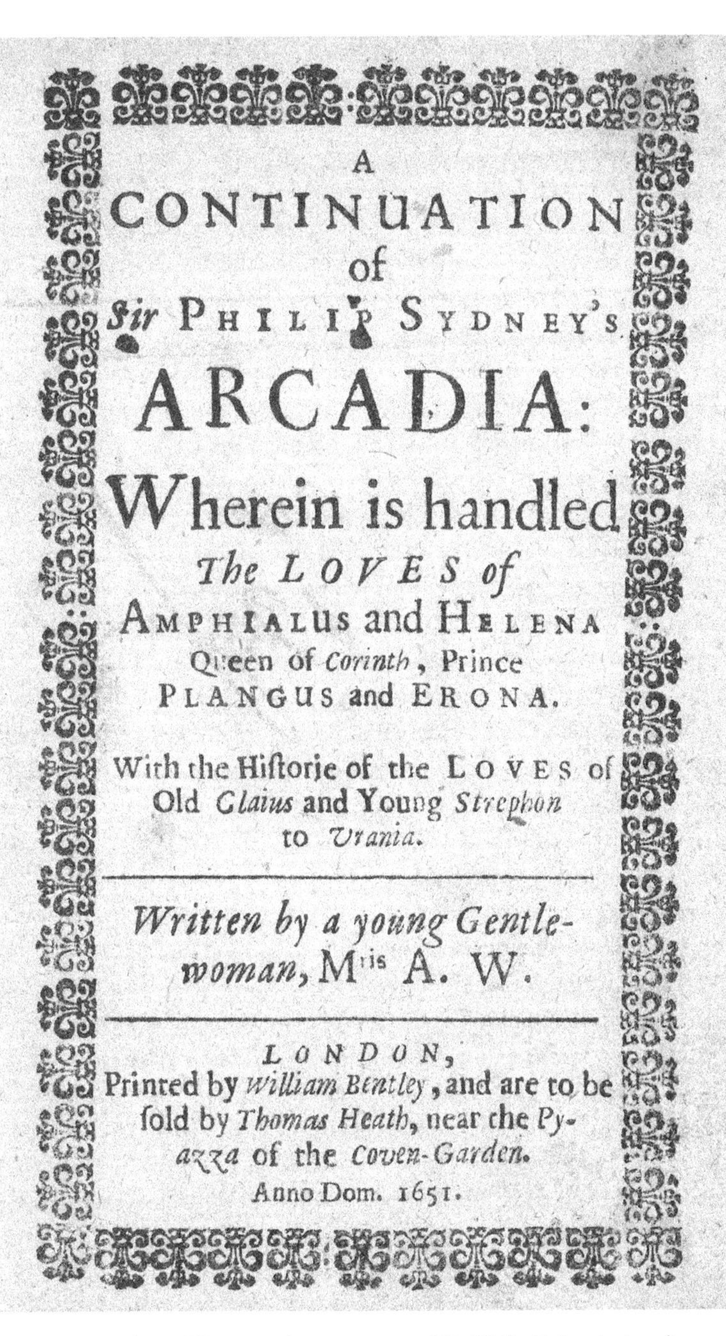

A
CONTINUATION
of
Sir PHILIP SYDNEY'S
ARCADIA:
Wherein is handled
The LOVES of
AMPHIALUS and HELENA
Queen of *Corinth*, Prince
PLANGUS and ERONA.

With the Hiſtorie of the LOVES of
Old *Claius* and Young *Strephon*
to *Vrania*.

Written by a young Gentle-
woman, Mris A. W.

LONDON,
Printed by *William Bentley*, and are to be
ſold by *Thomas Heath*, near the *Py-*
azza of the *Coven-Garden.*
Anno Dom. 1651.

Figure 6. Anna Weamys, *A continuation of Sir Philip Sydney's Arcadia: wherein is handled the loves of Amphialus and Helena Queen of Corinth, Prince Plangus and Erona. With the historie of the loves of old Claius and young Strephon to Vrania* (1651). RB79545, The Huntington Library, San Marino, California.

At one level, R. H. subordinates their voice to Bacon's: the narrator remains the same across the texts, even mentioning characters like Joabin from the original. However, they also underscore how Bacon's own actions hindered his completion of the ideal state, the language of being "diverted" by natural history functioning as a veiled critique, to be corrected in this "continued" work. By repeating Rawley's words that explain the trajectory of Bacon's late writings, R. H. establishes the necessity of their own subsequent work. *New Atlantis* is too important not to be "perfected." The statist visions that remain latent in Bacon's utopia ostensibly animate the later fiction, but it is R. H.'s adoption of the Baconian stance of simultaneous hubris and humility that ensures they can "*set forth*" (title page) the important work of explicating the political institutions and structures of the "best state or mould of a common wealth."

The title of Bacon's *New Atlantis* affords a particular kind of cultural capital, but seventeenth-century innovators seem to have located more broadly in Bacon's utopia an adaptable form with which to imagine intellectual as well as political reform.[68] For instance, Gabriel Plattes's *A Description of the Famous Kingdome of Macaria* (1641)—initially attributed to Hartlib—adapts the ethos of Bacon's work, as characters declare their desire to "learne knowledge" and state their concerns for "authority,"[69] into a sparse dialogic form. Discussing Macaria, a land mentioned in More's *Utopia*, the two interlocutors in Plattes's text (Traveller and Schollar) focus on concerns important to the Hartlib circle, which they lay out as a catalog of "Councell[s]": of "Husbandry," "Fishing," "Trade by Land," "Trade by Sea," and "new Plantations" (3). The text makes explicit that it outlines an achievable project. *Macaria* deploys its assertions of actual existence—a key trope of utopian fiction—not by narrating accounts of travels but by dismissing previous utopias as "impossible" (2) models. The Schollar states: "I have read over Sr. *Thomas Mores Vtopia*, and my Lord *Bacons New Atlantis*, which hee called so in imitation of *Plato*'s old one, but none of them giveth mee satisfaction, how the Kingdome of England may be happy, so much as this discourse, which is briefe and pithy, and easie to be effected, if all men be willing" (9). *Macaria* asks a question at the heart of the Baconian project—"how the Kingdome of England may be happy"—but one never fully articulated in *New Atlantis*. This separation of "my Lord *Bacons New Atlantis*" and the "Kingdome of England" allows Plattes to embrace the desire of the original, but he can claim that only a more pragmatic work can represent a "government" that is "very possible, and withall very easie" (3). We could consider *Macaria* as an opposition to Baconian fiction, where the latter is the zone of the "impossible." Or, we could treat it as the logical end point of *New Atlantis*—a desire to overlap the impossible ideal of "elsewhere" with the "Kingdome of England."

Texts like *Macaria* reveal the paradox of mid-seventeenth-century projects of improvement: they are utopian in content but not in form. By this I mean they advocate the desire for idealized governments, harmonious systems of power, and exemplary institutions of learning, even as they reject the formal elements of the utopian genre, including its immersive discussions of travel and transcultural exchange. As a result, utopia as a trope pervades even non-utopian writing on projects. For instance, William Petty's *The advice of W.P. to Mr. Samuel Hartlib. For the Advancement of Some Particular Parts of Learning* (1648), a text whose title invokes Bacon's *Advancement*, contrasts its own focus with the indirect methods of utopias: "Wherefore being not at leasure to frame Utopias, we shall onely speak of the Number and Salary of Ministers, the time of their service with their qualifications in generall, and Duties in particular, which are to be employed in this Nosocomium Academicum."[70] Petty casts utopia as the textual form (or imagined space) of extravagance, offering a modified version of the oppositions Humphrey Gilbert had outlined, as we saw in Chapter 1. But even as works such as Petty's discard the more obviously "literary" aspects of utopian fiction, in their instrumentality they show what is at the heart of the fictional form: the desire for transformation of current conditions. These texts, then, gesture to the ways in which aspects of imaginative work—from defining it to deploying it in pragmatic ways—became vital to natural philosophy in the immediately post-Baconian world.

These glimpses into the immediate afterlife of Baconianism invite us to pause at historical moments, agents, and documents that precede the experimenters of the Royal Society, who are usually studied as the intellectual heirs to Bacon's provocations for empirical methods of natural inquiry. Although historians of science have often pointed out disjunctions between Bacon's method and the probabilistic experiments of the Royal Society,[71] it is still commonplace to see the scientific institution as a descendant of Solomon's House—as if this lineage is inevitable. This stance is, of course, promoted by early advocates of the Society, as they invoke Bacon's name in prose and poetry, and as works such as Sprat's *History of the Royal-Society* use Bacon's image (frontispiece; Figure 7). Indeed, the gathering of experimental philosophers in the early Society supposedly reflects the Baconian ethos that "whatever deserves to exist deserves also to be known," as they envision an institution devoted to an equally capacious task: its "purpose is," claims Sprat, "to make faithful *Records*, of all the Works of *Nature*, or

Figure 7. Thomas Sprat, *The History of the Royal-Society of London: For the Improving of Natural Knowledge* (1667). RB706394, The Huntington Library, San Marino, California.

Art, which can come within their reach."[72] But as the above short excursions into mid-seventeenth-century Baconianism show us, there is no direct line from Bacon to the Royal Society. Instead, we need to think of the Society's work not as an obvious continuation of Baconian method but as one of its many digressions. We might better read the early practices of the Royal Society as one powerful attempt to transform the Baconian method into a generator of stabilized matters of fact.

One of Bacon's important legacies for Royal Society practitioners, as we saw in the Introduction, was his distinction between words and things, a distinction that the Society claims to implement: Sprat writes that its members have "been most rigorous in putting in execution, the only Remedy, that can be found for this *extrauagance*: and that has been, a constant Resolution, to reject all the amplifications, digressions, and swellings of style."[73] By predicating natural knowledge on the premise that things precede or exceed in value the force of words, Bacon's inheritors supposedly solve the problem that haunted their "guide." Yet, as I have shown, the homologies between Spenserian romance and Baconian method—and more broadly the imaginative techniques undergirding Baconian science—reveal how the "endlesse worke" of words drives the study of things. And even as Bacon's self-proclaimed intellectual descendants embrace the superiority of *res* over *verba*, other seventeenth-century writers claim unparalleled authority for *poiesis*. Indeed, two contemporaries of the early Royal Society, Margaret Cavendish and John Milton, suggest that imaginative writing is essential to addressing one of Bacon's key beliefs: that "whatever deserves to exist deserves also to be known, for knowledge is the image of existence." In the next chapter, I turn to Cavendish, who, like Bacon, embraced utopian fiction toward the end of her writing career. However, unlike Bacon—who sets out to define the parameters of natural philosophical knowledge at the expense of other arts, and who, in *New Atlantis*, deploys fiction in the service of his natural philosophical agenda—Cavendish refuses to sever fiction and philosophy. Instead, as we will see, her imaginative worldmaking insists that theories of the physical world must be accommodated to the forms of poesy.

CHAPTER 4

Margaret Cavendish's Physical Poetics

But my *Ambition* is such, as I would either be a *World*, or nothing.

—Margaret Cavendish, *Poems, and Fancies*
("*To Naturall* Philosophers," 1653)

In the last chapter, we saw how various technologies—from the aphorism, to the middle axiom, to the concept of "error"—overwhelmed Bacon's aspirations for "certain and demonstrable knowledge" with an endless profusion of possibilities. While seventeenth-century naturalists, including experimental philosophers in the Royal Society, eagerly tried to actualize such potentialities, one of the Society's prominent contemporary critics, Margaret Cavendish, afforded some of the most incisive ruminations on key aspects of Baconian thought—the relation between "Nature" and poesy,[1] for instance, or the appropriate methods of engagement with the natural world. The Duchess of Newcastle was one of the most prolific—and most ambitious—writers of the seventeenth century. From 1653 to 1668, she published poetry, essays, prose fiction, orations, and plays on the topics of politics, literature, natural philosophy, and moral philosophy. While Virginia Woolf famously lamented that "nobody ever reads" Cavendish's "torrents of rhyme and prose, poetry and philosophy," scholars in the last few decades have celebrated her expansive corpus and capacious thinking, locating in this "wild, generous, untutored intelligence"[2] not only an astute critic and commentator but also a producer of seventeenth-century philosophical thought. Cavendish's thinking engages the facet of possible knowledge that most explicitly refuses to sever "poetry and philosophy." Indeed, her "torrents of rhyme and prose" realize one radical promise of *poiesis* that Sidney had begun to theorize in the *Defence*, where he put pressure on the relations between poetic and natural creations: instead of imitating objects in "Nature," or even engendering "forms such as never were in nature," Cavendish's literary-thought experiments "grow in effect another nature" by making and remaking the matter of fictional worlds.

Celebrating the philosophical affordances of "fiction," which she theorizes in *The Blazing World* as an "issue of man's fancy, framed in his own mind, according as he pleases," Cavendish locates in literary modes, such as lyric poetry and utopia, ways of knowing that cannot be produced by the faculty of "reason," which "searches the depth of nature, and enquires after the true causes of natural effects" ("To the Reader," 123). A rich body of scholarship has approached Cavendish's principles of literary creation through rubrics of subjectivity, feminist debates, political exclusions, or philosophical networks—that is, through details of her biography, and in relation to the works of contemporary writers and thinkers.[3] Critics have also highlighted how Cavendish's authorial ambitions and intellectual aspirations work in conjunction with her conflicted sociopolitical condition: an elite Royalist, she negotiates between her advantageous position as noblewoman and her marginalization as a woman writer and philosopher, and she often constructs her persona by privileging her aristocratic status over her gender, her proto-feminism existing in tension with her imperialist leanings and support of absolute monarchy.[4] These various strands of scholarship mark the persistent connections between Cavendish's life and her work. In this chapter, I approach her "torrents of rhyme and prose" not only as products shaped by such political, social, or intellectual discourse but also as a logical limit-case of possible knowledge. Her poetry and philosophy are co-constituted from within an absolute "ambition" that cleaves authorial existence to the worlds she creates: Cavendish's desire to be a "World, or nothing"[5] is instrumental to the creation, sustenance, and even destruction of the material realms of her "fiction."

I focus on Cavendish's earliest and last non-dramatic "fiction," where she "figur[es] forth," to borrow Sidney's term, distinct kinds of material worlds within the parameters of specific literary forms. First, I study Part 1 of the first edition of *Poems, and Fancies* (1653), often termed the atomic poems, in which she creates worlds out of atomist physics.[6] In particular my interest here is a triad of lyrics that advance a coherent theory of microscopic worldmaking—of making worlds within worlds—only to expose the unbridgeable gaps between the "poetress" and the infinitesimal realms she creates.[7] These poems dramatize the inability of humans to perceive the atomic scale, producing a kind of epistemological uncertainty that results in the dissolution of the poetic worlds themselves. This disjunction between human knowledge and atomic existence raises the question whether a theory of the physical universe that connects, rather than separates, the author and the material world she creates might be more suited to ensuring the durability of the possible worlds of poesy. To test this idea, I then turn to her final imaginative work, *The Blazing World* (1666). This utopian prose fiction offers a "poetical" (183) enactment of Cavendish's vitalist physics; here, "self-moving" (154) material Nature becomes the generative principle that is not separable from but emerges within the mind of the "Authoress" (224) or "creatoress" (124). These literary endeavors mark the furthest poles, chronologically and conceptually, of the co-constitutive development of Cavendish's material worldmaking and her materialist theories of the physical world, and collectively reveal how the potential sustainability of a specific *kind* of fictional world can calibrate the value of a specific *physics*.

I term such worldmaking strategies, which are predicated on the making and remaking of material realms, "physical poetics." This phrase underscores that Cavendish's theories of literary making are the ground of her theories about the material world, and it acknowledges that the pleasures of making "fiction"—the "issue of man's fancy"—have the potential to reshape one's understanding of physics. Cavendish connects the concept of "fancy" to imaginative writing as early as 1653, when she conjoins it to "Poetry" in *Poems, and Fancies* ("To All Noble, and Worthy Ladies," sig. A3r); she will go on to more explicitly theorize the links between "fancy" and "fiction" in *The Blazing World*'s prefatory "To the Reader." She also advocates for the generative power of the author's inherent "natural phancies," and its intimacy with her "natural reason," in *The Philosophical and Physical Opinions* (1655, "To the Reader").[8] These phrases are part of a cluster of human faculties—including "natural eyes," "natural eloquence," and "natural wit" (*Blazing World*, 141, 160, 161)—that Cavendish invokes in her writings in order to articulate core elements of her philosophy, which privileges essence over appearance and natural capacities over artifice. I propose that, together, "fancy" and the "natural" faculties (which are framed by her understanding of Nature) are critical to Cavendish's formulation of physical poetics: Cavendish naturalizes her acts of creation by constructing theories of Nature as a variable and changeable entity, and she forges different *kinds* of imaginative worlds by aligning these multifarious conceptions of Nature with the author's "natural phancies." By postulating that the creating self (variously termed "authoress," "creatoress," and "poetress") is *naturally* predisposed to incorporate the ontological affordances of a changeable natural world into her mechanics of worldmaking, Cavendish suggests that the continued existence of a fictive world that is modeled on Nature's mechanisms hinges on a particular theory of the universe—a theory that is ultimately inseparable from the author's "fancy." In this chapter, then, I address an issue crucial to our understanding of possible knowledge: How does the "brazen" world of "Nature," to recall Sidney's words, interact with the possible worlds of fiction? Binding epistemology to ontology, Cavendish reveals that the durability of material fictive realms is contingent on the specific physics from which they are constructed.

Although Cavendish begins her writing career as an atomist, she significantly revises her physical theories of the universe over its course, so much so that by the time she composes her final work, *Grounds of Natural Philosophy* (1668), she is an avowed vitalist. The exact contour of this transition is a well-trodden path in Cavendish scholarship,[9] and here I offer one possible reason for this major shift: she reconceptualizes her materialist philosophy for the sake of her *fancy*, which, she argues in *Natures pictures drawn by fancies pencil to the life* (1656), is "not an imitation of nature, but a naturall Creation, which I take to be the true Poetry."[10] By the time she is writing *The Blazing World*, Cavendish believes that both fancy and reason are rooted in the shared physical substance: both originate in the "rational parts of matter" ("To the Reader," 124) but diverge in their methods and aims. Reason aims at "truth" and drives the work of natural philosophy, while fancy, from which fiction originates, is "voluntary" and "creates of its own accord whatsoever it pleases" ("To the Reader," 123). The unconventional features that scholars locate in Cavendish's atomist and vitalist physics are not arbitrary but rather deliberate constructions of her evolving theory of "fancy," especially its relation to "reason" and to "nature." Cavendish's search for appropriate models of durable, material worldbuilding propels her to compose new versions of Nature, rather than to accept extant theories about the physical world.

Cavendish's ideas of form, in other words, have implications for her physics. We know that she ascribes different intellectual values to different modes of writing. In 1653, she defends speculating about an atomic universe in verse by stating, "the Reason why I write it in *Verse*, is, because I thought *Errours* might better passe there, then in *Prose*; since *Poets* write most *Fiction*, and *Fiction* is not given for *Truth*, but *Pastime*; and I feare my *Atomes* will be as small *Pastime*, as themselves: for nothing can be lesse than an Atome" (*Poems, and Fancies*, "*To Naturall* Philosophers," sig. A6r). The minute scale of the atomic realm is the ontological counterpart to the mode of "*Pastime*," the less serious disposition of "*Verse*." Prose, by contrast, is the name she gives to intellectual exercises associated with "*Truth*" and is more suited to overcoming "*Errours*." I understand Cavendish's later choice of prose for vitalist worldmaking in *The Blazing World* through this rubric, rather than primarily as a problem of genre.[11] Prose, for her, is not synonymous with a literary form—essay, utopia, romance—but instead represents a constellation of forms that are conducive to philosophical work. It thus seems intuitive that she turns to prose narrative when she corrects her original philosophical stance, that is, when she becomes a vitalist. Certain literary genres seem to lend themselves better to the intellectual-creative questions that she addresses through the concept of "fancy" and to the more coherent theories she develops later in her career. At its most radical, this early musing about the difference between verse and prose indicates that writing a prose narrative about building durable material worlds requires a revised physics.

This chapter proposes that the literary form Cavendish chooses is a *cause*, and not merely representational vehicle, of the characteristics attributed to her physics. My emphasis on "poetics" as the governing force behind physics—rather than understanding physics as the catalyst for poetic production—challenges the notion that fictional works that engage with cosmology, materialism, and "Nature" primarily reflect existing, or conventional, scientific theories. Rather than reading literary works as vehicles to represent physical theories of the universe, physical poetics recovers how creative impulses alter one's understanding of physics. This chapter is thus the obverse of the relationship between literary making and knowledge production traced in Chapter 1, which examined the poetic *methods* underlying Spenser's allegorical epic-romance in order to highlight the ways in which a kind of fiction thinks and thereby generates its own theories of knowledge. For Cavendish, a specific physics (atomism or vitalism) functions as the apparatus, rather than the telos, of her fictive creations. My focus on physical poetics also illuminates how Cavendish's obsession with the authorial self—a consistent concern of modern criticism—is not about *herself*. Instead, it is a key component of an evolving literary theory that sees the creator as both instrument and agent of a changeable and pluralist natural world.

Physical poetics thus yokes a persistent authorial (and scholarly) concern—how does a self relate to the realms she creates?[12]—with questions about the structure and function of the cosmos. I read the variety of Cavendish's imaginative forms as literary-philosophical encapsulations of her belief that when natural philosophy's ability to explain the physical realm has been curtailed by its dependence on artificial technologies (like the telescopes that famously feature in *The Blazing World*),[13] one must turn to other modes of knowing and representing the world. To overcome this problem of dependency, Cavendish proposes acts of worldmaking that are founded on understanding Nature as a creative force and predicated on the author's "natural phancies." Underlying this evolving poetics is a feminist argument: Cavendish refuses to accept the commonplace that women's propensity to change is a liability. Instead, she perceives it as the vital ingredient that makes possible female writers' organic connection to Nature's infinite variety, the sign that they are best suited to know, represent, and mobilize its workings. In earlier chapters I explored the knowledge production that occurs in gaps between "might" and "should" in Spenser's poetry, in the rift between prediction and prophetic resolution in Shakespeare's tragedy, and in the chasm between processes and goals in Baconian induction. Here, in Cavendish's case, possible knowledge names the simultaneous creative and intellectual effort of a self that is perennially striving to emulate the dynamism of Nature as this self is pulled between the polarizing states of "World" and "Nothing."

The Authoress Theorizes Nature

Cavendish's varied ruminations on "Nature" culminate in the definition that it "is a perpetually self-moving body, dividing, composing, changing, forming and transforming her parts by self-corporeal figurative motions."[14] While its precise definition changes in the course of her writing career, certain features—Nature's materiality, creativity, variety, indestructibility, and freedom—persist across her works. Cavendish scholarship has focused extensively on the accuracy and coherency (or lack thereof) of her natural philosophy, and the continuing debates about her atomism, when she embraced vitalism, and which philosophers might have shaped her thinking are just a few facets of this strain of research. Rather than tracking the intricacies of these theories, or even pinpointing when she revised her physical theories of the universe, I ask what the continual fascination with Nature enables her, as an authoress who considers her state of being as linked to the material world, to do. Approaching Cavendish as a writer who theorizes Nature even as she constructs it enables us to better understand how the evolution of her physics is inextricable from her methods of worldmaking. To this end, I survey her discussions of Nature before turning to her imaginative writing in more detail. By recognizing how she imagines Nature as both source and vehicle of literary creativity, we can grasp the consistent logic that underpins Cavendish's mechanics of worldmaking *despite* the shift from atomism to vitalism. Since she perceives that processes governing matter-in-motion in the natural world correspond to modes of literary making, the primary difference between atomism and vitalism is not what the world is made of but *how* its making entails different relations between creator and created. The form and function of the natural world are inseparable from notions of invention, and the self that relies on her "natural phancies" is the ideal mediator between physical and imaginary worlds.

Cavendish recognizes Nature's variety and changeability as sources of imaginative production from her earliest publication, *Poems, and Fancies*. Noting that "first this World she [Nature] did create" ("*Nature calls* a *Councell, which was* Motion, Figure, matter, *and* Life, *to advise about making the World*," line 1), she associates the "*Free*" and "*uncurb'd*" style of writing with freedoms in the natural world ("The Claspe" between Parts 2 and 3, Untitled ["Give *Mee* the *Free*, and *Noble Stile*"], lines 1, 2). In *Philosophicall Fancies* (1653), she underscores the perpetuity of Nature, which "is *Infinite*, and *eternall*," elaborating, "There can be no *Annihilation* in *Nature:* not *particular Motions*, and *Figures*, because the *Matter*, remaines that was the *Cause* of those *Motions* and *Figures*."[15] Nature is a ceaseless organizing and operative principle, the source of the change that generates endless motions of matter: "*NAture* tends to *Unity*, being but of a kinde of Matter: but the *degrees* of this Matter being thinner, and thicker, softer, and harder, weightier, and lighter, makes it, as it were, of different kinde, when tis but *different degrees*: Like severall *extractions*, as it were out of *one* and the same thing; and when it comes to such an *Extract*, it turnes to *Spirits*, that is, to have an *Innate motion*" (9–10). Nature's "*Innate motion*" is emphasized here, even though there are tensions between "*Unity*" and differences that are of "degree" but might seem to be of "kinde." Its tendency to compositional unity wrestles with the "*Motions* and *Figures*" of its parts.

Cavendish dispels any indication that these activities are purposeless in *Philosophicall Fancies*: "there is not a *Confusion* in *Nature*, but an orderly Course therein" (10). And even though it is not omnipotent or omniscient—it cannot produce new matter but only arrange and reshape extant matter after God has created it—Nature's power lies in its ability to freely order and compose, serving as the foundation for inventions such as poetic production.[16] The poem that begins *Philosophical Fancies* marks this boundless potential:

> For *Nature's* unconfin'd, and gives about
> Her severall *Fancies*, without leave, no doubt.
> Shee's infinite, and can no *limits* take,
> But by her *Art*, as good a *Brain* may make.
> ("A Dedication to FAME," lines 7–10)

This stanza insists on the unrestrained extensibility of Nature—"unconfin'd" and "infinite," it "can no *limits* take." Her celebration of "unconfin'd" Nature's superiority as a creative force hints at her ongoing preoccupation with the value of natural creation, anticipating the art-nature debates that pervade subsequent works.[17]

In her later works, Cavendish more fully theorizes the notion that Nature—and the "sensitive and rational" (*Blazing World*, 176) matter of which her vitalist cosmos is composed—is volitional, self-moving, and changeable from within. Adapting her earlier ideas about its ordering capacity, she celebrates the constructive aspects of a "nature [that] is wiser than we or any creature is able to conceive: and surely she works not to no purpose, or in vain; but there appears as much wisdom in the fabric and structure of her works, as there is variety in them" (*Observations*, 60). Nature "has an infinite wisdom to order and govern her infinite parts; for she has infinite sense and reason, which is the cause that no part of hers, is ignorant, but has some knowledge or other; and this infinite variety of knowledge makes a general infinite wisdom in nature" (*Observations*, 85). Shaped by the spontaneity of Nature's movements, parts and whole mutually reinforce each other in the vitalist cosmos: "by the virtue of its self-motion, [Nature] is divided into infinite parts"; these parts, "being restless, undergo perpetual changes and transmutations by their infinite compositions and divisions" (*Blazing World*, 154). Constant change is the key feature of this ordering and reordering: Nature's "particulars are subject to infinite changes and transmutations by virtue of their own corporeal, figurative self-motions; so that there's nothing new in nature, nor properly a beginning of any thing" (*Blazing World*, 152–53). Nature is undeniably materialist—there is "one universal principle of nature, to wit, self-moving matter, which is the only cause of all natural effects" (*Blazing World*, 154)[18]—and its movements set the limit-case for order: "nature being in a perpetual motion, is always dissolving and composing, changing and ordering her self-moving parts as she pleases" (*Observations*, 55). The defining characteristics—"ordering," "self-moving," and "composing"—seem to describe an entity whose actions model creativity.

By predicating her mature materialist philosophy on an animate Nature, Cavendish enters contemporary debates about the composition of the universe and appropriate methods to study it. Her vitalist materialism, Sarah Hutton notes, shared commonalities with Thomas Hobbes's mechanism, which "'purported to account for the natural world in terms of the motion, rest, and position of corporeal particles in various structural combinations,'" but, Hutton continues, unlike mechanist philosophers who locate the source of motion as external to inanimate matter, Cavendish argues that motion arises from within and is internal to rational and sensitive matter.[19] These distinctions have implications for scientific methods as well as ontology. Not only does her self-motivated and volitional vitalism challenge externalist sources of motion proposed by the mechanist philosophies of Hobbes and Descartes, but her different understanding of the composition of the world also demands a different method for studying it—we witness this in her criticisms of what she perceives to be the passive and externalist practices of Royal Society members such as Hooke.[20]

We are better positioned to explain Cavendish's investment in crossing the boundaries of "reason" and "fancy" by recognizing that, for her, Nature's volitional, ordering, and self-directed actions underpin the author's natural abilities. "Nature being a wise and provident lady," Cavendish states in *Observations*, "governs her parts very wisely, methodically, and orderly" (105). Words such as "lady" hint at the constant negotiations of conflicting identity positions, especially status and gender, at work in this theorization of Nature. Perpetually in motion, Nature shapes worlds into being through the "rational and sensitive motions [that] do *figure* or *pattern* out something" (55, emphasis mine). This idea of "Nature" is undoubtedly a construction of authorial artifice; what interests me here is not whether such theories are accurate or even the rationales that Cavendish offers to justify them but rather the practices they enable her to adopt and promote. Nature functions as the concept to which she hinges her developing theory of literary making, enabling her to consistently naturalize—or essentialize—both the process of creation and the created entity, as well as the relation between them. We see an instance of such naturalization in *Philosophical and Physical Opinions*, where she declares that "natural reason is a better tutor then education" and argues that "a scholer is to be learned in other mens opinions, inventions and actions, and a philosopher is to teach other men his opinions of nature, and to demonstrate the works of nature"; unlike the scholar and philosopher who rely on training and education, her work emerges from her "head," which "was so full of my own natural phancies, as it had not roome for strangers to boord therein" ("To the Reader," sig. B1v–B2r).[21] Basing her authorial self-representation on the claim that those who actualize the potential of their "natural phancies"—rather than relying on artifice and acquired knowledge—are superior creators, she links the ideal poetic methodology to Nature's processes. The natural author, Cavendish suggests, is better able to embrace Nature's activities because she is uniquely attuned both to the composition of the physical world and to *how* it operates.[22]

Cavendish's writings enact the theory of creation for which Nature—governing, composing, patterning, and ordering—offers a blueprint. In an essay titled "Of Invention" in *The Worlds Olio* (1655), she argues that "invention comes from nature, and imitation from painful, and troublesome inquiries."[23] She aligns fancy—a key concept for her literary theory—with this notion of invention when she declares in *Natures Pictures* that "there is as much difference between fancy, and imitation, as between a Creature, and a Creator" ("To the Reader," sig. c3v). Invention, as she notes in *Worlds Olio*, adds "substance," while "an imitator adds nothing to the substance or invention, only strives to resemble it." These different origins and outputs signify distinct methods: "invention takes his own wayes, besides, invention is easie because it is born in the brain" whereas "imitation is wrought and put into the brain by force" (26). This contrast also demarcates appropriate forms of authorial labor. Invention "is easie" but imitation arises from "painful, and troublesome inquiries." Moreover, imitation is insubstantial; the inventor, by contrast, shapes and transforms the "substance." Invention is also pleasurable because it "comes from nature"; controlled by external forces, the imitator undergoes perpetual pain. At its most extreme, imitating is a violent act. By reflecting a limited understanding of imitation, at odds with more expansive versions described by writers like Sidney and Puttenham, Cavendish's words on invention turn the notion of perfection on its head: "an imitator can never be so perfect, as the inventor, if there can be nothing added to the thing invented; for an inventor is a kinde of a creatour; but most commonly the first invention is imperfect" (26).

Her argument celebrating imperfection echoes the Baconian and Spenserian projects that implore readers to "more inquyre." But she takes the argument to its logical limit when she makes epistemological lack a *necessity* of pleasurable creation. Such reversals of value are logical extensions of Cavendish's celebration of inconstancy: "NAture hath not onely made Bodies changeable, but Minds; so to have a Constant Mind, is to be Unnatural; for our Body changeth from the first beginning to the last end, every Minute adds or takes away: so by Nature, we should change every Minute, since Nature hath made nothing to stand at a stay, but to alter as fast as Time runs; wherefore it is Natural to be in one Mind one minute, and in another in the next; and yet Men think the Mind Immortal" (*Worlds Olio*, 162). The changeable mind, a product of Nature's dialectic of divergence and unity, is the site of human creativity. The phrase "by Nature" connects the interior of one's mind to exterior worlds, harmonizing the author's protean cognitions with changes in "Time." Cavendish's words are striking in their repudiation of constancy, a trope that many early modern women writers deployed to authorize their creative endeavors—Mary Wroth's *Urania*, for instance, combines the constant woman and the ideal poet in the figure of Pamphilia.[24] By contrast, Cavendish claims it is "Unnatural" to possess a "Constant Mind," thereby radically altering the meaning of women's characteristic changeability. Instead of treating inconstancy as a liability, she argues that it is natural, indeed enticing; since mental and corporeal changes are universal, those who deride women's fluctuating natures wish to reside in an anomalous state of being.

To challenge the idea that constancy is the norm, or even desirable, Cavendish predicates her worldmaking on the symbiosis of Nature's dynamism and the mind's tendency to change. *Poems, and Fancies* poeticizes how "GReat *Nature* by *Variations* lives, / For she no constant course to any gives" (Part 3, "Natures *Exercise*, and *Pastime*," lines 1–2) and also notes the mobility and vibrancy of cognition: "My *Thoughts* did travell farre, and wander wide" ("*The* Motion *of* Thoughts," line 4). *Natures Pictures* more forcefully transports change and variety to the figure of the author: "my Ambition is restless, and not ordinary; because it would have an extraordinary fame" ("An Epistle to my Readers," sig. c1r). The restless expansiveness of thoughts cannot be dismissed as instruments of "vain" wandering, even though they run exponentially ahead of the body—"I have not spoke so much as I have writ, nor writ so much as I have thought" (367). The poetic dialogue between "Reason" and "Thoughts" in *Philosophical Fancies* furthers the case that the corporeality of uncontrollable thoughts is vital to invention. Reason cautions thoughts to "run not in such strange phantastick waies," else "The *World* will think you mad, because you run / Not the same *Track*, that former times have done" ("Reason, *and* the Thoughts," lines 1, 5–6). Thoughts, however, claim that such restlessness is both productive and pleasurable since it is authorized by Nature: "wee do goe those waies that please us best. / *Nature* doth give us liberty to run / Without a Check, more swift far then the *Sun*" (lines 10–12). This propensity to wander enables them to access the unknown and realize the "phantastick," creating their own paths instead of following the "*Track*, that former times have done." Reason's slow pace and meticulousness—much like the imitator—negatively impact innovation. Where the mere presence of uncurbed thoughts provoked Bacon's anxieties, Cavendish longs for their proliferation.

These examples also intimate that *poiesis* can be deeply physical even when it is not about physics and the natural sciences. We see a vivid example of this in "The Claspe" that connects Part 1 to the rest of *Poems, and Fancies*; the physicality of the word "claspe" anticipates the corporeal aspect of the content and the extreme effects on the speaker that are documented in the poem. In the opening poem of this section, the speaker begins by highlighting the physical consequences of composition: "WHEN I did write this *Booke*, I took great paines, / For I did walke, and thinke, and breake my Braines" (Untitled, lines 1–2). The effects of her actions require that she let her thoughts rest—they "downe would lye, / And panting with short wind, like those that dye" (lines 3–4)—so that "When Time had given *Ease*, and lent them *Strength*, / Then up would get, and run another *length*" (lines 5–6). This poem captures the cycles of impact and recovery the author must undergo so that her thoughts "might run agen with swifter speed, / And by this course *new Fancies* they could breed" (lines 9–10). Once thoughts circulate in the world, her "*Braine* is more at ease" (line 12). It is imperative for her "ease" that thoughts "might run agen." More importantly, this kind of mobility is essential for the production of "*new Fancies*." Even as it exposes the negative effects on the writer, this poem exemplifies the necessity of a physical poetics. The writer who relies on Nature's "uncurb'd" motions to realize her "extraordinary" ambitions must bear the brunt of this force, as embodied thoughts determine the forms and ends of her capacities for invention.

The release of thoughts into the world is precisely what the writer wants. Given Cavendish's disdain for the "Constant Mind," it is unsurprising that she aligns such acts of creation with what she identifies as women's innate malleability. In the process, she turns stereotypical criticisms of female changeability into a virtue. She inaugurates her career by arguing that "*studying* or *writing* Poetry, . . . is the Spinning *with the* braine" (*Poems, and Fancies*, "The Epistle Dedicatory," sig. A2r) and declares fancy—and its product, poetry—to be the proper provenance of the woman writer: "*Poetry*, which is built upon *Fancy*, *Women* may claime, as a *worke* belonging most properly to themselves: for I have observ'd, that their *Braines* work usually in a *Fantasticall motion*; as in their *severall*, and *various dresses* . . . and thus their *Thoughts* are imployed perpetually with *Fancies*. For *Fancy* goeth not so much by *Rule*, & *Method*, as by *Choice*" ("To All Noble, and Worthy Ladies," sig. A3r). Cavendish makes the association between actor and labor an intrinsic, or *natural*, feature of her ideal poet. This kind of work is predicated on lack of "*Rule*, & *Method*." Hinging Fancy on "Choice" also distances it from painful force. As a woman writer who associates "spinning" with the spiraling of thoughts, and who values change and mobility, she seems perfectly suited to be "imployed perpetually with *Fancies*." Cavendish offers a theory of poetic production as ambitious as any we have seen so far. But as she rethinks fiction's relation to the natural world, elaborates on poetic labor, and considers a writer's responsibility to truth and actuality, she proposes that the "right poets," to recall Sidney, are female.

Atomist Poetics and Precarious Worlds

Cavendish first explores how the author can mobilize the changeability of the natural world in the atomic poems, the opening series of lyrics in *Poems, and Fancies*.[25] This work depicts an unconventional atomism where the four figures of atoms (sharp, long, round, and square) correspond to the four elements (fire, air, water, and earth). The worlds of poesy, however, do not merely represent a theory of matter. Instead, Cavendish atomizes creation: as fragmentary lyrics coalesce into a poetic universe composed of atoms, they reveal the cosmological scale of her imagination. Cavendish's linkage of atomism and poetics recalls Lucretius's *De Rerum Natura*, even though her claims to lack knowledge of Latin complicate attempts to seek a line of direct influence. Jessie Hock provides a useful way of understanding this resonance: "the broad similarities between *P&F* and *DRN* are indisputable" not because Cavendish directly appropriates *De Rerum Natura*'s "atomist physics" but because she embraces the broader "Lucretian ideas about the relation between poetry and philosophy."[26] Here, I propose that this "relation" manifests as an evolving theory of literary making that shapes works as varied as *Poems, and Fancies* and *The Blazing World*.

The opening poem of *Poems, and Fancies*, "*Nature calls* a *Councell, which was* Motion, Figure, matter, *and* Life, *to advise about making the World*," foreshadows key aspects of the atomic poems. While she does devote time to the work of individual atoms—for instance through descriptions such as "The *Atomes sharpe* hard *Mineralls* do make, / The *Atomes round* soft *Vegetables* take" ("A World made by *foure Atomes*," lines 5–6)—the poems collectively privilege the composed and ordered entities that result from cohesion. This focus on the whole is intimated by titles like "A World made by *foure Atomes*" and lines such as "*foure Atomes* the *Substance* is of *all*; / With their *foure Figures* make a *worldly Ball*" (lines 9–10). Individually, poems function as vehicles to explain the origins or operations of a range of things (from sickness to death, sun to stars).[27] Together they produce a universe.[28]

Like atoms that join and disjoin into infinitesimal realms, the poems generate a universe of "change and variety." (*The Blazing World*, 201). But even as atomic worldmaking models the composition of coherent entities from distinct elements, these created realms become embodiments of a potentially unbridgeable rift. Their minute scale makes them imperceptible to both author and reader, and this epistemological limit has ontological implications. Since perception is constrained by the available technology of the senses, the existence of microscopic domains is always in doubt. Atomic poetics exists as a paradox.

How might one overcome skepticism about worlds that are impossible to apprehend physically? Does the inability to experience their parts make entire worlds unreachable from our scale of existence? Can you preserve something you do not know exists?

To address these issues, Cavendish often subordinates detailing the features of individual atoms to larger conjectures on worldmaking. In an early poem in the series, she explores how atoms coalesce:

> SMall *Atomes* of themselves *a World* may make,
> As being subtle, and of every shape:
> And as they dance about, fit places finde,
> Such *Formes* as best agree, make every kinde.
> ("*A World made by* Atomes," lines 1–4)

The poem begins by gesturing to the volition of atoms, hinting at the possibility of a full description of these entities that are "of every shape" and that "make every kinde." But it almost immediately swerves toward exploring how an entire realm can be composed. While the modal verb "may" intimates that atoms are worlds *in potentia*, Cavendish postpones outlining their individual features till later in the series. Instead, she focuses on the cosmic scale, exploring how individual elements combine into a coherent system and spotlighting the world the poem itself constructs and becomes:

> For when we build a house of Bricke, and Stone,
> We lay them even, every one by one:
> And when we finde a gap that's big, or small,
> We seeke out Stones, to fit that place withall.
> For when not fit, too big, or little be,
> They fall away, and cannot stay we see.
> So *Atomes*, as they dance, finde places fit,
> They there remaine, lye close, and fast will sticke.
> Those that unfit, the rest that rove about,
> Do never leave, untill they thrust them out.
> Thus by their severall *Motions*, and their *Formes*,
> As severall work-men serve each others turnes.
> And thus, by chance, may a New *World* create:
> Or else predestinated to worke my *Fate*.
> (lines 5–18)

By the time the chiastic inversion from "fit places finde" to "finde places fit" occurs, the poem's investment in constructing an analogy of worldbuilding is clear. Atomism provides a blueprint for creative organization (atoms "may a New *World* create") through the trope of architecture. The language of "dance" also suggests that atoms perform choreographed motions, even as words like "rove" and "thrust" indicate these movements might be unsystematic. If initially the architect follows these atomic motions, by the end of the poem atoms seem to imitate the processes of building that "we" perform. The movements of atoms transfer the act of making onto the builder, indicating symbiosis between natural and human compositions. Yet, a disjunction persists.

Despite the fantasy of perfect alignment, random atomic motions are not the same as systematic architectural endeavors. The human maker does not exactly replicate Nature's volition; they plot and organize raw materials, or fit them into pre-planned structures. The discrepancy might be one cause for the strains of skepticism scholars locate in Cavendish's works.[29] The poetress makes worlds while doubting their existence; she can at best conjecture about potential harmony between the creating self and created world. This epistemic fragility is especially problematic for an author whose authority is based on the claim that her faculties mirror Nature. At its most extreme, this disjunction ruptures her intimacy with the natural world, raising the question of whether certitude requires physical access. While atomism provides a theory of pluralist worldmaking, the minute scale introduces instability to the worldmaker's ability to maintain the cosmological systems she creates.

Instead of writerly paralysis, however, skepticism about one's ability to acquire absolute knowledge acts as a catalyst for atomic creations. Cavendish makes epistemic lack central to her poetic methodology, as the indeterminate nature of cosmological knowledge fuels a kind of conjectural worldbuilding. A triad of lyrics toward the end of Part 1 models nested worldmaking, forcefully enacting this poetics by putting pressure on the question: Can worlds beyond our perception exist and persist? In these poems, the speaker engages in an evolving series of conjectures about states of being that are distinct from the realm she occupies. The title of the first poem, "*It is hard to beleive, that there are other* Worlds *in this* World," immediately registers doubt about the existence of microcosmic realms. But the speaker speculates about such ostensible "*Impossibilities*" (line 2), reminding readers that what "impossible to us appeare[s]" (line 3) results from constrictions of the senses: "For many things our *Senses dull* may scape, / For *Sense* is *grosse*, not every thing can *Shape*" (lines 7–8). In a peculiar echo of Spenser's Proem to Book 2—where the narrator defends Faerie Land's potential existence by questioning the general belief that "nothing is but that which he hath seene"—Cavendish's skepticism does not halt poetic production. Rather, it paves the way for new conjectures about entities of a different scale ("So in this *World* another *World* may bee" [line 9]) and that are beyond sensory experience ("That we do neither *touch, tast, smell, heare, see*" [line 10]). The poem employs a series of analogies to imagine the possibility of such realms:

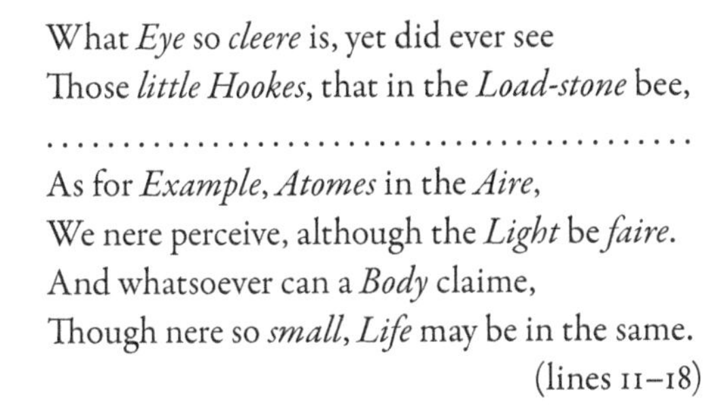

> What *Eye* so *cleere* is, yet did ever see
> Those *little Hookes*, that in the *Load-stone* bee,
> . . . .
> As for *Example*, *Atomes* in the *Aire*,
> We nere perceive, although the *Light* be *faire*.
> And whatsoever can a *Body* claime,
> Though nere so *small*, *Life* may be in the same.
> (lines 11–18)

These lines demonstrate how the atomic scale tests the limits of human experience. Imperceptible worlds might be concealed within reach. But if one accepts the possibility of such spaces, one also accepts their likely inhabitability, and a poetic realm can come into existence in the space of a few lines—"all without our *hearing*, or our *sight*, / Nor yet in any of our *Senses* light" (lines 29–30). Such worlds are dynamic systems emerging from our ability to accept that "probably may *Men*, and *Women* small / Live in the *World* which *wee* know not at all" (lines 21–22).

The poem progressively debunks the notion that sensory perception is necessary to establishing actual existence. The speaker's increasing insistence on what "may be" dramatizes an emergent strategy of conjectural worldmaking that highlights the failures that can result if we cling to verifiability as the basis of knowledge:

> And other *Stars*, and *Moones*, and *Suns* may be,
> Which our *dull Eyes* shall never come to see.
> But we are apt to laugh at *Tales* so told,
> Thus *Senses* grosse do back our *Reason* hold.
> Things against *Nature* we do thinke are true,
> That *Spirits change*, and can take *Bodies* new;
> That *Life* may be, yet in no *Body* live,
> For which no *Sense*, nor *Reason*, we can give.
> As *Incorporeall Spirits* this *Fancy* faines,

> Yet *Fancy* cannot be without some *Braines*.
> If *Fancy* without *Substance* cannot bee,
> Then *Soules* are more, then *Reason* well can see.
> (lines 31–42)

The imagined world straddles a dual ontology—beyond verification but cognitively within grasp if one accepts the premise of nested worlds—that places entire astronomical systems out of reach without negating their claims to existence. The "*Senses* grosse" not only mislead, they hinder rational knowledge of the complexity of the universe. Yet, the lack of evidence activates conjectures that oddly heighten the possibility that "other" realms "may be." The impossibility of verification opens space for potential states of being, as the poem professes that worlds originating in fancy are probable *physical* spaces of habitation. "*Fancy*" itself is substantial and related to the material brain. This poem thus attempts to reconcile a larger issue haunting the first part of *Poems, and Fancies*: skepticism about achieving certitude exists in tension with the conviction that *poiesis* can overcome evidentiary lack and produce its own forms of knowledge. Atomist theory is an ideal vehicle to examine this conundrum. As material entities that cannot be observed, atoms test the bounds between the real, potential, and nonexistent. These foundational elements underwrite why one might imagine states of being distinct from the actual realm one occupies. Yet, this kind of existence is always plagued by a sense of privation, since the author can only hypothesize about the minute scale of being.

The "If" in the concluding couplet also exemplifies how the poem uses suppositions about the cosmos to invite further speculation. Such conjectural tools are profoundly generative, as the breadth of Cavendish's atomic corpus shows. Indeed, this strategy is the foundation of physical poetics in poems such as "*If* Infinite Worlds, Infinite Centers":

> IF *Infinites* of *Worlds*, they must be plac'd
> At such a distance, as between lies waste.
> If they were joyned close, moving about,
> By justling they would push each other out.
> (lines 1–4)

The repeated "If"s may seem to deauthorize the actuality of this world. But as recurrent instruments of conjecture, each "If" demands an implicit "then" through which the author explicates the positions, activities, and interactions of multiple worlds. As its frequent presence in several of the atomic poems reveals, the "if" is a key grammatical technique of Cavendish's methodology, one of the linguistic vehicles through which her works facilitate further inquiry and imaginings.[30] We see comparable strategies across her works. For instance, in *Philosophicall Fancies* she states that "I *Could have inlarged my* Booke *with the* Fancies *of the severall* Motions" (72) and provides a list of things that do not yet exist in the published work: they range from movements and effects of the planets and sun, to "Motions [that] *make* Civil Wars," to questions on "Physicke," to the sympathy between creatures (72–77). Although she does not elaborate on these topics, the act of naming objects and procedures she *might* have examined hints at the possibility of augmentation. Mentioning potential avenues of inquiry, this list lays out a research agenda that reappears in different forms in her later works. Conjecture—in the form of the "If," "may," or "Could Have"—triggers the author's fancy. Remaking the persona of the Baconian guide, Cavendish ends by stating that she does not conclude her writing, since "that which I have writ, will give my Readers so much Light, as to guesse what my Fancies would have beene at" (77). While Bacon hoped to propel readers into the world to fulfill his *desiderata*, Cavendish suggests they may know more when they look inward into her "Fancies."

The next poem in the triad in *Poems, and Fancies*, "*Of many* Worlds *in this* World," presents different scales of existence as a universe of nested boxes:

> JUST like unto a *Nest* of *Boxes* round,
> *Degrees* of *sizes* within each *Boxe* are found.
> So in this *World*, may many *Worlds* more be,
> Thinner, and lesse, and lesse still by degree;
> Although they are not subject to our *Sense*,
> A *World* may be no bigger then *two-pence*.
> *Nature* is curious, and such *worke* may make,
> That our dull *Sense* can never finde, but scape.
> For *Creatures*, small as *Atomes*, may be there,
> If every *Atome* a *Creatures Figure* beare.
> If foure *Atomes* a *World* can make, then see,
> What severall *Worlds* might in an *Eare-ring* bee.
> For *Millions* of these *Atomes* may bee in
> The *Head* of one *small*, little, *single Pin*.
> And if thus *small*, then *Ladies* well may weare
> A *World* of *Worlds*, as *Pendents* in each *Eare*.
>
> (lines 1–16)

This poem is composed of elements now familiar to us: celebrations of Nature, imperceptible realms, confusions of scale, and conjectural techniques (moving from speculation to declaration, the shift from "If" to "thus" seemingly proves a hypothesis). The multiple suppositions bolster the idea that plural worlds are plausible. The analogies, however, reiterate disjunctions between the various tiers of existence. Worlds within worlds are "like" nested boxes but not identical to them. The poem also blurs the distinctions of scale by confusing the status of the container and the contained: a small "*two-pence*" comes to hold an entire world. Even the analogy of the "*two-pence*" denotes an imperfect comparison; the coin, unlike "many *Worlds*," is both visible and tactile. The container theory of worlds adheres to Cavendish's claims elsewhere about Nature's efficiency ("curious" Nature is cognizant that entities must be organized in proper configurations) and its composing power ("degrees," "thinner," and "less" point to different aspects of ordering). Yet, phrases like "*World* of *Worlds*" make the realms seem innumerable, spilling out of the contained structures with which the poem began.

But the most striking feature of the poem is its insistence on the potentiality of these worlds composed of invisible and intangible particles. The poem's operative word for this ontology is also familiar to us: the repetition of "may" highlights how microscopic worlds are predicated on their distance from actuality. The modal term (along with "might") denotes more than the possibility of "many *Worlds* more": it also gestures to a constant threat of loss, disappearance, even destructibility of these microscopic realms. The creatures, absent from view and impossible to touch, are as "small as *Atomes*," effectively nonexistent at our scale of perception. The poem does not resolve the issue of whether the distance between the authorial self and the worlds she creates can ever be eliminated. But this precarious relation raises the question: Are these created worlds sustainable?

To answer this, Cavendish expands on her speculation that "severall *Worlds* might in an *Eare-ring* bee." In the poem that follows, she personalizes worldbuilding on a microscopic scale by transporting readers to an individual organ of hearing. Physical poetics is also explicitly feminized, as she links this act of worldmaking to her broader interest in extravagant clothing and ornaments as instruments of self-presentation. Her most anthologized poem, "*A* World *in an* Eare-Ring," is a culmination of several issues with which earlier lyrics have been grappling.

AN *Eare-ring round* may well a *Zodiacke* bee,
Where in a *Sun* goeth round, and we not see.

> And *Planets seven* about that *Sun* may move,
> And *Hee* stand still, as *some wise men* would prove.
> And *fixed Stars*, like *twinkling Diamonds*, plac'd
> About this *Eare-ring*, which a *World* is vast.
> That same which doth the *Eare-ring* hold, the *hole*,
> Is that, which we do call the *Pole*.
>
> (lines 1–8)[31]

The repetition of what "may" be emphasizes the distance of this realm from the speaker's scale of existence. She cannot escape the terminology of actual cosmology ("*Zodiacke*," "*Sun*," "*Planets*," and "*Pole*"), even as she defamiliarizes readers through shifts in scale, and by extension perspective, between the micro and macro. A small "Eare-Ring" becomes a "vast" world that encompasses an atomic universe that "we [do] not see."

Yet, the speaker's tone increases in urgency as she confronts the lady's absolute distance from the world in her earring. The speaker offers a thick description of this realm:

> [In the World *in an* Eare-Ring] *nipping Frosts* may be, and *Winter* cold,
> Yet never on the *Ladies Eare* take hold.
> And *Lightnings*, *Thunder*, and great *Winds* may blow
> Within this *Eare-ring*, yet the *Eare* not know.
> There *Seas* may *ebb*, and *flow*, where *Fishes* swim,
> And *Islands* be, where *Spices* grow therein.
> There *Christall Rocks* hang dangling at each *Eare*,
> And *Golden Mines* as *Jewels* may they weare.
> There *Earth-quakes* be, which *Mountaines* vast downe sling,
> And yet nere stir the Ladies *Eare*, nor *Ring*.
> There *Meadowes* bee, and *Pastures* fresh, and *greene*,
> And *Cattell* feed, and yet be never seene:
> And *Gardens* fresh, and *Birds* which sweetly sing,
> Although we heare them not in an *Eare-ring*.
>
> (lines 9–22)

These natural resources are instruments of worldbuilding. They are also physically inaccessible. The rich details attempt to overcome an unbridgeable epistemic schism, as do descriptions of man-made entities and events, from "*Cityes*" (line

27) to "*Churches*" (line 29), "*Markets*" (line 33) to "*Battels*" (line 36). But the world remains out of reach, the "*Eare*" a synecdoche for the self who fails to detect the presence of inhabitable realms in close proximity. Several lines note the lady's inability to see, hear, or perceive the world in the earring ("yet the *Eare* not know" [line 12], "yet nere stir the Ladies *Eare*, nor *Ring*" [line 18], "yet not tidings to the *Wearer* bring" [line 38]). More problematically, this failure is contagious: "we heare them not in an *Eare-ring*," the speaker admits. This distance, Lara Dodds argues, is "the most important structuring principle of this poem."[32] It produces the kind of epistemic lack we witnessed earlier—for example, in the recurring phrase "yet the *Eare* not know" (lines 12, 32). But the poem takes to its limit the relation of existence and knowledge, as it documents hindered responses (events there "nere stir the Ladies *Eare*, nor *Ring*"), impoverished experience ("Yet never on the *Ladies Eare* take hold"), and oblivion ("yet the *Eare* be not disturb'd at all" [line 42] by potential "*dancing* all Night at a *Ball*" [line 41]). While the earlier poems also acknowledged gaps between actuality and potential elemental worlds, this lyric suggests that such rifts might be permanent—isolated from human experience, microscopic worlds are markers of perpetual, rather than temporary, ruptures. As the word "yet" becomes a refrain for denoting the impossibility to relate, the speaker's oppositions of existence and experience do not merely intimate limits of knowledge but foreshadow the possible loss of the worlds themselves.

Unable to sense the rich details of the world in the earring, "we" miss the processes that bring it into existence and thereby might be indirectly facilitating its subsequent disappearance:

> [In the World *in an* Eare-Ring] *Rivals Duels* fight, where some are slaine;
> There *Lovers mourne*, yet heare them not complaine.
> And *Death* may dig a *Lovers Grave*, thus were
> A *Lover* dead, in a faire *Ladies Eare*.
> But when the *Ring* is broke, the *World* is done,
> Then *Lovers* they in to *Elysium* run.
>
> (lines 43–48)

The disappearance results from the failure to empathize, a product of the inability to apprehend the world; the reader, like the lady, cannot identify with the "*Lovers* [who] *mourne*," or even feel the loss of their "death," because of the inability to perceive their world. Separation and unknowability reinforce each other, leaving us unable to respond to an active world calling out to be known, to be translated from "may" to "is." The difference in scale, then, exacerbates the gaps between Sidney's "bare *Was*" of history and poetry's "may be" and "should be." The destruction cements the distance between mundane and atomic scales. What human senses perceive as the breaking of a "*Ring*" is the erasure of an entire sphere of existence.

This demolition results from an eternal perceptual failure. The Lady's, and reader's, inability to reach a world that "may" exist, coupled with the speaker's inability to transfer her cognition to other minds, makes this world impossible to sustain. There is a perverse symbolism in using the lyric—meant to be sung to the tune of the lyre—to dramatize the failures of hearing. The earring is a detachable ornament, both part of and distinct from the one wearing it—its liminal status marks the limit between self and world. It is thus an ideal vehicle to highlight that while the poet's "fancy" can create atomic worlds, one creative mind cannot overcome the fragility of an unreachable realm that requires engagement. One shake of the head causes the ring to fall. The destruction of the world in the earring exposes the problematic side of potentiality: a microscopic realm exists in a precarious state, on the brink of dissolving into the impossible.

The collapse of the world exposes the tenuousness of the links of mind and matter that Cavendish attempts to establish through her unconventional atomism. The break between being and knowing also exposes a problem of worldmaking on the infinitesimal scale: atomist philosophy is at odds with her conception of Nature as an ordering, governing, knowledgeable body that undergirds *poiesis*. This physics might be enough to create worlds, but the disconnect between micro and macro levels of existence limits its value as a creative principle. The lack of communication and empathy between creator and created also challenges Cavendish's idea of a harmonious Nature tending toward unity. This disjunction between knowledge and existence belies the Baconian aphorism that "whatever deserves to exist deserves also to be known, for knowledge is the image of existence." *Poiesis* becomes a zero-sum game—or to transport Cavendish's dualistic ontology that structures the authorial ambition, a struggle between "a *World*" and "nothing." As scholars including John Rogers and Karen Detlefsen have shown, atomist theories are often predicated on principles that privilege individual and autonomous entities.[33] "*A* World *in an* Eare-Ring" provides a radical instantiation of both the possibilities and the failures of this form of worldbuilding. Dodds reads the distance "as a comic exaggeration of the disdainful lady of the Petrarchan tradition."[34] Here, I offer a different conclusion: the poetics of atomism finds its logical end in tragic annihilation.

The Durable Worlds of Vitalist Poetics

When skepticism leads to world annihilation, how does one stabilize fictional worlds? To overcome the disconnect between the creating self and created world—as well as her fear that she will be "*Annihilated* to nothing" (*Poems, and Fancies*, "*To Naturall* Philosophers," sig. A6r)—Cavendish requires a new physics. Her aspiration to link imaginative production to the creativity of the natural world finds its most harmonious focus in *The Blazing World*, which enacts how fictional realms constructed out of a vitalist philosophy (in which Nature is volitional and "self-moving," composed of sensitive and rational matter) are more compatible with the authorial self with which she identifies. By applying a different philosophy of Nature to create more durable fictional universes, Cavendish seems to resolve a key question underlying her literary ambitions: How might the materialist cosmos best serve the "natural" author in her act of making "fiction" from her "fancy"?

The Blazing World is physical poetics at its most ambitious, and perhaps most successful. The changeable natural world offers both matter and method for creation; modeling fictive realms on Nature's variability results in a hybrid form in which we witness the evolution of utopia and to which we can trace the origins of science fiction.[35] The culmination of Cavendish's prolonged efforts to harmonize the author's relation to Nature, *The Blazing World* reflects a realization that shifts in *poiesis* demand an adjustment of physics. Earlier in this chapter I proposed that Cavendish needs a revised physics only when she is writing in prose. Based on our study of the rifts between self and world that occur in *Poems, and Fancies*, we can now refine this claim: she must renounce atomism for the sake of creating durable fictional worlds.

As she foregrounds the epistemological underpinnings of *poiesis* in her later works, Cavendish yokes together the realms of words and things that English natural philosophers had been trying to separate. In the paratextual materials of *The Blazing World*, she underscores the distinct work of "fancy" by theorizing the entanglements of truth and fabrication. She highlights the epistemic values of fancy by connecting its origins to that of reason: "The end of reason, is truth; the end of fancy, is fiction: but mistake me not, when I distinguish *fancy* from *reason*; I mean not as if fancy were not made by the rational parts of matter;

but by *reason* I understand a rational search and enquiry into the causes of natural effects; and by *fancy* a voluntary creation or production of the mind, both being effects, or rather actions of the rational parts of matter" ("To the Reader," 123–24). Fancy is as substantial as reason, and they both represent activities rather than fixed entities. They originate in the "rational parts of matter," but the mind's activities aim toward different ends. Thus, instead of distinguishing fact from fiction along an axis of truth and falsehood, we need to recognize "truth" and "fiction" as distinct end points of the exercise of "rational parts of matter." Fancy emerges from within the creator's mind and mirrors Nature as a repository of "voluntary creation." Since it can "help" to "recreate the mind, and withdraw it from its more serious contemplations" (124), there are distinct pleasures in labors of volition. *The Blazing World* seizes on such circulating fancy as *the* way to actualize a "self-moving" materialist physics. The prose narrative offers the author an elastic form in which to actualize pleasurable invention by extending Nature's powers to the "*creatoress*" who makes "a world of [her] own creating" (124).

These entanglements of fancy and reason are crucial to Cavendish's late works, where the fictional worlds are now constructed from "sensitive" and "rational" parts of a vitalist cosmos rather than from atoms. The liberatory potential she repeatedly ascribes to fancy and invention she also locates in reason, which typically has more intellectual legitimacy ascribed to it. In *The Blazing World*, the Empress voices the author's belief that "art does not make reason, but reason makes art; and therefore as much as reason is above art, so much is a natural rational discourse to be preferred before an artificial: for art is, for the most part, irregular, and disorders men's understandings more than it rectifies them, and leads them into a labyrinth whence they'll never get out, and makes them dull and unfit for useful employments" (161). One of the many instances when Cavendish criticizes artifice and artificial methods, this declaration echoes ideas expressed in the exchange between the Empress and the bear-men in which she asks them to destroy their telescopes and rely on their "natural eyes." Moreover, by making the declaration that "as much as reason is above art, so much is a natural rational discourse to be preferred before an artificial," the Empress desires that her logicians and orators follow their "natural wit" and "natural eloquence" respectively, rather than pursuing "formal argumentations" or "artificial periods, connexions and parts of speech" (160–61). The text also sets a limit case for appropriate natural knowledge as the spirits in *The Blazing World* give voice to the idea that "Natural desire of knowledge . . . is not blameable, so you do not go beyond what your natural reason can comprehend" (178–79). The repeated dismissals of artificial modes of inquiry further bolster the connections between reason and fancy: Nature underwrites both faculties, and both are equally substantial and valuable. As a result, reason's intellectual function as an ordering principle is not separable from the theory that informs the fantastic naturalism of *The Blazing World*'s worldmaking practices.

This theory manifests in the text's plural worlds that fragment the actual into multiple possibilities. There are two explicit levels of being in the text: the "numerous, nay, infinite worlds" (184) that exist alongside the Blazing World and are inhabited by characters, and the immaterial worlds created by the Duchess and the Empress. The three "mentioned" (184) inhabited worlds are differently scaled fragmentations or "gross material" (186) realizations of political, social, and physical conditions in the authorial world, and they serve as particular sites for exploring the issues of corporeality, sensory knowledge, and materiality that undergird many of Cavendish's works. The Blazing World, ruled by the Empress, ostensibly presents the image of a "peaceful society, [with] united tranquility, and religious conformity" in its abundance of riches and "well ordered" government (189). The Duchess is the most explicit surrogate for the author in the fiction; her world, with its political "factions, divisions and wars" (189) and its "ruined and destroyed" (193) estates, dramatizes the economic paucity and political failures of the author's world. These two realms serve, in some ways, as models for how actuality *could* be and how it should not be, respectively.

There is no "passage out of the Blazing World into [the Duchess's world]" (216), and this absolute physical rift between the two realms might make readers wonder: are we approaching the disconnected world in the earring? *The Blazing World*, however, ensures worlds interact despite physical barriers. Narration promises a way to transgress boundaries: at one point, the Duchess entertains "some of her acquaintance" (221–22) in her world with "discourse" (221) about the Empress and the Blazing World. In doing so, she translates an inhabited space into a narrated one that facilitates pleasurable conversation, and she expands knowledge about a physically inaccessible realm. The affordances of narrative model extra-physical boundary crossings, as do the traveling souls of the Duchess and the Empress. Moreover, the physical barriers are not universal; there are also worlds that can be physically traversed. The "world which [the Empress] came from" (184) is termed ESFI and nominally represents a "nationalist acronym" for England, Scotland, France, and Ireland.[36] It is joined to the poles of the Blazing World: "the Poles of the other world, joining to the Poles of this, do not allow any further passage to surround the world that way," such that "if any one arrives at either of these Poles, he is either forced to return, or to enter into another world" (126). In Part 2, ESFI becomes the site for testing possibilities of military control and absolute power that are unrealizable in the other worlds.

But a different kind of world also comes into existence within the narrative as characters build worlds by channeling their natural faculties. These worlds are the products of the characters' "natural phancies," to transport a phrase from Cavendish's earlier writing. This act reflects the larger ethos of the text, which repeatedly privileges the natural by relating characters' occupations to their states of being. For instance, the text associates the Empress's sudden acquisition of "absolute power to rule and govern all that world as she pleased" (132) with both her inner virtue and external beauty.[37] And when she institutes numerous changes and establishes "schools" and "societies" to perfect learning and administration in the Blazing World, she assigns the beast-men work in areas such that "each followed such a profession as was most proper for the nature of their species" (134).[38] The identification of the ontology of the beast-men with their areas of inquiry equates proper function with natural states of being. While not all the species organically signify their fields of inquiry (there is no intrinsic explanation why bear-men are natural philosophers), the language of the "nature of their species" posits identification of kind and profession, reinforcing the idea that one's natural faculties are the best indicators of the endeavors one is ideally suited to pursue.

Of course, the category of the "natural" is also conceived through a singular authorial imagination, an instrument to serve the authoress's larger creative and intellectual goals. This contrivance is most apparent when, in time, the Empress perceives the transformations she had instituted as a series of failed experiments, since "the world is not so quiet as it was at first." She realizes such changes had led to "contentions and divisions between the worm-, bear- and fly-men, the ape-men, the satyrs, the spider-men, and all others of such sorts." Fearing that "they'll break out into an open rebellion, and cause a great disorder and the ruin of the government" (201), she resolves to overturn the changes. Yet, these reversals do not signify failure.[39] Instead, they dramatize an unusual manifestation of physical poetics. In order to rationalize her initial desire and subsequent reversals, she appeals to another naturalizing trope now familiar to us: "this [Blazing] world was very well and wisely ordered and governed at first, when I came to be Empress thereof; yet the nature of women, being much delighted with change and variety, after I had received an absolute power from the Emperor, did somewhat alter the form of government from what I found it" (201). The "nature of women," like Nature itself, thrives on "change and variety," and thus the reversals the Empress proposes are also a form of *natural* occurrence.[40] As Cavendish recuperates a perceived flaw of "women"—their changeability—into her larger project about matter, she ties *The Blazing World*'s physical poetics to an issue with which she had grappled throughout her career. The Empress's position recalls Cavendish's declaration in *Worlds Olio* that "to have a Constant Mind, is to be Unnatural." The character's reversals naturalize the connection between voluntary individual action and the self-moving, internal vitalism of Nature, resolving a problem that had long occupied the author. By linking her actions to an apparently inherent quality, this episode of sudden change is presented as a positive act, authorized by Nature itself.

These issues find their most dramatic instantiation when the Empress and Duchess turn to worldmaking as the source of ultimate authority. The Duchess expresses her "desire to be Empress of a material World" (186), prompting the Empress to ask the immaterial spirits "whether there be not another world, whereof you [the Duchess] may be Empress as well as I am of this" (184). The spirits inform the Duchess, however, "you can enjoy no more of a material world than a particular creature is able to enjoy, which is but a small part." Surprisingly, the Empress's sovereign status exemplifies the impossibility of enjoying a realm in its totality: "although she possesses a whole world, yet enjoys she but a part thereof; neither is she so much acquainted with it, that she knows all the places, countries and dominions she governs." The spirits take a counterintuitive position, stating that "a sovereign monarch has the general trouble; but the subjects enjoy all the delights and pleasures in parts" (186). Physical limits make it impossible to access the entirety of extant worlds, even for the absolute monarch. Fuller experience, knowledge, and pleasure arise not from conquest but from creation.

When the immaterial spirits distinguish "celestial" worlds "within" the mind from the "terrestrial" worlds "without," the Empress inquires "can any mortal be a creator?" In response, the spirits suggest that creating an "immaterial world fully inhabited by immaterial creatures, and populous of immaterial subjects, . . . and all this within the compass of the head or scull" can uniquely fulfill the Duchess's ambition (185). Through her act of *poiesis*, she can surpass the Empress's "absolute" power, since, as the spirits reveal, "by creating a world within yourself, you may *enjoy* all both in whole and in parts, without *control* or opposition, and may *make* what world you please, and *alter* it when you please, and *enjoy* as much pleasure and delight as a world can afford you" (186, emphasis mine). The "may" that has been part of Cavendish's poetic toolbox since the atomic poems now becomes a call to action.

This exchange results in a vitalist fiction-making as the Duchess creates worlds "within" her mind by employing her "sense and reason." The text finally actualizes the forms of framing and composing that originate in the creator's mind, demonstrating how created worlds might survive without sensory verification. The mind's "pure, that is, the rational parts of matter" ("The Epilogue to the Reader," 224) correspond to and mimic the self-moving entity of the physical world. It brings possible worlds into existence by imitating the creativity of a changeable Nature, "one infinite self-moving, living and self-knowing body, consisting of the three degrees of inanimate, sensitive and rational matter" (176). While earlier Cavendish had marked distinctions between the creator and imitator, she now erases the distance by appropriating Nature's *activities*, thereby solving some of the central problems that had haunted her atomic poems.

Cavendish now makes the durability of fictive realms central to her physical poetics. In "The Epilogue to the Reader" she declares that anyone "may" make any number of worlds, from those "full of factions, divisions and wars," to those of "peace and tranquility," but that she chooses to create the particular worlds in her fiction because "[she esteems] peace before war, wit before policy, honesty before beauty" (224). However, she also advocates that one create worlds most suited to one's own ambitions: "if any should like the world I have made, and be willing to be my subjects, they may imagine themselves such, and they are such, I mean, in their minds, fancies or imaginations; but if they cannot endure to be subjects, they may create worlds of their own" (224–25). Within the prose narrative, the immaterial spirits echo the egalitarian principle underlying such claims, informing the Empress and Duchess that any "human creature" "may alter that world as often as he pleases, or change it from a natural world, to an artificial; he may make a world of ideas, a world of atoms, a world of lights, or whatsoever his fancy leads him to" (185–86). There are associated benefits of such creative acts: "since it is in your power to create such a world, what need you to venture life, reputation and tranquility, to conquer a gross material world" (186). By incorporating such ideas into her acts of *poiesis*, the fictional Duchess can surpass the Empress's "absolute" power over the world she inhabits.

The Duchess enacts this theory by imitating Nature's rationality and its "change and variety," continuing to create till she "had brought [her] worlds to perfection" (186). Her actions model physical poetics at its clearest: adopting different theories of the universe, she offers the most vivid example of how one's creative priorities drive one's choice of physics. First, she unsuccessfully tries to appropriate ancient world systems. She rejects Thales's cosmology because the "demons" there "would not suffer her to take her own will" and renounces

Pythagoras's world because "she was so puzzled with numbers, how to order and compose the several parts, that she [had] no skill in arithmetic." She discards Platonic "Ideas, [which] having no other motion but what was derived from her mind, whence they did flow and issue out, made it a far harder business to her, to impart motion to them, than puppet-players have in giving motion to every several puppet." In an implicit rejection of Cavendish's earlier atomism, the fictional Duchess also dismisses the world of Epicurus, because "the infinite atoms made such a mist, that it quite blinded the perception of her mind; . . . the confusion of those atoms produced such strange and monstrous figures, as did more affright than delight her, and caused such a chaos in her mind, as had almost dissolved it." She finds Aristotle's world impossible to materialize: "her mind, as most of the learned hold it, was immaterial, and that according to Aristotle's principle, out of nothing, nothing could be made; she was forced also to desist from that work" (187). The atomic poems had exposed how sensory limits led to the collapse of worlds; *The Blazing World* refines this idea to show why characters cannot construct realms that do not align with their theory of the universe. Ancient cosmologies are too distant from her own ontology for the Duchess to appropriate.

The physical impact becomes more pronounced as she continues her efforts. Worlds based on modern mechanist physics—with externalist and reactive theories of motion—are possible but unsustainable, the Duchess learns through trial. She experiences the impact of mechanistic systems that Cavendish criticizes in her philosophical works. Descartes's cosmology of "ethereal globules" makes the Duchess's mind "so dizzy with their extraordinary swift turning round, that it almost put her into a swoon; for her thoughts, by their constant tottering, did so stagger, as if they had all been drunk" (188). It drives her away from ordered composition and destabilizes the mind. Terms like "dizzy" and "tottering" accentuate her lack of control; the matter she deploys controls her, while she desires the opposite. This moment anticipates, in a peculiar way, the effects that physical poetics can have: the Duchess's experiences are reflected in readers' experiences with *The Blazing World* itself. As Marjorie Hope Nicolson famously declared, her "head . . . still spins" after a single venture "in the pages of [Cavendish's] ponderous tome."[41]

Hobbes's mechanist cosmology literalizes the effects of adopting a physics of force and reaction.[42] The Duchess successfully makes a world according to "Hobbes' opinion," but

> when all the parts of this imaginary world came to press and drive each other, they seemed like a company of wolves that worry sheep, or like

> so many dogs that hunt after hares; and when she found a reaction equal to those pressures, her mind was so squeezed together, that her thoughts could neither move forward nor backward, which caused such an horrible pain in her head, that although she had dissolved that world, yet she could not, without much difficulty, settle her mind, and free it from that pain which those pressures and reactions had caused in it. (188)

The violent impact of its "parts" as they "press and drive each other" makes this "imaginary" world unsustainable, the hunting metaphors underscoring the threat of destruction implicit in the motions. "Reaction," the basis of the external movements, affects both parts of the world and the "mind" that creates it. Such motions control and, at their most extreme, destroy the entity that generates them. Even after the Duchess dislodges the "pressures and reactions" from her mind, the "pain" of containing them lingers. Pain replaces pleasure, the primary incentive of "invention." The Duchess's experience with the Cartesian and Hobbesian worlds suggests that while it is logically and materially possible to create worlds by appropriating foreign "patterns" (188), the creator can more pleasurably sustain the world that emerges from her natural kind. A creator should make a world suited to her state of being instead of inhabiting another's. In this moment, the larger problem we encountered in the atomic poems persists: a disconnect threatens the created world's durability. But this moment also foregrounds a starker side of physical poetics. While the lady with the earring remains oblivious to the destruction of a microscopic world, here the creator, too, experiences repercussions. Her attempts might destroy both her and the world she desires to bring forth.

To overcome these issues, the Duchess formalizes the naturalizing trope that was applied to the beast-men—that "each" act in a manner "most proper for the nature of their species"—when she realizes her own ontological status in "a world of her own invention" (188). She is introduced in *The Blazing World* as "a plain and rational writer, for the principle of her writings, is sense and reason" (181), as one who "writ sense and reason" (181), and one who prefers "poetical or romancical" (183) cabbalas over philosophical, moral, or scriptural ones. This description presents her as the human embodiment of Nature's "corporeal, figurative self-motions," especially its "sensitive and rational" matter. Thus, the Duchess's self-reflecting world—her "own" world—is perfectly suited to give imaginative form to Cavendish's philosophy of matter. The Duchess finally creates a world "composed of sensitive and rational self-moving matter; indeed, it was composed only of the rational, which is the subtlest and purest degree of matter" (188). She thus enacts Cavendish's claim in the Prologue that "fancy" voluntarily creates from the mind's rational matter; this world re-creates the matter in motion on which it is built. The Duchess's actualization of a "celestial" world not only is a poetical representation of the self but also harmonizes this self with both natural and imagined worlds.

Her actions resolve several problems haunting Cavendish's physical poetics from its earliest manifestation. This world "appeared so curious and full of variety, so well ordered and wisely governed, that it cannot possibly be expressed by words, nor the delight and pleasure which the Duchess took in making this world of her own" (188). This world, uniquely "her own," actualizes the author's vitalist philosophy; within the fiction, it also replicates its creator's state of being. Full of "variety," perfectly "ordered and wisely governed," it gives ontological form to Cavendish's *poiesis*. The orderliness, stability, and variety produce "delight and pleasure," making the Duchess wish for a proliferation of self-modeled worlds where there is perfect correspondence between world and self: she now instructs the Empress to create a world in "her own mind" rather than inhabiting the Duchess's, because "your Majesty's mind is full of rational corporeal motions" (189). In this way, she settles the Empress's doubts about the nature of the worlds that one might produce, control, alter, and enjoy.

Cavendish's fictional worlds, then, give form to her natural philosophy. Sharing the same concepts *and* the same matter, the value of one can only be explained by reference to the other. While innumerable worlds are logically, even materially, possible, the creator's ability to control, enjoy, and sustain created worlds is predicated on their relationship to matter and the ability to manipulate it—poetics is physical in both source and effect. The form of the fictional realm is inseparable from the matters out of which it arises, as Cavendish triangulates the creator's state of mind with the world she occupies and the realms she creates. This identity between mind and world offers an alternative to the unsustainable existence in the earring. *The Blazing World*'s physical poetics produces possible worlds that are also durable.

The Blazing World realizes Cavendish's "ambition" "not only to be Empress, but Authoress of a whole world." Indeed, the perfect alignment of "the Blazing and the other Philosophical World" ("The Epilogue to the Reader," 224) with the self allows one to sustain and enjoy one's original creations. She continues in the Epilogue that all "may create worlds of their own, and govern themselves as they please": "let them have a care, not to prove unjust usurpers, and to rob me of mine; for concerning the Philosophical World, I am Empress of it myself; and as for the Blazing World, it having an Empress already, who rules it with great wisdom and conduct, which Empress is my dear Platonic friend; I shall never prove so unjust, treacherous and unworthy to her, as to disturb her government, much less to depose her from her imperial throne, for the sake of any other; but rather choose to create another world for another friend" (225). The author who began her career by oscillating between extreme binaries—desiring to be either a "world" or "nothing"—now envisions communities for the writing subject. She argues that worlds predicated on a naturalized relation between writer and matter can accommodate other ambitions: anyone "may create worlds of their own." Cavendish's writing overcomes the destructive potential of the world in the earring because the Duchess's self-realized world affords pleasure a central place in creation, a desire that goes as far back as *Worlds Olio*, where Cavendish had noted the "painful" inquiries of imitation. The Duchess's dramatization of actualizing possibility reveals the intimate connections of pleasure, creation, and learning. Pleasure from original creation (a product of Nature's tendency to creative motions), rather than the pain of imitation or the fear of oblivion, drives this pluralist authorial enterprise, converting Cavendish's individual ambition into a theoretically egalitarian project that labels all as potential creators, so they "may make what world [they] please."

Physical poetics marks one logical end point of the ambitions for poesy that Sidney articulated in the *Defence*. The "authoress," like Sidney, relies on the figure of the poet-maker to describe her acts of creation and invokes what "may be" to authorize her worldbuilding. It is likely that Cavendish is not beginning with a clear one-to-one relation between literary genre and physics. Instead, as she tries to address how her fiction can best sustain what she perceives as natural links between self and world, she offers tenuous relations between materialist philosophy and genre: *Poems, and Fancies* is comprised of individual lyrics, each of which is analogous to a discrete atom that coheres into a cosmos. The prose utopia, on the other hand, connects parts and whole into one narrative. By yoking literary production and the motions of Nature, Cavendish also revises crucial elements of early modern poetic theories. We see her complicated relationship to literary history reflected in William Cavendish's praise for her "creating Fancy," which can "make [her Blazing] World of Nothing, but pure Wit."[43] By locating her ability to "make a world" in "Nothing, but pure Wit," William seems to situate her literary endeavors within a tradition of English poetic making that can be traced back to the Sidneian poet who "rang[es] only within the zodiac of his own wit." Cavendish's writings, however, belie William's claim that her creations arise from "Nothing." The matter of fiction here is not predicated on what Sidney would identify as "forms such as never were in nature." Instead, Cavendish's lyrical and utopian realms emerge from the productive, creative potential of the natural world itself.

Indeed the concept of Nature is vital to Cavendish's "endeavour to be *Margaret* the *First*" (*Blazing World*, "To the Reader," 124): it is both ontological and rhetorical marker of how a changeable physical world can authorize the ambitions of a female writer at the margins of male intellectual networks and institutions. Even more radically, Cavendish reclaims "Nature" for the domain of poesy. In doing so, she both challenges contemporary experimental philosophers who declare the natural world to be their purview and explicitly rejects the mechanistic physics proposed by male philosophers such as Hobbes. Even as Cavendish's theories about the physical world, and her approaches to knowing it, distance her from thinkers who would have been allies in the Royalist cause, they unexpectedly align her with Milton, the "mighty poet" whose opposition of the monarchy would sever him from public life in the Restoration.[44] While scholars have marked the affinities between Cavendish's vitalism and Milton's monist materialism, and have noted how these philosophical ideas mobilize both their poetics and politics, Milton does not directly challenge or satirize the methods and theories of contemporary natural and experimental philosophers. Instead, he enacts poetry's capacity to "furnish such a vast expense of mind" (in Andrew Marvell's words),[45] and in the process lays bare the fallacy at the heart of the ultimate ambition of natural philosophers from Bacon onward: that natural philosophy was the route to a pre-fallen, original state of human existence. It is to this connection between Milton's and Cavendish's understandings of the epistemic power of *poiesis* that I turn in the final chapter. By dramatizing how experiential knowledge, grounded in the fallen natural world, cannot restore paradise, Milton reveals why the natural philosopher's knowledge can never match the poet's "vast design."[46]

CHAPTER 5

John Milton's Evental Poetics

> And as at first, mankind *fell* by *tasting* of the forbidden Tree of Knowledge, so we, their Posterity, may be in part *restor'd* by the same way, not only by *beholding* and *contemplating*, but by tasting too those fruits of Natural knowledge, that were never yet forbidden.
>
> —Robert Hooke, *Micrographia* (1665)

Satan's temptation of Eve marks the turning point in *Paradise Lost*, condensing into an unrepeatable event the epic's core subject: "man's first disobedience, and the fruit / Of that forbidden tree, whose mortal taste / Brought death into the world, and all our woe, / With loss of Eden."[1] The event is the fulcrum of the poem, instantly disrupting the continuum of experience in Eden and propelling readers, with Eve, into the narrative's fallen world, a state of existence so fundamentally altered that even "Earth felt the wound, and Nature from her seat / Sighing through all her works gave signs of woe" (9.782–83). Eve retrospectively describes her actions as her "sad experiment" (10.967), apprehending with certainty how this moment of "disobedience" produces the absolute difference of the future. She becomes keenly aware that she must disavow the very modes of experiential learning that have enabled her to know about the future in an instant. Eve's "sad experiment" has proved fatal to her—and, in Milton's epic, for the world (and readers) to come.

Eve's usage of the term "experiment" is unique in Milton's poetic corpus, and it raises a question that continues to haunt scholarship on this scene: Is Eve experimenting? In this final chapter, I follow the invitation embedded in Eve's phrasing to read Milton's representation of this pivotal incident in relation to an event-centered enterprise that was becoming increasingly influential in the 1660s:

experimentation in the early Royal Society. A year before *Paradise Lost*'s first publication, Milton's contemporary "authoress" Margaret Cavendish distinguished the scope of fictional worldmaking from what she perceived as the "artificial" working of experimental practice and—as discussed in the prior chapter—evoked possible knowledge in terms of a physical poetics. Here, I propose that Milton's "advent'rous song" (1.13) enacts how the labors of *poiesis* can challenge fundamental aspirations underlying the practices of experimental philosophy. Such aspirations are explicated in texts like Robert Hooke's *Micrographia* (1665), which declares that a prelapsarian state "may be in part *restor'd*" by "tasting too those fruits of Natural knowledge, that were never yet forbidden."[2] Sprat's *History of the Royal-Society*, as we saw in the Introduction, extends the parameters of such aspirations by connecting the restoration of a pre-fallen existence to liberation from the trappings of rhetorical eloquence; the "return back to the primitive purity" is inextricable from the return to a shared Adamic language.[3] The methods of experimental philosophers, suggests Sprat, are intimately linked to their desire to abandon the "colours of *Rhetorick*, the devices of *Fancy*, or the delightful deceit of *Fables*." It is within this intellectual matrix that Milton composes his "great argument" (1.24).

Proposing that his epic will "justify the ways of God to men" (1.26), Milton indicates that it is the art of poesy—whose apparatus consists of "the colours of *Rhetorick*" and "devices of *Fancy*"—that will fulfill the seemingly impossible task of making the state of "primitive purity" apprehensible to fallen readers. Unlike Cavendish's *The Blazing World*, which directly challenges the Royal Society's pursuits when the Empress orders the destruction of instruments that facilitate experimental philosophy, *Paradise Lost* does not so pointedly engage with the work or purpose of the Society.[4] Yet, the ways of knowing that Milton anatomizes in Eve's irreproducible transgression expose a deeper problem with the method of experiment itself: designed to study a "Nature" whose "signs of woe" reveal that it, too, has been dissociated from its prelapsarian state, experimental philosophy cannot "return" practitioners to paradise.

Beyond underscoring Milton's theological beliefs and his commitments to reason, Eve's "disobedience" stages with particular clarity the epistemology that animates *Paradise Lost*—what I call an "evental poetics" that foregrounds how unique incidents shape our understanding of prelapsarian states of being. Epic events—the distinct and delimited occurrences that shape the logic of action in *Paradise Lost*—stage singularity and unrepeatability to produce moments of certitude, and evental poetics delineates how this unit of epic poetry becomes an instrument of knowing. The singular and anomalous event that disrupts continuity to reveal what is true also stands apart from repeated experiences that promise regularity and invite analogy and prediction. We can define evental poetics, then, as the intellectual habits underlying Milton's poetic method, in which epic narration turns toward instances of rupture, contingency, and uniqueness to generate a form of epistemic certainty about a realm that is inherently beyond common experience.

I understand "event" both as occurrence—"Something that happens or takes place, esp. something significant or noteworthy; an incident, an occurrence"—and as result: "The outcome of an action or occurrence; a result, a consequence."[5] Milton's evental poetics enacts how such logics of "eventness" govern incidents of absolute change in *Paradise Lost*. Jacques Lezra argues that the "'event' is a moment marked out as especially meaningful; an event is always 'special.' But it is also what 'merely happens.'"[6] The event is thus both noteworthy and potential consequence of unremarkable occurrences. The battle in heaven, for example, or the falls of Adam and Eve, follow an evental structure in which the event realizes an unanticipated future. In such moments, Milton isolates the "special" aspect of eventness by underscoring the epistemic effects of finite occurrences bound in time: in actualization, the event simultaneously ruptures previous trajectories of action and changes what was supposed to occur.[7] If, according to Witmore, events "simply happen of themselves" while actions "are directed from without,"[8] the absence, or at least the dispensability, of an active agent suggests the event's emergence is always unpredictable. Events, then, are defined by the unexpected turns that distinguish them from other kinds of experiences they seem to echo. While characters such as Eve and Satan might perceive themselves as masters of their own actions, they are unable to fully direct how matters unfold in time. Confronted with the unprecedented, they come to recognize the contingency of incidents that propel them to act, or react, in unexpected ways.

Eve's retrospective terminology of "sad experiment," however, points to the difficulty of isolating the distinctive epistemological insights of unique events from questions regarding repeatable experiments that were used to know the physical world. In its early years, the Royal Society championed a method of experimentation in which the repetition of an experience within controlled conditions would generate probabilistic knowledge and matters of fact. The Society's practices redefined what "experiment" meant in a specific way: a procedure that couples singular "events" with habitual "experiences" to produce repeatable and generalizable results. As Milton poeticizes ways of knowing in the *unrepeatable* actions of Eve, he decouples this linkage of habitual experience and unique event. Instead, his epic poetry provides an alternate, evental epistemology predicated on interpreting the singular, unrepeatable, and ultimately inaccessible conditions of prelapsarian Eden.

By delinking the modes of event and experience that are conjoined in the practice of experiment, the notion of evental poetics invites us to revisit the commonplace in Milton scholarship that Satan and Eve are experimenters. Stanley Fish's claim that "Satan proceeds to initiate Eve into the mysteries of empirical science" has long invited scholars to historicize this scene as one of scientific experimentation.[9] Critiquing Fish, Karen L. Edwards dismisses Eve as a naïve experimenter: "Eve ought to have been more, not less, of an empiricist."[10] While Fish condemns Eve for succumbing to reason, Edwards suggests Eve falls because she does not experiment enough: "holding fast to her own reading of nature's ordered ways would have enabled Eve to see the talking snake . . . for the monstrosity it is."[11] Although Shannon Miller has challenged such readings of Eve as an inept experimenter, Edwards's account of Eve's failures is a powerful one.[12] Joanna Picciotto draws on Edwards's argument to dismiss Eve as a "zealous but incompetent natural philosopher," equating Adamic productivity with experimentalism while reading Eve as one manifestation of Adam's "sensitive body and private fancy."[13] In Picciotto's reading, experimentalist authors, including Milton, demonstrate a "commitment to collaborative empiricism," "replicate the crucially prosthetic, collective, and processual character of experimentalist insight," and turn their readers into "virtual witnesses."[14] *Paradise Lost* enacts "a formal manifestation of experimentalist progress, combining the copious accumulation of experimental results with a modest deferral of certainty about their meaning," and emerges as "a literary counterpart to experimentalist observation."[15] This experimentalist Milton is methodologically aligned with Royal Society experimenters, who embraced cumulative inquiry and virtual witnessing—the technology that, in Steven Shapin and Simon Schaffer's formulation, "involves the production in a *reader's* mind of such an image of an experimental scene as obviates the necessity for either direct witness or replication."[16]

My claim that Milton's poetics is animated by an *evental* epistemology differs both in focus and method from these explorations of "experimentalist literature," which persuasively espouse an "Adamic epistemology" to link "innocence and experience" but fail to accommodate an equally innocent Eve.[17] Instead of equating Eve's fall with a Royal Society experiment, or trying to classify Eve as a perfect (as Miller argues) or imperfect (as Fish, Edwards, and Picciotto differently propose) experimenter, I employ the concept of "event" to read this scene as a crucial example of the larger epistemic problem of singularity and unrepeatability that Milton raises throughout the epic. Stressing the differences among the categories of experience, experiment, and event, I demonstrate that Milton employs varied ways of knowing to make intelligible states of being in prelapsarian Eden. While experimental method joins habitual experiences and singular events in a fallen world, Milton disentangles them by foregrounding the evental nature of Eve's fall. Experimentalism cannot account for unrepeatable and distinct scenes such as Satan's temptation of Eve because experiments were contrived to study the fallen natural world through techniques of collaboration, probability, and processual repetition. Indeed, Eve's fall stages the impossibility of experiment in Paradise as it irreversibly fractures her prelapsarian existence and pivots her to the unknown future. Eve's belated terminology captures her recognition of her error; in a fallen state, she can label her capitulation to the lure of replication as an "experiment." When she eats the fruit, she mistakes an event—singular, disruptive, and irreversible—for an experiment and attempts an impossible duplication.

Evental poetics thus uncovers the habits of thought shaping Miltonic poetics without aiming to establish that the poet is experimenting or suggesting he specifically critiques a Royalist scientific enterprise. By demonstrating that poetry theorizes knowledge practices in a manner distinct from experiments, this approach compels us to rethink how we can reorient scholarship on Milton and science. Partially to counter Milton's perceived backwardness to developments in natural philosophy (as Kester Svendsen argues),[18] scholarship has predominantly focused on recovering connections and exploring objects of knowledge—the natural world, the structure of the universe, materialist philosophy, mathematics—in Milton's corpus.[19] This growing body of work has drawn some rare but stringent criticism: William Poole, for instance, warns readers against proposing close connections—"especially at the level of methodology"—between Milton and the new science.[20] Without claiming commonalities between the methodologies of Milton and his contemporary experimental philosophers, in this chapter I model a critical attention to method or to ways of knowing—not to objects that appear to connect or separate Milton and contemporary naturalists—that can effectively unearth the relations between poetic and natural knowledge. Picciotto's study offers one particularly rich example of how a focus on method can situate Milton's poem within a long tradition of tracing productive and intellectual labor to the figure of Adam: she demonstrates that "by reinventing the innocent Adam as an agent of innocent curiosity, experimentalists made him a model member of the knowledge-producing public of which he was also the embodiment."[21] My focus on Eve's "sad experiment," on the other hand, demonstrates that Milton's poetics is not primarily concerned with knowing the natural world. Instead, I propose that it is because Milton is deeply invested in questions of certitude and probability that were vital to his times (and not because he replicates a specific experimental method that emerged out of natural philosophical practices) that his poetry enables us to comment on epistemological shifts that are also discernable in the Royal Society's practices.

My reorientation from object to method reveals that the epic poem is raising a philosophical and theological, rather than a scientific, problem: What is the state of being before the Fall? At the very moment when natural philosophers had begun to privilege probabilistic knowledge, Milton's evental poetics dramatizes the expansive epistemic scope of poesy by linking certain knowledge with questions about being. In *Paradise Lost*, Milton is concerned with prelapsarian ontology, but the poet can only appeal to postlapsarian epistemologies to make intelligible an unfallen world that fallen audiences can never, in theory, fully grasp. As such, the issues about ways of knowing that drive the epic narrative are not only epistemic, they reveal the ontology of Miltonic poetics. The question of what the inaccessible prelapsarian world *is* remains inseparable from *how* one knows it. Events serve as narratological units of certainty that explicate prelapsarian ontology; they are both educative and transgressive because they enact what is disruptive and unparalleled in prelapsarian Eden. By simultaneously drawing attention to unique moments of knowing and their results, events also underscore that concerns about epistemology—of theories of knowledge—cannot be separated from the methods the poet employs.

Because I situate this chapter in a long tradition of scholarship in early modern literature and science, it offers the most acute example of the critical tendencies that *Possible Knowledge* has aimed to push against—tendencies that suggest we can prove Milton's modernity by reading his poetry as reflective of, or directly engaging with, scientific developments. By reading Milton's poetics with the assumption that it echoes developments in natural history and philosophy, we subscribe to the notion that approaching poetry in terms of science—especially scientific disciplines and philosophical traditions—is key to establishing imaginative writing as forward-looking. By extension, we also accept the evaluative categories that mark scientific knowledge as the standard through which other forms of knowledge-making must be judged. My focus on evental poetics, however, elucidates a dimension of possible knowledge that reveals why *poesy* can serve as a powerful medium to raise questions about—and to test the limits of approaching—truth and certitude. *Paradise Lost* stages moments of transformation that take readers to the brink of certitude, even as the vanishing nature of these occurrences reminds them that such certainty exists in a state of being from which they are forever severed. Only through unprecedented, exceptional, and always-disappearing incidents, suggests Milton, can fallen readers begin to apprehend "that happy state" of "our grand parents" (1.29).

Experimental Philosophy: Joining Experiences to Events

Experiential ways of knowing had long governed how one understood the natural world, and in the beginning of the seventeenth century, the terms "experience" and "experiment" signified similar practices.[22] Both designated generalizable processes in which the accumulation of many instances could lead to facts. "Experiment," as Elizabeth Spiller demonstrates, could refer to a range of practices across varied fields, from operations in alchemy and natural magic to controlled trials, from the creation of model worlds to cognitive and imaginative exercises broadly understood as "thought experiments."[23] The meanings of the terms "experiment" and "experience" began to diverge when early modern naturalists came to understand the former as a controlled method that would provide new insights about the natural world; Peter Dear and Thomas Kuhn, among others, document the differences between the mathematical and non-mathematical (or Baconian) experiments that formed the cornerstones of the "Scientific Revolution."[24] In the mid-seventeenth century, English natural philosophers increasingly considered "experiment" a regulated and contrived event produced within formal or informal laboratories.[25] Such experimental practice also redefined the aims of knowing: as Dear argues, an experiment corresponded to "*how something had happened* on a particular occasion," whereas the term "experience" was still primarily used in the Aristotelian sense of "*how things happen* in nature."[26] The distinction was not yet absolute, however. When Milton's nephew Edward Phillips wrote *The New World of English Words*, he defined both terms under one entry: "*Experience* or *Experiment*, (lat.) proof, trial, or practise."[27]

The term "experience" was central to an academic tradition that had been inherited from Aristotle; in this convention, Dear shows, "an 'experience' was a universal statement of how things are, or how they behave."[28] Predicated on habit and on the expectation of regular behavior, experience did not account for, or even consider, the unique and the anomalous as suitable objects of investigation. One gained an Aristotelian experience through the repeated perception of regularities and by holding them in memory. An experience was meant to explain things that were already "'common knowledge.'"[29] In *Posterior Analytics*, Aristotle elaborates how an experience is developed: we understand through "perception," and "although you perceive particulars, perception is of universals."[30] Aristotle states that experience emerges "from pre-existing knowledge": "from perception there comes memory, as we call it, and from memory (when it occurs often in connection with the same item) experience; for memories which are many in number form a single experience. And from experience, or from all the universal which has come to rest in the soul (the one apart from the many, i.e. whatever is one and the same in all these items), there comes a principle of skill or of understanding—of skill if it deals with how things come about, of understanding if it deals with how things are."[31] Experience in this formulation represents a many-to-one function. Perceiving many "particulars" enables one to constitute a single "experience" from various "memories" of "the same item." An experience is thus gained with the aid of "pre-existing knowledge," if one can perceive this "same item" "often." The process provides "understanding" of "how things are," and it depends on regularity. As the culmination of acts of perception, experiences are built on the Aristotelian premise that "'there is nothing in the mind which was not first in the senses.'"[32]

This notion of experience is also related to what early modern thinkers would classify as probable knowledge, which was based on the credible testimony of authority.[33] Aristotle had demarcated, in the *Topics*, demonstrative knowledge as the domain of certainty and dialectical knowledge as the realm of probability. Demonstration concerns itself with "premisses from which the reasoning starts" that "are true and primary," and the "dialectical" names "those opinions [that] are 'generally accepted,'" that is, those that are "accepted by every one or by the majority or by the philosophers—i.e. by all, or by the majority, or by the most notable and illustrious of them."[34] Works such as John of Salisbury's twelfth-century *Metalogicon* would adopt the divisions listed in the *Organon*, declaring that probable logic "is concerned with propositions which, to all or to many men, or at least to the wise, seem to be valid."[35] Many early modern thinkers continued to echo Aristotle's definitions, linking probable knowledge with the credible testimony of authoritative figures. Blundeville, in *The Arte of Logick*, states, "Things probable, according to *Aristotle*, are these that seeme true to all men, or to the most part of men, or to all wise men, or to the most part of wise men, or else to the most approued wise men."[36] Thomas Spencer's *The Art of Logick* (1628) also announces its debts to Aristotle and declares that an "Axiome is probable which seemes so to all, to many, or them that are wise, by certaine frequent notes, and cleerenes."[37] We must consider the shifting relations of "experience" and "experiment" in the seventeenth century within this nexus of credible testimony, authority, and probable knowledge.[38] Based on *explaining* how things exist in their natural state, the Aristotelian experience was probabilistic and depended on the authority of knowers; in the absence of demonstrable knowledge, one could appeal to "the nature of experience" that "depended on its embeddedness in the community; the world was construed through communal eyes."[39]

The new experimental philosophy in Restoration England, too, operated on the premise that probable knowledge was sufficient for making knowledge claims about "Nature," but it shifted attention from explanation to "successful prediction and control."[40] As Shapin and Schaffer have shown, this experimental philosophy was institutionalized through interconnected conventions: collective witnessing, prediction, repetition, and standardization.[41] Experiments were composed of singular events that were meant to represent universal facts. Experimenters claimed these localized events could be repeatedly experienced and generalized. As the Royal Society modified aspects of Baconian method, the term "experiment" became associated with "the notion that what nature can be made to do, rather than what it usually does by itself, will be especially revealing of its ways."[42] As Dear argues, the "singular experience could not be *evident*, but it could provide *evidence*."[43] I understand this "singular experience" as the *evental* aspect of repeatable experimental practice. The early Royal Society experiment was a communal form of knowledge production, but it differed from the "common knowledge" of the habitual Aristotelian experience. Experiments depended on the agreement of a (supposedly) disinterested body of witnesses who collectively worked to reveal nature's secrets.[44] They functioned within a culture of trust, where the "identification of trustworthy agents," as Shapin has demonstrated, was vital to reliable knowledge production.[45]

Apologies for the experimental method, including Sprat's *History*, outline its main protocols. Witnesses need not have specialized knowledge as long as they display sincerity and bring their labors to the enterprise: "we find many Noble Rarities to be every day given in, not onely by the hands of Learned and profess'd Philosophers; but from the Shops of *Mechanicks*; from the Voyages of *Merchants*; from the Ploughs of *Husbandmen*; from the Sports, the Fishponds, the Parks, the Gardens of *Gentlemen*."[46] The disinterested witness is defined in opposition to "*perfect Philosophers*"; Sprat accepts "plain, diligent, and laborious observers," who might not bring "much knowledg" but bring "their hands, and their eyes uncorrupted" to the task.[47] Sprat does not acknowledge, however, what Robert Hooke—the Royal Society's "commoner assistant"—recognizes: the importance placed on disinterestedness ensures that experiments have, as Michael McKeon argues, "socially leveling implication[s]."[48] Sprat's discomfort with the "equal Balance of all Professions" becomes apparent when he stresses the indispensability of certain social groups:

"though the *Society* entertains very many men of *particular Professions*; yet the farr greater Number are *Gentlemen*, free, and unconfin'd."[49] Sprat's words, perhaps inadvertently, expose that the witnesses who seem disinterested could counterfeit for their own interested purposes; by marking extant hierarchies (he privileges "*Gentlemen*, free, and unconfin'd") he echoes the division of labor between the philosopher and the "factors and merchants" that Bacon had made. Even though the various proponents of the "New Science" distance their work from Aristotelian paradigms, Sprat's words are haunted by the specter of credible testimony and probable knowledge that early moderns had inherited from Aristotle.[50]

While ostensibly rejecting the regular and habitual scholastic experience in favor of a collective and event-based experimental philosophy, the Royal Society implicitly translates the *repeated* perception of the Aristotelian experience into a different kind of probabilistic and cumulative epistemology: a diverse body of witnesses that observes the same "experiment" replaces multiple acts of perception. Witnessing could consist of repeated acts of perception, but some would merely read about the event and even this act of virtual witnessing was considered a reliable form of replication. Sprat's attempts to define collective witnessing through contrived, repeatable experiences underscore a paradox in the aspirations of the Royal Society: "experiment" needs multiple disinterested witnesses but also presumes their diversity will cancel individuated interests. The Society aims to translate numerous particulars into a unified whole by suggesting that experimental method authorizes the effacing of the plurality, difference, and interests of individual witnesses. In the very practices in which Hooke sees the potential for social leveling, Sprat locates the erasure of particularity. Yet in redefining experiment as a series of repeatable and replicable instances, Sprat also couples the experience—habitual and common—with the event: singular, delimited, and disruptive.

Eve's "sad experiment": Staging Unrepeatability in Paradise

Although the terms "experience" and "experiment" remained partially interchangeable in the mid-seventeenth century, in what follows I employ "experience" to discuss regularities, habits, and common modes of perception that make things evident, and I use "experiment" to refer to an enterprise that combines event and "experience," that aims at repeatability and prediction, and that seeks to provide evidence. In *Paradise Lost*, Milton disconnects the evental structure of experiments—their historically specific, localized, and singular nature—from the common and habitual experience. He associates these different modes of knowing with the poem's evolving ideas about ontology to suggest that while replicable experiments are sufficient to explain conditions in the fallen world, only singular events can reveal with any degree of certainty the unique states of being in prelapsarian Eden.

Satan initially applies an experiential strategy to tempt Eve but soon reverts to an experimental approach. Eve, however, continually appeals to an evental epistemology until she succumbs to the enticement of replication. Eve's trial dramatizes the eventness of the Fall in the epic's narrative logic, as she aims to address an ontological problem through an empirical epistemology: whereas the repeatable structure of experimental knowledge is suited to the sociable world of the Royal Society as a way of understanding the natural world, it remains unsuitable for Eve, who can only draw on and modify her own individual knowledge, and whose current state of existence can be altered by a single act of transgression. As Satan and Eve enact the promise and dangers of these varied modes of knowing, the temptation captures the entire "loss of Eden" in a unique event that irreversibly separates the fallen Adam and Eve from their innocent states.

Satan models an experiential mode of knowing in the beginning of his encounter with Eve.[51] His tactic is to make apparent the habitual craving of "all other beasts" for the fruit and to produce in Eve a "like desire" for it (9.592). He does not mark an intrinsic difference between the serpent and other creatures; they are connected by "like desire":

> I was at first as other beasts that graze
> The trodden herb, of abject thoughts and low,
> As was my food, nor aught but food discerned
> Or sex, and apprehended nothing high.
> (9.571–74)

While the serpent's experience is personal, it is not necessarily a unique event, Satan implies. He hopes to tempt Eve not with the evidence of singular transformation but by appealing to the possibility of a common, unplanned experience. The inner change in being is incidental to this act of consumption:

> Sated at length, ere long I might perceive
> Strange alteration in me, to degree
> Of reason in my inward powers, and speech
> Wanted not long, though to this shape retained.
> (9.598–601)

Even when Satan speaks about the "strange alteration" he "perceive[s]" in himself (as the serpent), he does not present himself or the tree as proof. This change in "inward powers, and speech" leads him to "speculations high or deep" (9.602) and culminates in this encounter: her singular "divine / Semblance" has "compelled" him "to come / And gaze, and worship [her] of right declared / Sov'reign of creatures, universal dame" (9.606–12). Satan presents this enhanced cognitive faculty as a by-product of desire and suggests to Eve that a similar experience is possible for all. At this moment, he is not "falsifying experimental data," as Edwards argues.[52] He tries to make "evident," not to provide evidence, that the serpent speaks as a result of eating the fruit.

Satan is primarily interested in articulating the possibility that all creatures can undergo this common experience. To do so, he stresses its unpremeditated nature, using indefinite articles and appeals to contingency—"on a day roving the field, I chanced / A goodly tree far distant to behold" (9.575–76)—to prevent the reader, and Eve, from attaching his description to particular examples or locations. Unlike planned experiments that necessitate the explication of "singular events, explicitly or implicitly located in a specific time and place," Satan's experience, he implies, could occur anywhere, anytime.[53] Moreover, if an experience is an occurrence in which "there is nothing in the mind which was not first in the senses," Satan's recounting of the serpent's supposed transformation is experiential in its saturation with sensory details. The tree "pleased [his] sense" (9.580); its "savory odor" drew him to "smell" it and "gaze"; it produced an "appetite" and finally propelled him to "tasting" it (9.578–85). He repeatedly ate till he had his "fill," and "spared not" (9.595–96). The tree habitually creates "hunger and thirst . . . , / Powerful persuaders" (9.586–87) in all animals that approach it, and who look on "Longing and envying" when the serpent is the only creature that can "reach" it (9.593). Satan's description, which mostly lacks particulars (although he does once label the fruits "fair apples" [9.585]), is about the habitual, common, and partially equalizing force of sensory perception rather than about particular knowledge claims.

Eve, however, displays a cautious curiosity from the moment she witnesses the talking serpent.[54] She interrogates this natural anomaly, or "monstrosity," to borrow Edwards's term:

> Not unamazed she thus in answer spake.
> "What may this mean? Language of man pronounced
> By tongue of brute, and human sense expressed?
> The first at least of these I thought denied

> To beasts, whom God on their creation-day
> Created mute to all articulate sound;
> The latter I demur, for in their looks
> Much reason, and in their actions oft appears."
> (9.552–59)

Eve initially draws on her memory, noting that the "language of man pronounced" and "human sense expressed" by the serpent falsify her past experience of creatures, "mute to all articulate sound." But she almost immediately turns to her unique cognitive skills and focuses on what she *believes*: "The latter I demur." She separates what she "thought," or her past knowledge, from what exceeds perception: that beasts might have reason. By dissociating this specific instance from the common knowledge that exists about the capacities of all creatures, she avoids the mistake of equating seeming with being. Eve is not so "'overwhelmed by wonder,'" as Fish suggests, that she cannot question deviations from nature.[55] The double negative, "Not unamazed," intimates her suspension between skepticism, wonder, and curiosity. The speaking serpent could well be one of the "deviating instances" that Bacon calls on naturalists to study, one of the wonders and marvels that fascinated early modern thinkers. In answering the "articulate" serpent, she responds to something unique and anomalous that demands explanation. Eve's interest would thus be recognizable to many of Milton's contemporaries in the context of the study of wonders. Her seeking the cause—"What may this mean?"—becomes a query about signification as well as being. This focus on causes prompts a series of questions about process and ontology: "How cam'st thou speakable of mute, and how / To me so friendly grown above the rest / Of brutal kind, that daily are in sight?" (9.563–65).

Instead of accepting the serpent's conclusion that his transformation results from a habitual experience, Eve interprets this alteration as an anomalous event. In other words, Milton makes Eve inaugurate an *evental* mode of knowing. Eve does not succumb to tasting the fruit, partially because she thinks the unique ontological status of the serpent is a specific, worthy object of inquiry. Adam and Eve are defined, from creation, as beings who rely on local details to understand the world. For instance, Eve's detailed exposition of her waking moment draws on the particulars of her surroundings. While phrases such as "That day" and "I first awaked" (4.449–50) enable her to locate the self in a particular time, descriptions like "Under a shade on flow'rs" (4.451), "the green bank" (4.458), and the "clear / Smooth lake" (4.458–59) associate this temporal immediacy with a specific place. Moreover, Eve's interrogations, of "*what*

[she] was, *whence* thither brought, and *how*" (4.452, emphasis mine), demand specific responses, much like Adam's initial interrogations about his identity: "*who* I was, or *where*, or *from what* cause" (8.270, emphasis mine). Scholars have long marked Eve's particular association with the garden, but even more broadly, both Adam and Eve define and know themselves and the place they inhabit by associating their singular identities with local entities and bounded markers.[56] Eve repeats this process of localization in her interaction with the serpent.

Eve focuses on the unique event that enabled the serpent's transformation and "more amazed unwary," reiterates her cautious curiosity: "Serpent, thy overpraising leaves in doubt / The virtue of that fruit, in thee first proved: / But say, where grows the tree, from hence how far?" (9.614–17). Eve demands answers, seeking the specific site of consumption—"where" the tree is, and "how far" it is from their current location—where she can unravel the truth of the serpent's claims and locate the cause and meaning of his transformation. She does not dismiss the efficacy of the tree ("For many are the trees of God that grow / In Paradise, and various, yet unknown / To us" [9.618–20]), but the serpent's "overpraising" heightens rather than lessens skepticism. Eve, and not Satan, seeks *evidence* for the "virtue of that fruit," which she acknowledges has been "proved" by the latter without accepting that this anomalous event is replicable.

Since Eve's cautious response indicates she might not succumb to a purely sensory experience or repeat Satan's dissimulated act, he transforms his narration of consumption into an experiment: a historically singular incident that is repeatable and therefore provides generalizable evidence of change. Satan now deploys localized particularities to present himself as a witness, initiating his foray into the experimental method. He proposes, "if thou accept / My conduct, I can bring thee thither soon" (9.629–30), countering Eve's wariness about his "overpraising" by asking her to "accept" his "conduct" and toning down his eagerness through the conditional "if." The suppression of desire and personal investment suggests disinterestedness and transfers the burden of action onto Eve. While she continues to define her interlocutor as an anomaly she must investigate, Satan transforms the serpent into an agent of knowledge who will lead Eve to the imagined laboratory where the tree will be subject to scrutiny. By directing her to one particular tree, "the way is ready, and not long, / Beyond a row of myrtles, on a flat, / Fast by a fountain, one small thicket past / Of blowing myrrh and balm" (9.626–29), Satan translates an act of consumption into a dissimulated experimental event. Refusing to objectify the serpent into a piece of evidence, he slowly subjects the tree, and Eve, to trials.

When Satan points her to the particular tree, Eve forgoes the possibility of conducting a trial herself, refusing his experimental strategy, which invites her to replicate his action:

> "Serpent, we might have spared our coming hither,
> Fruitless to me, though fruit be here to excess,
> The credit of whose virtue rest with thee,
> Wondrous indeed, if cause of such effects.
> But of this Tree we may not taste nor touch;
> God so commanded, and left that command
> Sole daughter of his voice; the rest, we live
> Law to ourselves, our reason is our law."
>
> (9.647–54)

Stressing her inability to "taste" or "touch" the fruit of "this Tree" because "God so commanded," Eve discounts a repeatable trial. Consumption is a transgression and a violation of divine law. Therefore, the serpent's altered ability can still be interpreted as the sign of a singular event. Yet as the serpent's "credit" and "virtue" become evidence of the tree's qualities, this transference of credit transforms him from object of inquiry to subject of knowledge. Although Eve still employs conditional reasoning (inquiring "if" the tree were the "cause of such effects"), and although she echoes a distinctly Miltonic argument ("reason is our law"), her increasing trust in the serpent's words obscures her initial focus on his anomalous state of being. Eve stabilizes her original object of study—cause of "language of man pronounced"—into a fact when she describes his change as an effect and accepts the tree as the "cause" of this effect. In accepting the serpent as a reliable witness, she also accepts his anomaly as an established fact. By designating him as a credible witness—one whose unique act provides new understanding—she forgoes his status as an object of inquiry and replaces one singularity with another: the tree for the serpent.

It is Eve who educates Satan about the value of local knowledge when she privileges the event over the possibility of its replication and directs him away from a purely sensory experience toward its epistemic stakes. An epistemology evolves, her words suggest, out of the eventness of a specific occurrence: Where and when does it occur? Which particulars are worth investigating? This focus on Eve's initiating role helps reverse the strain of argument that suggests that "rather than to 'make experiment' of the serpent's claims, [Eve] chooses to accept the experience he offers."[57] Such a line of thinking inevitably defines Eve as a passive recipient, automatically (though often implicitly) associating her fall with her failure to experiment within the parameters defined by Milton's contemporary English naturalists. Yet it is Eve who, through her demands for specific answers, makes Satan revise his strategies. Eve does not "abandon open-mindedness" when she lets credibility "rest" with the serpent,[58] but she does relinquish her focused object of inquiry: How and why does the serpent speak? The constant revision of objects, experiments, and aims enables Satan to "[abuse] the potential of the new experimental philosophy," as Edwards argues.[59] Eve accepts the serpent's word only because she reads the animal as evidence of an unrepeatable event. But since Satan's aim is to make Eve eat from the tree, he must force replication. He has to convince Eve that the serpent is both witness and evidence of a generalizable, not unique, transformation.

To do so, he redirects Eve's attention back to the serpent's irregular ontology, indicating that the serpent speaks because of a distinct change in being. Satan newly stresses the possibility of trial by coupling Eve's concept of the singular event with the repeatable experience: his experimental strategy, he now promises, will transform Eve. Depersonalizing his fabricated experience into an experiment, Satan presents his alteration as an occurrence that actualizes a generalizable process of ontological change. While he had earlier stressed *his* perceptions through personal and possessive pronouns ("I," "me," "my" [9.575, 588, 580]), he now primarily directs attention away from the self. He employs analogy not to liken himself to other beings but to describe a singular event that might provide evidence, in the figure of the serpent, of universal change in being:

> [God] knows that in the day
> Ye eat thereof, your eyes that seem so clear,
> Yet are but dim, shall perfectly be then
> Opened and cleared, and ye shall be as gods,
> Knowing both good and evil as they know.
> That ye should be as gods, since I as man,
> Internal man, is but proportion meet,
> I of brute human, ye of human gods.
>
> (9.705–12)

Satan argues, that "since" he is able to become "as man / Internal man," by analogy Eve can rise to be "as gods." His current state is a proven instance that serves as evidence of what *she* will be. His heretical pronunciation of plural "gods" is masked by his promise that proportionate rise maintains a gap between the subject undergoing change and that toward which they tend. The repeated "as" reiterates this disparity and suggests that one is always engaged in a process of ontological change. What one "should be" is meted out in "proportion" and one only rises appropriate to one's proper position in the hierarchical order of being.

Satan's tactic points to the larger questions of singularity and repeatability that govern the poem's strategy of accommodation: How does one know and interpret the state of being that any entity occupies in Paradise? By redirecting attention to the serpent's anomalous status, Milton asks: Is one's mode of being unique and irreproducible, or is it part of a repeatable process? Accommodating knowledge of the unknown in the prelapsarian realm remains dependent on explicating ontological states using epic events; if one accepts conditions of being are unique and irreproducible, as Eve interprets at this stage in the scene, one can potentially avoid transgressive replications. But if one accepts Satan's appeal to repeatability and generalization, one moves closer to contemporary philosophers who promised to standardize knowledge by studying the natural world. In the narrated world of *Paradise Lost*, one also moves closer to the Fall. Satan succeeds because his experimental strategy combines two distinct modes of thinking—on being and knowing—and he can redirect Eve's initial query about the serpent's anomalous singularity into a generalizable process that is familiar to her.

Satan's argument is of particular interest to Eve because in her earlier dream she was visited by a figure "shaped and winged like one of those from Heav'n" (5.55), who promised she could "ascend to Heav'n, by merit thine" (5.80). More importantly, Satan echoes the epic's most forceful assertions about changes in being. In his tutorial to Adam, Raphael promises ontological leveling as the end of individual progression. Scholars typically read this exposition of proportional rise as proof of Milton's monist materialism in his later works.[60] Raphael states that all things are made of "one first matter" (5.472) and are

> Endued with various forms, various degrees
> Of substance, and in things that live, of life;
> But more refined, more spiritous, and pure,
> As nearer to him placed or nearer tending
> Each in their several active spheres assigned,
> Till body up to spirit work, in bounds
> Proportioned to each kind.
>
> (5.473–79)

Raphael frames being as a process of "nearer tending" to something, and Milton presents readers with an ontology of tendency, rather than fixity. This propensity to change creates "active" rather than static states of being in prelapsarian Eden. The narrator, too, terms the serpent "spirited" (9.613), suggesting that there occurs a change in his substantial proportions.[61] Satan echoes the narrator's and Raphael's promises of attaining "perfection" (5.472) when he describes his proportional rise in degree to "internal man." He undercuts Eve's initial emphasis on the serpent's unique ontology by arguing that this anomaly provides an indirect route to "nature's ordered ways" (to recall Edwards's phrase), in which the serpent tends toward godhead by achieving human speech and sense. By making a graduated understanding of individual change central to his narration, Satan convinces Eve that events might be replicable because they are connected to the "nearer tending" process of prelapsarian becoming.

Satan successfully converts Raphael's ontology of gradual progression into an epistemology of repeatable instances, demonstrating that knowing is dependent on apprehending states of being: his experiment, he proposes, will contrive the alteration in Eve's current state. Observing Eve's vacillation about action and ends, Satan foregrounds the tree—the "Mother of science" (9.680)—as the object under investigation and subordinates his own experience and supposed transformation:

> now I feel thy power
> Within me clear, not only to discern
> Things in their causes, but to trace the ways
> Of highest agents.
>
> (9.680–83)

By introducing the tree as a mediator that actualizes the possibility of a repeatable process, Satan suggests it produces experiments. The tree allowed him to see inherent "causes" and "trace the ways" to proper understanding. By stressing the tree's gifts, he also translates the serpent from a Baconian "deviating instance" to an exemplar of the methods that enable the constitution of the natural order of things. He now draws attention not to his craving but to his newfound agency and to the possibility of knowledge he embodies:

> Look on me,
> Me who have touched and tasted, yet both live,
> And life more perfect have attained than fate
> Meant me.
>
> (9.687–90)

The tree directs subjects toward godhead, leading them to what now seems like an inevitable course of ontological development; the experience of eating becomes secondary to the knowledge this consumption provides. The serpent claims to report an experiment that can be replicated because he embodies its effects: just as he has internally risen from bestiality to manhood by achieving the faculties of "reason" and "speech," so too humans can tend toward godhead. The implicit claim is, of course, about a generalizable fact: not only serpents or humans, the two examples he chooses, but *all* living creatures might attain "perfection." In this way, Satan combines Eve's initial focus on the singular event with his general experience to propose a perfectly replicable experiment.

Eve accepts his argument because it is predicated on structures of being she need not question; the serpent is merely stating how Raphael's logic of habitual change could operate. Satan's conversion of the ontology of tendency into a repeatable method also seems plausible because he maintains the hierarchy latent in Raphael's tutorial: there is similitude but never identity between two orders of being. To maintain her expected distance from the serpent—they can only be "as" each other—Eve has to cast herself into the role she initially assigned to the animal, that is, an entity undergoing change. Perceiving the serpent's words are "impregned / With reason" (9.737–38), Eve—Milton's creation—must explore the possibility this presents. But she makes the error of supposing her action will initiate the universal promise of proportional ascent.

She forgoes her focus on the event's uniqueness, giving "too easy entrance" (9.734) to the words she had resisted before:

> Pausing a while, thus to herself she mused.
> "Great are thy virtues, doubtless, best of fruits,
> Though kept from man, and worthy to be admired,
> Whose taste, too long forborne, at first assay
> Gave elocution to the mute, and taught
> The tongue not made for speech to speak thy praise."
> (9.744–49)

Earlier, Eve had marveled at the serpent's ability to speak and decided "such wonder claims attention due" (9.566). However, as Satan shifts the focus from the experience of the serpent to the virtues of the tree, she accepts the latter as the "Mother of science." Instead of examining the *specific* "wonder" she identified earlier, the "best of fruits" now "claims attention due." The serpent's "elocution" becomes a by-product of the tree's yet untried virtues. "Pausing a while," Eve can still "mus[e]," but eating becomes secondary to the universal knowledge and change replication promises—the idea that the tree will produce its generalizable effects in her. Her belief in the possibility of ontological progression persuades Eve to replicate the event she had earlier deemed unrepeatable.

If Eve's position corresponded to that of Royal Society experimenters, she could follow their footsteps and seek corroboration whether this "first assay" ever occurred, which might have, as Edwards suggests, "exposed Satan as a charlatan."[62] But Edwards's conclusion is based on an impossible equation between Eve and seventeenth-century natural philosophers. Eve cannot be an experimental philosopher. Her condition is very different from the sociable world of the Royal Society, whose principles of corroboration and replication remain unavailable to her: Eve's singularity, the unrepeatability of her act of transgression, and Eden's ontological distance belie the possibility of communal knowledge production. Eve is deceived by a dissembling Satan as she ventures beyond the laws set for her unique position and practices on herself. Her encounter, instead of instantiating her ontological progression, completely disrupts it. Her act reverts into a singular instance of consumption, a sensory event punctuated by the narrator's rueful words, "she plucked, she ate" (9.781). This moment generates a different universal experience to what Eve expects: "Earth felt the wound, and Nature from her seat / Sighing through all her works gave signs of woe, / That all was lost" (9.782–84). Through the lure of replication, Satan transforms Eve's evental epistemology into her "sad experiment."

The echoes of Eve's dream and Raphael's tutorial in her trial suggest that the prospect of repeatability lurks behind the most unprecedented of events. Yet the outcome of her fall also marks its absolute distinction. While Eve's dream enacts and Raphael narrates a regular progression toward a higher order of being, the Fall unexpectedly ruptures the continuity between her current and predicted states of existence. Eve's consumption halts all expectations by refuting the logic of predictability that underlay the outcomes of the dream or the tutorial. Taking into account the dual nature of the event—not only as a moment of actualization but also as an outcome—we can isolate the Fall's evental nature: in the very instant Eve acts, the poem demarcates her fall as an unparalleled event whose result erases the repeatable promise of those earlier narrative moments. By staging how the Fall irrevocably disrupts ontological progression, and by revealing how it forever distances Eve from the tendency to godhead, Milton shows that an event might approach repeatable experiences and experiments but is finally irreducible to them. Experiments, as postlapsarian ways of understanding, cannot fully accommodate the irrevocable shifts in prelapsarian ontologies. Forever severed from her own unfallen existence, Eve retroactively understands that experiment cannot take her back to prelapsarian Eden.

Angelic Experiences and the Failures of Repetition

Eve's trial provides an acute example of Milton's evental poetics, which models a poetic method to explore the epistemic potential of unrepeatability and by extension exposes the shortcomings of repeatable experiences. But concerns about the scope of replication and irreproducibility are deeply embedded in the poem's explorations of action, knowledge, and existence. *Paradise Lost* is broadly interested in questions of singularity and repetition, rather than in a specific critique of experimentation, and Milton employs an inverse logic to underscore the limitations of cumulative knowledge: when singular epic events offer stark moments of certitude, repeatable experiences might at best provide probable knowledge. Yet, since characters cannot dissociate their self-knowledge from a multiplicity of experiences until their very existence is overturned by an unprecedented event, the poet must continually grapple with the fallacy of experiential learning that is based on cumulative and analogical reasoning.

The relation of part to whole, and of singularity to recurrence, structures various aspects of the epic. *Paradise Lost* balances its vision of creation as a singular event alongside its hypotheses about a pluralistic cosmos—with Raphael's suspension of the reader between Ptolemaic and Copernican world systems, the narrator's invocations of new methods of observation through references to Galileo's telescope, and his espousal of a prelapsarian cosmology, the poem refuses to settle the question about the exact form of the universe. The creation of earth provides the clearest example of how what one perceives as a singularity might actually be part of a recurring yet changing pattern. The Argument to Book 1 mentions "a new world and new kind of creature to be created, according to an ancient prophecy or report in Heaven"; this description gestures both to Satan's speech about novelty and creation—"Space may produce new worlds; whereof so rife / There went a fame in Heav'n that he ere long / Intended to create, and therein plant / A generation" (1.650–53)—and to the creation of earth and the living beings that inhabit it.[63] Yet, these descriptions of newness exist alongside claims that the earth is "built / With second thoughts, reforming what was old!" (9.100–101). The claim that the "new" is predicated on the complete replacement—and thus erasure—of the old, as well as the repetition of the concept of the "new-created world" (3.89) in various forms, reframes the structure of repetition and repeatability that abounds in the poem in a unique epic event of creation.[64]

Paradise Lost also often defines its characters' forms of existence through complex analogies between particulars and universals. Eve identifies herself both as individual and as type. For instance, after eating the fruit, she contemplates her imminent death: "Then I shall be no more, / And Adam wedded to another Eve, / Shall live with her enjoying, I extinct" (9.827–29). While Eve's name lends itself to this interplay between exemplarity and universality, her imagination of "another Eve" raises the question not only of nomenclature but of ontology: What does it mean to *be* Eve?[65] The threat of being replaced by "another Eve" exposes her inability to know herself completely as either individual or type. The proper name signifies a unique poetic character, but it also refers to all individuals who might replace her. Eve's singularity, then, serves as an index of universality. Moreover, when Eve imagines the "extinct" "I" being replaced, she cannot predict the future without calibrating her existence in time: "Then I shall be no more." Yet her understanding of states of being does not accommodate disappearance; after all, she will be replaced by another *Eve*. As Eve's reading of the self as type is predicated on the erasure of an original individual, her dilemma exposes that experiential routes to self-knowledge are inseparable from states of being.

Satan, the universalizer par excellence, transforms such interpretations of individual being into an analogical mode of knowing. We see a forceful manifestation of this work when he appropriates Raphael's language to tempt Eve. Here, he employs analogy and resemblance to convince her (he addresses Eve as "Fairest resemblance of thy Maker fair" [9.538]) and calls on tropes of universality (Eve is, he states, "universally admired" [9.542]). Specifically, he circumvents the complexity that Eve faces in her paradoxical understanding of self by drawing analogies from past experiences to future possibilities, in order to link individual experiences to universal knowledge.[66] Satan's tendency to generalize from past experiences inevitably leads to the prediction of the future by appeals to angelic being:

> since by fate the strength of gods
> And this empyreal substance cannot fail,
> Since through experience of this great event
> In arms not worse, in foresight much advanced,
> We may with more successful hope resolve
> To wage by force or guile eternal war
> Irreconcilable, to our grand foe.
>
> (1.116–22)

Marking the relative immutability of angelic states of existence, Satan attempts to forecast a future. The modal verb "may" gestures to the "hope" or possibility of future success. He uses his past knowledge of material being—"substance"—and their recent "experience" as sufficient and linked proofs against unexpected effects. This analogical reasoning relies on the regularity of surroundings and the predictability of the past, as Satan imagines this "great event" has given him "foresight." Under Satan's influence, Beelzebub too draws resemblances between human and angelic states ("some new race called Man, about this time / To be created like to us, though less / In power and excellence, but favored more" [2.348–50]). Based on this likeness, he predicts an analogous transformation: the fall of Adam and Eve. As God informs his audience in heaven, however, there is a crucial difference between the angel's fall and that of man: "The first sort by their own suggestion fell, / Self-tempted, self-depraved: man falls deceived / By the other first" (3.129–31). As the fallen angels predict possible futures, they disregard how the war in heaven has ruptured the continuum of being and action, making impossible such analogies.

Raphael's ontology of tendency serves as a useful corrective to Satan's misreading: the "empyreal substance" is not immutable. Satan's surety of the past, based on his habitual experiences, cannot function as a precise indicator of future knowledge and does not guarantee the possibility of replication. The fall of the angels has brought about the degeneration of their material form, as Stephen Fallon has persuasively argued.[67] Zephon, who spots a disguised Satan whispering in Eve's ear, marks this ontological change: "Think not, revolted Spirit, thy shape the same, / Or undiminished brightness, to be known / As when thou stood'st in Heav'n upright and pure" (4.835–37). Satan fails to recognize the war as an exceptional event that has radically altered his state of existence, but Zephon's observation, like God's pronouncement, exposes the insufficiency of Satan's predictive analogy, which extrapolates the future by joining perceptions about repeatable experiences and a stable past to ideas of immutable being.

Such analogical processes, however, are not limited to the fallen angels. The unfallen angels also repeat actions based on prior experiences, even under exceptional circumstances. This tendency is most apparent when they unquestioningly follow Satan's rebellion. Their previous allegiance, as well as Satan's reputation, is sufficient to draw a "third part of Heav'n's host" (5.710): "all obeyed / The wonted signal, and superior voice / Of their great potentate" (5.704–6). The angels do not doubt the word of their "great potentate" when he invites them to perform the most transgressive act imaginable "for great indeed / His name, and high was his degree in Heav'n" (5.706–7). Only Abdiel, "than whom none with more zeal adored / The deity, and divine commands obeyed" (5.805–6), discards the codes of regularity that demand they unconditionally follow hierarchy. He rejects the angelic error of analogy and exposes the rebellion as an unprecedented event that fractures forms of existence in heaven.

Yet knowledge in heaven remains analogical even after the war, and the unfallen angels continue to predict the unknown through the habitual nature of their past experiences. Uriel, the "sharpest sighted spirit of all in Heav'n" (3.691), fails to recognize Satan, who, disguised as a "stripling Cherub" (3.636), persuades Uriel of his sincerity. Satan claims that an "Unspeakable desire to see, and know / All these his wondrous works, but chiefly man, / . . . / Hath brought me from the choirs of Cherubim" (3.662–66). Appealing to curiosity, he voices desire to know more of God's works ("Brightest Seraph tell / In which of all these shining orbs hath man / His fixèd seat" [3.667–69]) and tempers this zeal by expressing the appropriate end, "as is meet / The Universal Maker we may praise" (3.675–76). Since Satan draws on recent history and prior knowledge and performs the signs that Uriel's experience has trained him to expect and accept, the latter cannot detect this dissembling act. Sincerity is the default mode in heaven, and appearances usually correspond to inner truth ("goodness thinks no ill / Where no ill seems" [3.688–89]); Uriel has no reason or need to expect the unexpected. He continues to trust the repeatable structure of habitual experience. He allows the "false dissembler" to pass by "unperceived" (3.681), believing the cherub's wish to "know" and "glorify" God's works "merits praise" (3.694–97).

Uriel does discern the imposter, however, when Satan's external expressions reveal his inner state of mind. Uriel, like Eve and Abdiel, learns from an anomalous event. Satan encounters the "Hell within him" (4.20) on witnessing earth. His "conscience" "wakes the bitter memory / Of what he was, what is, and what must be / Worse; of worse deeds worse sufferings must ensue" (4.23–26). His expressions reflect this descent:

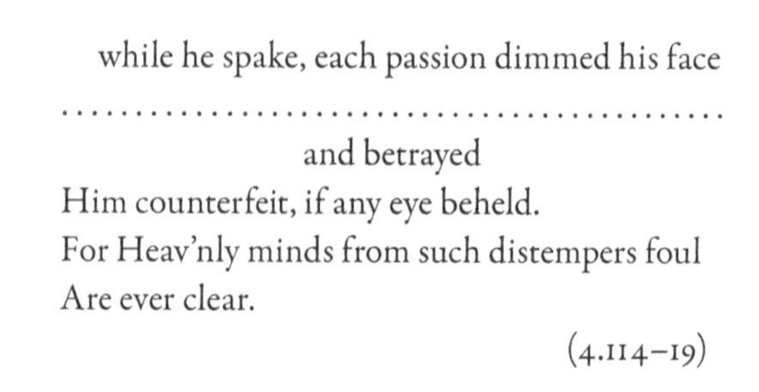
while he spake, each passion dimmed his face
. . . .
and betrayed
Him counterfeit, if any eye beheld.
For Heav'nly minds from such distempers foul
Are ever clear.

(4.114–19)

While Uriel's common experiences misdirect him, the anomalous nature of Satan's reaction—brought about by Satan's increasing grasp of his changed state—arouses suspicion. Because Satan

> Yet not enough had practiced to deceive
> Uriel once warned; whose eye pursued him down
> The way he went, and on th' Assyrian mount
> Saw him disfigured.
>
> (4.124–27)

Uriel admits to Gabriel that although Satan "seemed" (4.565) zealous, he "soon discerned his looks / Alien from Heav'n, with passions foul obscured" (4.570–71). Taught by experience to trust what seems to be true, Uriel's sole perception of irregularity ("once warned") exposes that regular, past experiences can be performed to deceive.

Uriel's initial mistake and subsequent realization of the truth from an exceptional observation demonstrate that, at most, repeatable experiences provide probabilistic knowledge about what is likely to occur in any anticipated moment. Every actualized event, such as Eve's trial and Uriel's correct reading of Satan, can potentially disrupt such regularity in an instant. Uriel and the "third part of Heav'n's host" are deceived when they embrace different forms of analogical prediction, in which habitual experiences misdirect from the "counterfeit," leading them to expect complete alignment of past, present, and future, and of seeming and being. The predictive analogy proposed by Satan and adopted by the angels falsely suggests to its practitioners that likelihood is certainty, until an anomalous event unmasks the gap between the two. By stressing the irreproducibility of events, which serve as units of surety in the epic, Milton rejects the possibility that particular instances will progressively lead to generalizable knowledge, a course that is differently promised by scientific experiments and habitual experiences. The unique and exceptional Miltonic events I explore reveal the impossibility for cumulative experiential methods, including experiments, to serve as instruments of certainty in *Paradise Lost*. The epic, then, continually enacts its promise of certainty through unprecedented and unrepeatable events, emphasizing that states of being in Eden are not fully accessible through the repeatable structures of experience and experiment that will come to govern ways of knowing in the fallen world.

The poem does seem to offer a more appropriate mode of analogical learning, when Raphael employs likelihood (rather than prediction) in his conversation with Adam. Since Adam remains unable to grasp events in heaven, Raphael undertakes a strategy of accommodation: "What surmounts the reach / Of human sense, I shall delineate so," he explains, "By lik'ning spiritual to corporal forms, / As may express them best" (5.571–74). Although the Earth is "the shadow of Heav'n, and things therein / Each to other like, more than on Earth is thought" (5.575–76), Raphael acknowledges that he cannot narrate the war in heaven in plain terms and wonders "what things / Liken on Earth conspicuous" can "lift / Human imagination to such highth / Of godlike power" (6. 298–301). His solution is "to set forth / Great things by small" (6.310–11). The analogy of "lik'ning" serves as a corollary to the ontology of tendency he espouses: it retains an incommensurable gap between the two entities being compared, and it only allows an imperfect accommodation. Raphael models the impossibility of fully grasping the unknown because experience is, at best, "like" the unexperienceable.

Poetry and Certainty in a Fallen World

These moments in *Paradise Lost* reinforce the unbridgeable gap between cumulative experiences and certain knowledge. As the poem emphasizes why collective, replicable, and repeatable processes remain insufficient to achieve certitude, it also raises questions about the possibility of learning in the realm of "evil days" and "evil tongues" (7.26) occupied by the poet and the reader: How does one learn in the fallen world of *Paradise Lost*? How do repeatability and eventness work after the Fall? I propose that Milton locates certainty not in collectives of actors who accumulate knowledge but in individual voices who can translate the complete unknown of unfallen and divine language into something that fallen audiences might understand. Unlike the principles of repeatability and replication championed by contemporary experimenters, Milton turns to an alternate method to accommodate the inaccessible event in prelapsarian Eden: prophetic insight.

As Milton's epic elevates the vatic figure's singular role, it also resolves a larger question of access to truth and certitude that had haunted his earlier writing. In the autobiographical section of *The Reason of Church-Government Urg'd Against Prelaty* (1642), he identifies himself as one of the "selected heralds of peace and dispensers of treasure inestimable" (836). After outlining his ambitions to compose a national epic, he argues that he must write this tract because he is one of the "selected" few: "These abilities, wheresoever they be found, are the inspired gift of God rarely bestowed" (841). This claim about individual abilities exists in tension with the inspiring ideas of collective prophesying he imagines in *Areopagitica* (1644), where actors learn through a process of predictive analogy: they "[search] what [they] know not by what [they] know, still closing up truth to truth as [they] find it" (956). This "brotherly search after truth" (958) is a collective exercise, resulting in a "noble and puissant nation rousing herself like a strong man after sleep and shaking her invincible locks" (959). Indeed, "wise and faithful laborers" are sufficient to "make a knowing people a nation of prophets, of sages, and of worthies" (957). These labors will ultimately "unite those dissevered pieces which are yet wanting to the body of Truth" (956). Truth represents the sum of all knowledge; since it existed in "perfect shape" in a distant past, "the sad friends of Truth" have undertaken the laborious task of reassembling it by "gathering up limb by limb still as they could find them" (955).

In the 1640s, the revolutionary Milton predicts that England will become a land of prophets through its engaged collective of readers and writers and accordingly defines the public role of the individual—he argues, "to sequester out of the world into Atlantic and Utopian polities which never can be drawn into use will not mend our condition, but to ordain wisely as in this world of evil, in the midst whereof God hath placed us unavoidably" (943).[68] *Areopagitica* offers a potential method of fulfilling Milton's self-representation as one of the "dispensers of treasure inestimable" by translating the logic of individual "abilities" into a cumulative and collective search for Truth. But by the time he published *Paradise Lost* in Restoration England, "this land" had changed, with limited or no public role left for him, and in the epic he fully embraces his belief that only a select few, and primarily the prophetic poet, can access divine truth. In this changed environment of "evil days" and "evil tongues," a "nation of prophets, of sages, and of worthies" will not materialize through the unison of "wise and faithful laborers." *Paradise Lost* therefore supplants *Areopagitica*'s optimistic vision of collective prophetic experiences with individual voices, such as the narrator and Michael.

In the last two books of the epic, the archangel Michael, who "future things canst represent / As present" (11.870–71), stands at the "top / Of speculation" (12.588–89) and uses examples from approaching times that fully "show" Adam "what shall come in future days" (11.357).[69] The latter inquires, "say where and when / Their [the son and serpent's] fight, what stroke shall bruise the victor's heel" (12.384–85). To Adam's demand for specifics Michael responds: "Dream not of their fight, / As of a duel, or the local wounds / Of head or heel" (12.386–88). Adam must "learn, that to obey is best" (12.561) by "fulfilling that which [he] didst want / Obedience to the law of God" (12.396–97). Michael's "prediction" of the approaching future must rely on particulars because the "seer blest" can only "[measure] this transient world" (12.553–54).[70] Yet the archangel repeatedly warns Adam to reject the "local," the "where and when," and adhere to a universal law of obedience.

Michael suggests that Adam's acceptance of the local is at best fragmented and at worst a complete misrecognition of what constitutes understanding. Adam adopts an immediate mode of knowing that enables him to situate his experiences within an evolving temporality. Michael repeatedly corrects Adam's responses, which are based on his understanding that he is foreseeing "nature . . . fulfilled in all her ends" (11.602). This tutorial, suggests Michael, is not intended to reveal the "bent of nature" (11.597) and "delight" (11.596) in it but to provide a larger understanding of one's place in the universe:

> Judge not what is best
> By pleasure, though to nature seeming meet,
> Created, as thou art, to nobler end
> Holy and pure, conformity divine.
> (11.603–6)

Toward the end of his instruction, when Adam acknowledges that he has learned by "example" (12.572), Michael—the "prophet of glad tidings" (12.375)—corrects him:

> thou hast attained the sum
> Of wisdom; hope no higher, though all the stars
> Thou knew'st by name, and all th' ethereal powers,
> All secrets of the deep, all nature's works,
> Or works of God in heav'n, air, earth, or sea,
> And all the riches of this world enjoyedst,
> And all the rule.
> (12.575–81)

Michael recognizes that Adam's learning by "example" leads to the "sum / Of wisdom." But Michael also reminds Adam that accumulation and experience—the forms of knowing that the angels adopt, that Satan promises to Eve, and that had been the ideal form of action proposed in *Areopagitica*—are not the ends of learning. What lies beyond is "higher." One attains the "sum" of "wisdom" to move away from it and act: "only add / Deeds to thy knowledge answerable, add faith / Add virtue, patience, temperance, add love" (12.581–83). "Deeds" might supplement "knowledge," but experiential actions do not precede and provide direct routes to certainty. Adam's particular responses show his grasp of the distinct character of each revelation and his ability to apprehend the physical world, but Michael exposes these "local" experiences of the future to be secondary to "higher" knowledge.

Embedded in this theological exchange on obedience, law, and action, we find the duality that governs ways of knowing immediately after the Fall, an absolute disjunction between experiential instruments of instruction (the "local wounds") and its universal goals (to "learn" that "to obey is best").[71] While Michael's typological exegesis directs attention from the present toward a future in which one gains complete understanding at the end of history, the fallen Adam can only articulate an experiential mode that extrapolates knowledge from distinct examples.[72] Adam, like Uriel and the fallen angels, relies on former instances to predict the lessons that might be embedded in each new vision. Michael's corrections explicate the failures in understanding that arise when one replicates the teachings of past experiences. He underscores that Adam can learn through an exposure to multiplying examples only if he forgoes analogical predictions and refuses to accept that diverse, unique, "local" instances will aggregate into universal truth. Michael's separation of prophetic and experiential knowledge thus forcefully negates the possibility of direct relationships between singular instances and certain ends that was imagined in *Areopagitica*, that generates probable knowledge in the epic, and that remains the promise of the repeatable experiment.[73]

While Michael instructs an individual character to inquire whether cumulative experiences can produce absolute knowledge, the narrator—another prophet figure who operates in the poem's postlapsarian world—transports the disjunctions between habitual perceptions and unpredictable instances into the realm of poetic production itself. The narrator is the most explicit figuration of Milton's long association of poetry with "universal insight into things" (*Of Education* [1644], 978), but this figure remains both within and outside the world of the epic. His moments of identification with readers are few, but they break any illusion of complete separation from them. His simultaneous distance from and identification with the reader is apparent from the poem's opening lines, when he shifts from announcing the topic of "man's first disobedience" to a lament about "all *our* woe" (emphasis mine).[74] Ultimately a "character" in the poem, he is surprised by the very events he narrates.[75] In this way, he recalls Spenser's Merlin, who also expresses surprise when he reveals the future to

Britomart. While the narrator's prophetic claims rely on an all-encompassing view, the certainty and omniscience he projects as a vatic figure are undercut by his immersion in the poetic realm. Unlike Michael, the narrator remains trapped in a world in which he must negotiate between event and experience.

The narrator enacts how one of the most repeatable and predictable experiences in the poem, the "nightly" (3.32, 7.29, 9.22) composition of the epic, is transformed into an unexpected, temporally bound event. As he ruminates on how the muse's repeated visitations provide him prophetic insights, he increasingly associates inner illumination with darkness: prophecy compensates for loss of external vision by providing access to inner truth. In his first invocation, he states: "what in me is dark / Illumine, what is low raise and support" (1.22–23). This interplay between darkness and light continues not only in the narrator's words but also in the increasing association of illumination with prophetic poetry. For instance, Michael initiates Adam's tutorial by removing the "film" (11.412) from Adam's eyes. Again in Book 3, the narrator walks a precarious line between lamenting his blindness and praising this loss as a gift from God. He misses the "holy light" (3.1), which "Revisit'st not these eyes" (3.23), but he uses the loss of sight to associate himself with other seers, "Blind Thamyris and blind Maeonides, / And Tiresias and Phineus prophets old" (3.35–36). The potential alteration of lack to "renown" (3.34) becomes a source of consolation, and he accepts this lack as a form of internal gain and, ultimately, as a motivation for creativity:

> celestial light
> Shine inward, and the mind through all her powers
> Irradiate, there plant eyes, all mist from thence
> Purge and disperse, that I may see and tell
> Of things invisible to mortal sight.
>
> (3.51–55)

By appealing to the "celestial light" to "plant eyes" inward, the prophet can "see and tell" what cannot be seen by all. Prophetic poetry in the form of a theological epic emerges as the specific, "unattempted" (1.16) answer to the questions of belatedness and untimeliness raised by Milton earlier in his career, in poems such as "Lycidas" and Sonnets 7 and 19. The lack that originally signified isolation emerges as a marker of talent—and one that is available only to the "selected" few.

Even as he grapples with the permanent "darkness" and "solitude" (7.27–28) that engulf him, the narrator acknowledges that the muse's regular visits activate his poetic insights:

> not alone, while thou
> Visit'st my slumbers nightly, or when morn
> Purples the east: still govern thou my song,
> Urania, and fit audience find, though few.
> (7.28–31)[76]

But it is only when he realizes that he "must change / Those notes to tragic" (9.5–6) that he undertakes a fuller exploration of the relationship between his "higher argument" (9.42) and the "nightly" inspiration that catalyzes his prophetic poetics:

> If answerable style I can obtain
> Of my celestial patroness, who deigns
> Her nightly visitation unimplored,
> And dictates to me slumb'ring, or inspires
> Easy my unpremeditated verse:
> Since first this subject for heroic song
> Pleased me long choosing, and beginning late;
> Not sedulous by nature to indite
> Wars, hitherto the only argument
> Heroic deemed
> . . . .
> Me of these
> Nor skilled nor studious, higher argument
> Remains, sufficient of itself to raise
> That name, unless an age too late, or cold
> Climate, or years damp my intended wing
> Depressed, and much they may, if all be mine,
> Not hers who brings it nightly to my ear.
> (9.20–47)

As the narrator again acknowledges the role of inspiration, he reveals his understanding of the prophetic poet as one who stands apart from his subjects as an interpreter, or as a translator who accommodates what the "celestial patroness" reveals. The muse makes possible the process as well as the product of poetic creation. This invocation offers the final take on the complex relationship between vision and insight that he has been attempting to map in the earlier invocations. Ultimately, the narrator recuperates hope as fate, making the act of poetic creation seem inevitable. Although "beginning late," in these lines he enacts the realization of his potential to compose the "higher argument." He replaces the uncertainty about potential by ascertaining that he was meant to construct this "argument." The composition of the poem becomes both the act and the product of fulfilling his vatic role.

Yet, the repetition of the visits is by no means predictable or guaranteed. The narrator's stress on the unexpected and uncategorizable nature of these visions ("unimplored," "unpremeditated"), and on his inability to direct how his poetry will be composed (the muse "dictates," "inspires," she "brings" the "higher argument" to him), introduces contingency and surprise into the most repeatable, and seemingly stable, aspect of his composition. As the narrator translates to "tragic" notes, he can only hope to "obtain" an "answerable style" to his heroic content. Poetic inspiration is not the active, collective prophesying imagined in *Areopagitica*; rather, it is the product of "unimplored" visits of the "celestial patroness, who deigns" to reveal to the narrator the subject of epic poetry. When the narrator questions his own capacity to act as anything more than a receptacle of his inspiration, he intimates how events lurk within routine actions and threaten to interrupt his regular mode of existence. His attempts to highlight his prophetic role bring him closest to a performative, evental mode of knowing. This invocation dramatizes how evental poetics works: each nocturnal visit becomes a unique and potentially unrepeatable event, out of the character's control, and always carrying with it the possibility of disappearance. As the only source of poetic insight, the distinct and unpredictable event becomes the sole unit that can provide certainty and make prelapsarian Eden intelligible to readers through a postlapsarian poetics.

In the figures of Michael and the narrator, the epic provides two different examples of how the prophet-poet, himself trapped in "evil days," instructs fallen audiences to look beyond the habitual and repeatable structure of experiential learning. Michael directs Adam away from cumulative and analogical learning, and the narrator exposes the unpredictability embedded within the most repeatable of experiences. Through such moments, as well as unique occurrences like Eve's fall, *Paradise Lost* makes claims to certitude, appealing to the singularity of events and dissociating unique prophetic insights from the "Adamic epistemology" of cumulative experimentation. These occurrences thus become crucial to foregrounding the conceptual underpinnings of evental poetics: epic poetry deploys events to unsettle the predictable trajectory of regular experiences in order to explicate an absolutely different state of being. As unrepeatable events disrupt regularity and in the process come to function as instruments of certainty, the poem also reveals the epistemic limitations of habitual and repeatable experiences.

Paradise Lost and *The Blazing World* were published within a decade of the formation of the Royal Society, an event that continues to serve as a touchstone for the institutionalization of natural philosophy in seventeenth-century England. Because Milton and Cavendish were writing in a very different intellectual climate from the one in which Sidney had marked the distinctions between poesy's "golden" worlds and the "brazen" worlds of "nature," their writing feels even more charged with concerns about how poetic knowledge can thrive—even survive—in a world that longs for a time, to recall Sprat's words, "when men deliver'd so many *things*, almost in an equal number of *words*." Yet, we might see Cavendish's celebration of the pleasures of "fiction" and Milton's "great argument" as different *defenses* of poesy, where their distinctive "consideration[s] of what may be and should be," to return to the phrase that sparked my own inquiries, reveal how the ontological affordances of literature can expand, even supplant, the knowledge of those who refuse to look beyond "Nature." In Cavendish's case, these manifest as vitalist worlds that actualize the author's "fancy" in ways that are impossible through the theories proposed by contemporary male philosophers. But for Milton, the inaccessibility of prelapsarian states of being has created an even more acute problem, one that underlies the impossibility of repeated experiences to restore the "perfect shape" of certainty through direct appeals to natural philosophy. Although experiential methods—including controlled ones like the Royal Society's experiments, which Cavendish explicitly criticizes—are necessary to grasp the fallen world, they remain untenable with a poetics that wishes to reveal a state of being that inevitably escapes translation. The unknown, unknowable, and singular ontology of prelapsarian Eden requires a poetic project that recognizes the incommensurability of experiential and imaginative routes of knowing: it demands events "unattempted yet in prose or rhyme" (1.16).

CODA

The Ethics of *Poiesis*

This book's chapters trace various vectors of the "possible" to document how poesy thinks. We have seen the literary epistemology of possible knowledge in construction within romance, tragedy, utopia, lyric, and epic. The formal apparatus of different genres of imaginative writing delimits the conditions of possibility within a fictive world and, in doing so, determines the kinds of knowledge a particular kind of fiction can generate. In different ways, the chapters uncover the linked epistemological and ontological affordances of literary form. Most broadly, then, *Possible Knowledge* is a book about literature itself. By focusing on the mechanics of fiction-making, on form as an engine of thinking, and on literature's ambitions to affect actuality, I have sought to theorize poesy as an aesthetic *and* a philosophical endeavor. Since *Possible Knowledge* studies literary knowledge-making through its relation to science, this book inevitably sits alongside scholarship on the history of disciplines and research that locates the origins of the contemporary "crisis of the humanities" in the early modern period.[1] Yet, I have suggested that we cannot defend humanistic knowledge merely by rehearsing the Renaissance commonplace that poetry is valuable because it can "teach and delight," by arguing for the value of literature through its proximity to "useful" or "practical" disciplines, or even by reading literary texts for their proto-scientific gestures. Instead, my argument locates early modern literature's knowledge-making potential in tendencies that made critics of fiction question its utility and value: its formal inclinations toward extravagance, dilation, and unruliness. In other words, *Possible Knowledge* defends the epistemological—and by extension, pedagogical—value of literature by excavating poesy's peculiar modes of being and thinking.

In this brief conclusion, I reflect on the promise and dangers of such thinking. In its ideal expression, early modern literature professes a kind of practical ethics, offering readers models of *praxis* that urge them to cultivate themselves in the image of figures they encounter in the worlds of poesy.[2] What might we, as twenty-first-century readers, make of such an invitation? How might we—indeed can we—incorporate such ethics into our own intellectual practices? To begin to think through these questions, I return to the text that laid the foundations for and that serves as a touchstone throughout this book. *The Defence of Poesy* liberates the poet from the "bare *Was*" that constrains the historian and celebrates the poet's propensity to affirm "nothing." At the same time, Sidney emphasizes that poetry's aim of "moving" (39) the reader is crucial to inculcating in them a desire for *praxis*. One of the ways in which the poet can spark the impetus for "well-doing," he suggests, is by acting as a "moderator" between philosophy and history; in this role, the poet "coupleth the general notion with the particular example" (31–32). By appropriating from the historian the use of particulars and combining them with precepts of philosophy, the poet models exemplars. Sidney provides the example of the figure of Cyrus, who is created through a process "not wholly imaginative, as we are wont to say by them that build castles in the air; but so far substantially it worketh, not only to make a Cyrus, which had been but a particular excellency as nature might have done, but to bestow a Cyrus upon the world to make many Cyruses, if they will learn aright why and how that maker made him" (24). This perfected Cyrus—surpassing the "particular excellency as nature might have done"—incorporates into a figure a cluster of ideas, values, or philosophical precepts that will compel the reader to emulate them. It is striking that as Cyrus emerges as a fully realized linkage between philosophy and history, this figure that poesy "bestow[s]" becomes increasingly prescriptive. Sidney's practical ethics is dependent on a reproduction, one in which the poet's act of making transfers the responsibility of further remakings onto the reader. Yet, in modeling the self into another Cyrus, the reader also forecloses other potential forms of being—the multiplicities of what "may be" fold into this singular figure. As Cyrus prescribes replication of this image as one necessary end of poetic production, the figure also exposes how poesy's ambitions to shape reality through readerly *praxis* exist in tension with its claims to *be otherwise*. Instead of merely standing in for the poetic ideal, Sidney's example also puts pressure on the notion that this methodology "nothing affirms."[3] Cyrus, in other words, exposes how poesy's ethical goal of "well-doing" depends on fictional worlds infiltrating the "bare *Was*" of history.

Or, we might read Cyrus as the logical end of poesy's fantasies about remaking history. Even as this figure serves as a pattern to instigate ethical action, this single instantiation occludes other unexpected ramifications—what might transpire when readers participate in the project of literary production through their own acts of remaking? Throughout this book, we have glimpsed how fictive existences might seep into readers' lived realities and implicate readers in literature's own processes of becoming: when Spenser's invitation to readers to "discoue[r]" Faerie Land collides with Europe's nascent colonial ventures; when Bacon's plan of seeking the "inner chambers" of "nature" results in the exploitation of natural resources; or even when Shakespeare stages the violent origins of a political dynasty. In her foundational work in literature/science studies, Mary Baine Campbell ruminates on poesy's collision with actuality, noting how the imagination of other worlds both holds the promise of scientific knowledge and becomes the impetus to unleash the horrors of colonization: "The longing for another world seems to have been a *real* pressure on the construction of the Edenic narrative of America," where the "efforts of the imagination to see and yearn past the bounds of the known and approved" is inseparable from the "terrible exploitations of people, nations, land, and resources."[4] More broadly, a rich body of scholarship in early modern literary studies has documented the role of imaginative writing and rhetorical practices in reflecting and shaping early modern discourses on race, colonialism, empire, and slavery.[5] In considering the ethics of literary making, we must confront the horrifying side of intellectual and aesthetic endeavors, which in the early modern period often thrived on tropes of racial othering, nationalistic pride, colonialist aspirations, and exploitation of natural resources.

This awareness seems especially important for readers as we embrace what I have elsewhere termed the "participatory readerly ethics" embedded in early modern poesy, a phrase that recognizes the ways in which "writers engage in acts of world-making with the tacit knowledge that their fictive worlds, incomplete and visionary, will be remade by various reading publics."[6] Sometimes, what early modern writers might have considered means of fostering the reader's "good endeuours," to recall Spenser's words, were also catalysts of terrible injustice. Patricia Akhimie has demonstrated, for instance, how the discourse of "cultivation"—what Spenser describes as the desire to "fashion" the reader—was a vital instrument of race-making.[7] Utopian fiction in particular (as we saw in Chapters 3 and 4) becomes a form of imagining alternate polities that inspired characters and readers to exert control over reality in radical ways. One dismaying account of such potential readerly remaking has been documented by Silvio Zavala, who chronicles how the Spanish in Mexico may have employed More's

Utopia as a model for organizing colonial settlement in the "New World."[8] Even if only circumstantially provable, Zavala's account exposes how literature could contribute to reconstructing the imagined (and the ironic, in the case of *Utopia*) into something dreadfully real, a form of *praxis* of remaking not merely the self but the world.

Since *Possible Knowledge* has been invested in tracing *how* literature thinks, instead of concluding by simply acknowledging the atrocities that unfold when actualization of possibility becomes the means of harming real people, I wish to push us one step further to ask what we—as scholars, teachers, and students—can *do* with the modes of thinking, knowing, and being that abound in early modern literary texts. Can thinking like poets—or, perhaps more pointedly, thinking through the aims poets set out for their readers—enable twenty-first-century readers to foster new forms of just engagement with their own spheres of existence? Can we embrace the ethical goals of worldmaking delineated by writers like Sidney and Spenser while rejecting the world-ending patterns that fantasies of poesy inevitably leave in their wake?[9]

Such questions about what we can do with the imagined and the fictional—terms that have complex associations with the false, the futile, and the ontologically empty—seem newly urgent in our own era of "truthiness" and "alternative facts,"[10] when truth and facts are questioned across media environments, and where climate change is dismissed as a hoax by appealing to constructionist arguments in science studies.[11] The questions I pursue in this book, completed during an ongoing global pandemic, are also haunted by another reality: the toll on human life that has been exacerbated by the relentless circulation of misinformation about science, medicine, and public health. Writing about the relation of scientific knowledge and fictionality seems especially fraught in such context. What do we learn by marking the imaginative practices underlying scientific knowledge, when the latter is already under assault as something fabricated? As I have shown in this book, such an approach makes us better attuned to how parameters of truth, knowledge, and existence are as much defined by imaginative discourse as by methods oriented toward the production of "facts." It also reminds us that to be fictional does not automatically entail being false, divorced from ethical obligations to the actual—these associations, too, are the constructs of thinkers who were carving out domains of scientific truth, and in the process investing both epistemic and ethical values in the categories we have come to associate with the terms "literature" and "science." While scholarship in history of science and science studies has underscored the constructivist nature of scientific practice—that is, this research reveals scientific knowledge and matters of fact to be constructed entities[12]—*Possible Knowledge*'s attention to imaginative modes also highlights that such constructions happened at the expense of the language arts and poesy. Much as early modern writers mobilized words and things to their own ends, whether for "searching into and discovering truth" or to "justify the ways of God to men," these concepts—and related categories of truth, fact, and fiction—continue to be activated in the service of personal or political interestedness, sometimes to devastating ends.

My primary goal, however, has not been to trace the formation of modern disciplinary boundaries or to elaborate the effects (positive or negative) of these institutional changes in our own time. Instead, *Possible Knowledge* aims to recover the methods by which literary worlds come into being. In this book's final pages, I extrapolate from my particular case studies and suggest that the value of literature—and in particular early modern literature—in *our* environment also lies in its indeterminacy and open-endedness as it makes and unmakes worlds, and as it reveals which forms of existence are sustainable. Early modern literature's palpable affiliation with the multivalent ontology of what is "possible," and its enactment of how things "may be and should be," is ultimately an invitation for further thought and action—that is, for new acts of worldbuilding. Poesy not only requires, but also assumes, that a reader must, like the Baconian philosopher, "enquire farther"—such inquiries could lead to generative worldbuilding, as we see in *The Blazing World*, or they might drive one, as Macbeth learns, toward a tragic end. In such ways, poesy invites readers into the co-construction of the conditions of possibility to which fiction gives form, a form that is necessarily incomplete.

Early modern literature's salience lies (to return to Andrew H. Miller's account of modalities of literary criticism) in its continual "performance or display of thinking." Miller finds this kind of thinking, which he terms "implicative," "more reliably in the writing of philosophers than in that of literary critics."[13] Implicative criticism "does not have ending as an end, is not justified by its conclusions, by any facts established, information conveyed, or judgments made; instead it is successful to the extent that it implicates, it enfolds its reader. To succeed it requires a reply."[14] This description also offers insight into the kinds of force imaginative writing exerts on its readers, and to his list of modern philosophers who revel in this modality of thinking, Miller could add some early modern names: Sidney, Spenser, Shakespeare, Bacon, Cavendish, and Milton. These writers flaunt their "performance or display of thinking" in capacious narratological and performative domains, as they simultaneously enact and theorize how alternate forms of existence can generate innovative ways of knowledge production.

Ultimately, the early modern literary texts I have studied in this book push readers to imagine radically diverging worlds, but they all do so without foreclosing the possibility that things *could be otherwise*. These works suggest that by engaging with a realm that is imagined by an author, readers and spectators might be able to expand—even give new form to—the kinds of worlds to which *they* desire access: Britomart's refusal to embrace Merlin's prophecy could have forestalled the machinery of the "endlesse" romance; Macbeth's refusal to embrace the witches' prophecy might have halted the march toward tragedy; the lady's ability to access microscopic worlds in her earring could have sustained an atomist cosmology. Early modern poesy invites its interlocutors, like the Governor in Bensalem, to "ask [it] questions." The future of early modern studies—perhaps of literary studies more broadly—requires not only that we "reply," as Miller advocates, but also that our replies keep imagining how criticism itself could be otherwise. When early modern poesy asks of us that we, like the Spenserian reader, "more inquyre," what we do to answer its calls tests our ethical commitments both to what we consider actual or probable and to what we believe is possible.

NOTES

INTRODUCTION

1. On the intellectual climate "when the settled Aristotelian, Galenic, and Ptolemaic accounts of how the universe worked began to fall apart and the new ideas that would replace them were still inchoate and in flux," see Mary Thomas Crane, *Losing Touch with Nature: Literature and the New Science in Sixteenth-Century England* (Baltimore: Johns Hopkins University Press, 2014), 2. On how there was no swift reorientation of knowledge or an immediate embrace of new philosophies in the period, see Katherine Eggert, *Disknowledge: Literature, Alchemy, and the End of Humanism in Renaissance England* (Philadelphia: University of Pennsylvania Press, 2015), esp. 2–3.

2. I use the terms "fiction," "literature," and "imaginative writing" interchangeably to refer to what early moderns term "poesy" or "poesie." I draw on Philip Sidney, who states that "it is not rhyming and versing that maketh a poet" but "feigning notable images." *A Defence of Poetry*, ed. Jan Van Dorsten (Oxford: Oxford University Press, 1966), 27. All citations of the *Defence* are from this edition and hereafter cited parenthetically by page number. Francis Bacon echoes this claim: while "verse is but a character of style," poesy is "feigned history or fables." *De Augmentis*, in *The Works of Francis Bacon*, ed. James Spedding, Robert Leslie Ellis, and Douglas Denon Heath, 15 vols. (London: Longman, 1857–64), 4:292.

3. Although I do not pursue a cultural keywords or critical semantics approach, "possible" does fit into the category of ordinary words that Roland Greene speculates can "produce a fresh and compelling version of the period." *Five Words: Critical Semantics in the Age of Shakespeare and Cervantes* (Chicago: University of Chicago Press, 2013), 12.

4. John Florio, *A VVorlde of Wordes, Or most Copious, and Exact Dictionarie in Italian and English* (London: By Arnold Hatfield for Edw. Blount, 1598), s.v. "possible," Early English Books Online (hereafter EEBO); Claudius Hollyband, *A Dictionarie French and English: Published for the Benefite of the Studious in that Language* (London: Imprinted at London by T. O. for Thomas Woodcock, 1593), s.v. "possible," EEBO. The *OED*'s first definition elaborates on this formulation: "That is capable of being; that may or can exist, be done, or happen (in general, or in given or assumed conditions or circumstances); that is in a person's power, that a person can do, exert, use, etc." *Oxford English Dictionary Online* (Oxford: Oxford University Press, March 2022), s.v. "possible, adj., adv., and n." For the definition of "possibility" as the "condition or quality of being possible," see *OED Online*, s.v. "possibility, n."

5. Steven Shapin deconstructs the term in the first line of *The Scientific Revolution* (Chicago: University of Chicago Press, 1996): "There was no such thing as the Scientific Revolution, and this is a book about it" (1).

6. For representative examples of Sidney's importance in literature/science studies, see Elizabeth Spiller, *Science, Reading, and Renaissance Literature: The Art of Making Knowledge, 1580–1670* (Cambridge: Cambridge University Press, 2004), 24–44; Henry S. Turner, *The English Renaissance Stage: Geometry, Poetics, and the Practical Spatial Arts, 1580–1630* (Oxford: Oxford University Press, 2006), 82–113; and Eggert, *Disknowledge*, 206–10. There is a rich body of scholarship on the philosophical influences and resonances in the *Defence*. For instance, whether Sidney's theories of poetry are Aristotelian or Platonic, or a mixture of both, has occupied critics for a long time. See S. K. Heninger Jr., *Sidney and Spenser: The Poet as Maker* (University Park: Pennsylvania State University Press, 1989); A. C. Hamilton, "Sidney's Idea of the 'Right Poet,'" *Comparative Literature* 9, no. 1 (1957): 51–59; John P. McIntyre, "Sidney's 'Golden World,'" *Comparative Literature* 14, no. 4 (1962): 356–65; and Morriss Henry Partee, "Sir Philip Sidney and the Renaissance Knowledge of Plato," *English Studies* 51, no. 5 (1970): 411–24.

7. For the golden world as an idea, see Michael Mack, *Sidney's Poetics: Imitating Creation* (Washington, DC: Catholic University of America Press, 2005), 124–26, 139–56. For the relationship of the golden world to England, see Spiller, *Science, Reading, and Renaissance Literature*, 24–45. See also Ronald Levao, "Sidney's Feigned Apology," *PMLA* 94, no. 2 (1979): 223–33, for the relationship between the ideal and the actual.

8. In *The Art of English Poesy*, ed. Frank Whigham and Wayne A. Rebhorn (Ithaca, NY: Cornell University Press, 2007), George Puttenham offers a slightly different understanding of the poet's relation to other thinkers. He elaborates on the topic: "How the poets were the first philosophers, the first astronomers, and historiographers, and orators, and musicians of the world" (98).

9. See also Sidney, *A Defence of Poetry*, 39.

10. Criticisms took various forms, ranging from Stephen Gosson's *The Schoole of Abuse* (London: By [Thomas Dawson for] Thomas VVoodcocke, 1579) to Theseus's declaration in *A Midsummer Night's Dream* that the poet "gives to airy nothing / A local habitation and a name" (5.1.16–17). *The Norton Shakespeare*, ed. Stephen Greenblatt et al., 3rd ed. (New York: W. W. Norton, 2016). All citations of Shakespeare's plays are from this edition and given by act, scene, and line number in the text.

11. I draw on Caroline Levine's definition of affordance as "potential uses or actions latent in materials and designs." *Forms: Whole, Rhythm, Hierarchy, Network* (Princeton, NJ: Princeton University Press, 2015), 6.

12. Throughout the book, I use the phrase "literature/science studies" to describe the subfield in early modern literary studies that has emerged out of engagements with—and ultimately significantly deviated from—scholarship in the history of science as well as science and technology studies.

13. See, for instance, Marjorie Hope Nicolson, *Science and Imagination* (Ithaca, NY: Cornell University Press, 1956).

14. Mary Baine Campbell, *Wonder and Science: Imagining Worlds in Early Modern Europe* (Ithaca, NY: Cornell University Press, 1999), 2. Works that have shaped the field of early modern literature/science studies include Denise Albanese, *New Science, New World* (Durham, NC: Duke University Press, 1996); John Rogers, *The Matter of Revolution: Science, Poetry, and Politics in the Age of Milton* (Ithaca, NY: Cornell University Press, 1996); Spiller, *Science, Reading, and Renaissance Literature*; Turner, *English Renaissance Stage*; Carla Mazzio, ed., "Shakespeare and Science," special issue, *South Central Review* 26, nos. 1–2 (2009); Joanna Picciotto, *Labors of Innocence in Early Modern England* (Cambridge, MA: Harvard University Press, 2010); Howard Marchitello, *The Machine in the Text: Science and Literature in the Age of Shakespeare and Galileo* (Oxford: Oxford University Press, 2011); and Juliet Cummins and David Burchell, eds., *Science, Literature*

and Rhetoric in Early Modern England (Burlington, VT: Ashgate, 2007). On mathematics, see David Glimp and Michelle R. Warren, eds., *Arts of Calculation: Quantifying Thought in Early Modern Europe* (New York: Palgrave Macmillan, 2004). On medicine, anatomy, and theories of the body, see Gail Kern Paster, *The Body Embarrassed: Drama and the Disciplines of Shame in Early Modern England* (Ithaca, NY: Cornell University Press, 1993); and Jonathan Sawday, *The Body Emblazoned: Dissection and the Human Body in Renaissance Culture* (New York: Routledge, 1995). On Shakespeare and cognitive science, see Mary Thomas Crane, *Shakespeare's Brain: Reading with Cognitive Theory* (Princeton, NJ: Princeton University Press, 2001); and Laurie Johnson, John Sutton, and Evelyn Tribble, eds., *Embodied Cognition and Shakespeare's Theatre: The Early Modern Body-Mind* (New York: Routledge, 2014).

15. Spiller, *Science, Reading, and Renaissance Literature*, 3. This scholarship responds to and complicates work in the history of science as well as science and technology studies, fields that have also been expanding what counts as an object of natural inquiry in the early modern period at least since Thomas Kuhn's distinction of experimental (Baconian) and classical (mathematical) sciences: "Mathematical Versus Experimental Traditions in the Development of Physical Science," in *The Essential Tension: Selected Studies in Scientific Tradition and Change* (Chicago: University of Chicago Press, 1977), 31–65. For representative works with which literary scholars have typically engaged, see Steven Shapin and Simon Schaffer, *Leviathan and the Air-Pump: Hobbes, Boyle, and the Experimental Life* (Princeton, NJ: Princeton University Press, 1985); Bruno Latour, *Science in Action: How to Follow Scientists and Engineers Through Society* (Cambridge, MA: Harvard University Press, 1987); Steven Shapin, "The Invisible Technician," *American Scientist* 77, no. 6 (1989): 554–63; Paula Findlen, *Possessing Nature: Museums, Collecting, and Scientific Culture in Early Modern Italy* (Berkeley: University of California Press, 1994); William Eamon, *Science and the Secrets of Nature: Books of Secrets in Medieval and Early Modern Culture* (Princeton, NJ: Princeton University Press, 1994); Peter Dear, *Discipline and Experience: The Mathematical Way in the Scientific Revolution* (Chicago: University of Chicago Press, 1995); Lorraine Daston and Katharine Park, *Wonders and the Order of Nature, 1150–1750* (New York: Zone Books; Cambridge, MA: Distributed by the MIT Press, 1998); Pamela O. Long, *Openness, Secrecy, Authorship: Technical Arts and the Culture of Knowledge from Antiquity to the Renaissance* (Baltimore: Johns Hopkins University Press, 2001); and William R. Newman, *Promethean Ambitions: Alchemy and the Quest to Perfect Nature* (Chicago: University of Chicago Press, 2004). For the importance of material changes to the emergence of ideas, see Elizabeth L. Eisenstein, *The Printing Press as an Agent of Change: Communications and Cultural Transformations in Early Modern Europe* (Cambridge: Cambridge University Press, 1979); and Adrian Johns, *The Nature of the Book: Print and Knowledge in the Making* (Chicago: University of Chicago Press, 1998).

16. Frédérique Aït-Touati, *Fictions of the Cosmos: Science and Literature in the Seventeenth Century*, trans. Susan Emanuel (Chicago: University of Chicago Press, 2011), 10.

17. Claire Preston, *The Poetics of Scientific Investigation in Seventeenth-Century England* (Oxford: Oxford University Press, 2015), 6. Several studies use imaginative forms, and the imagination writ large, to interrogate relations of literature and science. See Lawrence Lipking, *What Galileo Saw: Imagining the Scientific Revolution* (Ithaca, NY: Cornell University Press, 2014); Suparna Roychoudhury, *Phantasmatic Shakespeare: Imagination in the Age of Early Modern Science* (Ithaca, NY: Cornell University Press, 2018); and David Carroll Simon, *Light Without Heat: The Observational Mood from Bacon to Milton* (Ithaca, NY: Cornell University Press, 2018). See also Jenny C. Mann and Debapriya Sarkar, eds., "Imagining Early Modern Scientific Forms," special issue, *Philological Quarterly* 98, nos. 1–2 (2019), on how "early modern science is shaped by imaginative engagements with the problem of form" ("Introduction: Capturing Proteus," 2). For a

recent work that captures the various strands of scholarship in early modern literature/science studies, see Howard Marchitello and Evelyn Tribble, eds., *The Palgrave Handbook of Early Modern Literature and Science* (London: Palgrave Macmillan, 2017).

18. Preston, *Poetics of Scientific Investigation*, 20.

19. Tita Chico, *The Experimental Imagination: Literary Knowledge and Science in the British Enlightenment* (Stanford, CA: Stanford University Press, 2018), 5. See also 3–12.

20. Ibid., 9, 8.

21. Colleen Ruth Rosenfeld, *Indecorous Thinking: Figures of Speech in Early Modern Poetics* (New York: Fordham University Press, 2018), 3. See also Jenny C. Mann, *The Trials of Orpheus: Poetry, Science, and the Early Modern Sublime* (Princeton, NJ: Princeton University Press, 2021); Wendy Beth Hyman, *Impossible Desire and the Limits of Knowledge in Renaissance Poetry* (Oxford: Oxford University Press, 2019); and Andrea Gadberry, *Cartesian Poetics: The Art of Thinking* (Chicago: University of Chicago Press, 2020).

22. Both *De Humani Corporis Fabrica Libri Septem* and *De Revolutionibus Orbium Coelestium* were published in 1543.

23. On vernacular and practical epistemologies, and the maker's knowledge tradition, see Pamela H. Smith, *The Body of the Artisan: Art and Experience in the Scientific Revolution* (Chicago: University of Chicago Press, 2004); and Antonio Pérez-Ramos, *Francis Bacon's Idea of Science and the Maker's Knowledge Tradition* (New York: Oxford University Press, 1988).

24. See, for instance, Spiller, *Science, Reading, and Renaissance Literature*, esp. 5–16; and Turner, *English Renaissance Stage*, esp. 12–16.

25. Turner, *English Renaissance Stage*, 46–47.

26. For representative works on these topics, see Jane Gallop, "The Ethics of Reading: Close Encounters," *Journal of Curriculum Theorizing* 16, no. 3 (2000): 7–17; Jonathan Culler, "The Closeness of Close Reading," *ADFL Bulletin* 41, no. 3 (2011): 8–13; and Frank Lentricchia and Andrew Dubois, eds., *Close Reading: The Reader* (Durham, NC: Duke University Press, 2003).

27. Corey McEleney, *Futile Pleasures: Early Modern Literature and the Limits of Utility* (New York: Fordham University Press, 2017), 8. See also Heather Dubrow, *Echoes of Desire: English Petrarchism and Its Counterdiscourses* (Ithaca, NY: Cornell University Press, 1995), 13.

28. J. K. Barret, *Untold Futures: Time and Literary Culture in Renaissance England* (Ithaca, NY: Cornell University Press, 2016), 11. Barret continues, "In looking to literature's microstructures, we can work outward from the formal resources employed within a given text to investigate its conception of and experiments with larger philosophical and conceptual questions" (11). See also McEleney, *Futile Pleasures*, esp. 7–8.

29. Andrew H. Miller, "Implicative Criticism or the Display of Thinking," *New Literary History* 44, no. 3 (2013): 345–60, 347.

30. My ideas on the relations between early modern poesy and methods of literary criticism have been shaped by ongoing conversations with Colleen Rosenfeld. See also Colleen Ruth Rosenfeld, "Opening Remarks" (paper presented at the Renaissance Project Symposium [online], Pomona College, Claremont, CA, June 1, 2021).

31. On how the early modern culture of incertitude engenders "an enormous expansion of the realm of the probable and a contraction of the certain," see Barbara J. Shapiro, *Probability and Certainty in Seventeenth-Century England: A Study of the Relationships Between Natural Science, Religion, History, Law, and Literature* (Princeton, NJ: Princeton University Press, 1983), 4. See also Ian Hacking, *The Emergence of Probability: A Philosophical Study of Early Ideas About Probability, Induction and Statistical Inference* (Cambridge: Cambridge University Press, 1975), esp.

25; Douglas Lane Patey, *Probability and Literary Form: Philosophic Theory and Literary Practice in the Augustan Age* (Cambridge: Cambridge University Press, 1984); Lorraine Daston, *Classical Probability in the Enlightenment* (Princeton, NJ: Princeton University Press, 1988); Lisa Jardine, *Francis Bacon: Discovery and the Art of Discourse* (Cambridge: Cambridge University Press, 1974); and Walter J. Ong, *Ramus, Method, and the Decay of Dialogue: From the Art of Discourse to the Art of Reason* (Cambridge, MA: Harvard University Press, 1958). On the increasing belief that "certitude" was possible in "only mathematics and a few logical and metaphysical principles," see Shapiro, *Probability and Certainty in Seventeenth-Century England*, esp. 4–5. See also Peter Dear, "From Truth to Disinterestedness in the Seventeenth Century," *Social Studies of Science* 22, no. 4 (1992): 619–31.

32. The possible worlds of modern philosophy and logic convert (sometimes without acknowledgment) Gottfried Wilhelm Leibniz's metaphysics of "the best of all possible worlds" into tools of modal logic. For representative works on these topics, see David Lewis, *Counterfactuals* (Cambridge, MA: Harvard University Press, 1973; rev. printing, Oxford: B. Blackwell, 1986); and Nelson Goodman, *Ways of Worldmaking* (Indianapolis: Hackett, 1978). On modal logic and possible-worlds semantics, see Saul A. Kripke, *Naming and Necessity* (Cambridge, MA: Harvard University Press, 1980); David Lewis, *On the Plurality of Worlds* (New York: B. Blackwell, 1986); and Alvin Plantinga, "Actualism and Possible Worlds," *Theoria* 42, nos. 1–3 (1976): 139–60. On tenets and problems of possible-worlds semantics and modal logic, see A. C. Grayling, *An Introduction to Philosophical Logic*, 3rd ed. (Oxford: Blackwell, 1997), 49–71. Theories of possible worlds of fiction derive but ultimately depart from this philosophical discourse. For the relations of possible-worlds semantics and literary works, see Marie-Laure Ryan, *Possible Worlds, Artificial Intelligence, and Narrative Theory* (Bloomington: Indiana University Press, 1991); Lubomír Doležel, *Heterocosmica: Fiction and Possible Worlds* (Baltimore: Johns Hopkins University Press, 1998); and Thomas G. Pavel, *Fictional Worlds* (Cambridge, MA: Harvard University Press, 1986), 11–29. Particularly relevant to early modern literature is Keir Elam's description: dramatic possible worlds "are hypothetical ('as if') constructs, that is, they are recognized by the audience as counterfactual (i.e. non-real) states of affairs but are embodied *as if* in progress in the actual here and now." *The Semiotics of Theatre and Drama*, 2nd ed. (New York: Routledge, 2002), 91.

33. See Ryan, *Possible Worlds, Artificial Intelligence, and Narrative Theory*, 17. Scholars refer in particular to Aristotle's claim that "the poet's function is to describe, not the thing that has happened, but the kind of thing that might happen, i.e. what is possible as being probable or necessary." *Poetics*, in *The Complete Works of Aristotle*, ed. Jonathan Barnes (Princeton, NJ: Princeton University Press, 1984), 2322–23.

34. Early modern literary scholarship, on the other hand, embraces the terminology of possible worlds but dismisses the relevance of modern analytic philosophy to our field of study. See Simon Palfrey, *Shakespeare's Possible Worlds* (Cambridge: Cambridge University Press, 2014), esp. 16–17.

35. Jardine, *Francis Bacon*, 19.

36. Ong, *Ramus, Method, and the Decay of Dialogue*, 97. See also Jardine, *Francis Bacon*, 5, 13–14, 25, for these shifts, and 31 for how Agricola's work models a turn "towards the literary." On how the swing in intellectual gravity from certainty to probability was part of broader reforms in humanist education, see also Lisa Jardine, "Lorenzo Valla and the Intellectual Origins of Humanist Dialectic," *Journal of the History of Philosophy* 15, no. 2 (1977): 143–64, 164.

37. Jardine, *Francis Bacon*, 13. See also 4–26.

38. On Ramism and rhetoric, see Ong, *Ramus, Method, and the Decay of Dialogue*; Wilbur Samuel Howell, *Logic and Rhetoric in England, 1500–1700* (Princeton, NJ: Princeton University Press, 1956), 247–81; and Rosenfeld, *Indecorous Thinking*, esp. 25–40.

39. Mann and Sarkar, "Introduction: Capturing Proteus," 8.

40. Francis Bacon, *The Advancement of Learning*, in *Francis Bacon: The Major Works*, ed. Brian Vickers (Oxford: Oxford World's Classics, 2008), Book 1, 139. All in-text citations of the *Advancement* are from this edition and given parenthetically by book and page number.

41. William Gilbert, *De Magnete*, trans. P. Fleury Mottelay (New York: Dover Publications, 1958), l.

42. For recent work on description and natural history, see Roychoudhury, *Phantasmatic Shakespeare*, 164–90; and Vin Nardizzi, "Daphne Described: Ovidian Poetry and Speculative Natural History in Gerard's *Herball*," *Philological Quarterly* 98, nos. 1–2 (2019): 137–56. For an account of the main debates for plain style, see Jenny C. Mann, *Outlaw Rhetoric: Figuring Vernacular Eloquence in Shakespeare's England* (Ithaca, NY: Cornell University Press, 2012), 182–88.

43. Thomas Sprat, *The History of the Royal-Society of London: For the Improving of Natural Knowledge* (London: Printed by T. R. for J. Martyn and J. Allestry, 1667), 113, EEBO. One solution to what Turner terms this "dream of mimesis" (*English Renaissance Stage*, 114) was to turn to belief in the universality of mathematical signs. Galileo famously declares that "the book [of the universe] . . . is written in the language of mathematics, and its characters are triangles, circles, and other geometric figures without which it is humanly impossible to understand a single word of it." *The Assayer* (excerpts from), in *Discoveries and Opinions of Galileo*, ed. and trans. Stillman Drake (Garden City, NY: Doubleday, 1957), 237–38. This dream also motivates the perpetual search for universal language and character: Descartes's and Mersenne's discussions about a philosophical language give way to attempts in England by William Petty, John Wilkins, Francis Lodwick, and others (see Shapiro, *Probability and Certainty in Seventeenth-Century England*, 243–48; and Turner, *English Renaissance Stage*, 114–16). Since this book focuses primarily on the Baconian tradition of empirical philosophy up to the early years of the Royal Society, when prominent members focused less on mathematics, my discussion of mathematical principles and approaches is limited. In marking this distinction, I follow scholars like Peter Dear. Dear argues that even though early members of the Society such as John Wilkins imagined its work to involve "'Physico-Mathematicall-Experimentall Learning,'" prominent members like Robert Boyle "were indifferent at best to the mathematical sciences." Wilkins's vision is realized only "when the Royal Society gave itself up, at the end of the century, to the self-labeled 'mathematical' natural philosophy of Isaac Newton," and it is only in retrospect that we see that "Boylean experimental philosophy was not the high road to modern experimentalism; it was a detour." *Discipline and Experience*, 2–3.

44. Francis Bacon, *The New Organon*, in *The New Organon and Related Writings*, ed. Fulton H. Anderson (New York: Liberal Arts Press, 1960), book 1, aphorism 38. All quotations of the *New Organon* are from this edition and are cited by book and aphorism number in the text; the "Preface" is cited by page number. Bacon is not unique among natural philosophers in worrying about the scope of language. William Gilbert acknowledges that he must "sometimes employ words new and unheard-of, not (as alchemists are wont to do) in order to veil things with a pedantic terminology and to make them dark and obscure, but in order that hidden things which have no name and that have never come into notice, may be plainly and fully published" (*De Magnete*, l). Galileo is more skeptical about naming: "names and attributes must be accommodated to the essence of things, and not the essence to the names, since things come first and names afterwards." "Letters on Sunspots" (excerpts from), in *Discoveries and Opinions of Galileo*, 92. Robert Boyle offers a more complex perspective of how words are misused and how to rectify this shortcoming, when he laments

"that the word *Nature* hath been so frequently, and yet so unskilfully imploy'd, both in Books and in Discourse, by all sorts of Men, Learned and Illiterate. For the very great Ambiguity of this term, and the promiscuous use Men are wont to make of it, without sufficiently attending to its different Significations, makes many of the Expressions wherein they imploy it, (and think they do it well and truly) to be either not intelligible, or not proper, or not true" (30–31). He thus "heartily wish[es], that Philosophers and other Learned Men (whom the rest in time would follow) would by common (tho' perhaps Tacite) consent, introduce some more Significant, and less ambiguous Terms and Expressions in the room of the too licenciously abused word *Nature*, and the Forms of Speech that depend on it. Or would, at least, decline the use of it, as much as conveniently they can; and where they think they must imploy it, would add a word or two, to declare in what clear and determinate sense they use it (31–32). *A Free Enquiry into the Vulgarly Receiv'd Notion of Nature made in an Essay Address'd to a Friend* (London: Printed by H. Clark for John Taylor, 1686), EEBO.

45. See Preston, *Poetics of Scientific Investigation*, esp. 9–10. The body of scholarship on words and writing in Bacon's works is vast, covering rhetoric, figurative and allegorical language, universal language, poetry and fiction, etc. For representative works, see Brian Vickers, "Bacon and Rhetoric," in *The Cambridge Companion to Bacon*, ed. Markku Peltonen (Cambridge: Cambridge University Press, 1996), 200–231; Ronald Levao, "Francis Bacon and the Mobility of Science," in "Seeing Science," special issue, *Representations* 40 (1992): 1–32; Julie Robin Solomon, "'To Know, To Fly, To Conjure': Situating Baconian Science at the Juncture of Early Modern Modes of Reading," *Renaissance Quarterly* 44, no. 3 (1991): 513–58; Anthony Grafton, *Defenders of the Text: The Traditions of Scholarship in an Age of Science, 1450–1800* (Cambridge, MA: Harvard University Press, 1991), 30–34; Rhodri Lewis, "Francis Bacon, Allegory and the Uses of Myth," *Review of English Studies* 61, no. 250 (2010): 360–89; and James Stephens, "Bacon's New English Rhetoric and the Debt to Aristotle," *Speech Monographs* 39, no. 4 (1972): 248–59.

46. Sprat, *The History of the Royal-Society of London*, 62. Early modern writers often personify and feminize the concept of "nature," but they are not always consistent in how they describe the concept. Nature emerges as a constructed entity that both poets and natural philosophers deploy to their distinctive ends, even though the writers often *present* it as an external, extant thing that stands apart from art and artifice—as well as from human affairs. In this book, I thus understand "Nature" as a constructed and artificial entity.

47. Ibid., 113.

48. Francis Bacon, *The Great Instauration*, in *The New Organon and Related Writings*, 3 (Proem).

49. On early modern discussions about how an "ideal philosophical language would recapture at least some of the elements of the original Adamic tongue," see Peter Harrison, *The Fall of Man and the Foundations of Science* (Cambridge: Cambridge University Press, 2007), 176. See also 177–78, 192–96, 207–12.

50. W. Gilbert, *De Magnete*, l.

51. For futility and early modern literature, see McEleney, *Futile Pleasures*.

52. See Rosenfeld, *Indecorous Thinking*, 3; and Mann, *Outlaw Rhetoric*, 173, for the increasing links between the two.

53. Galileo, *The Assayer*, 237.

54. Sprat, *The History of the Royal-Society of London*, 112.

55. On how Shakespeare's appeals to the image-making faculty of "imagination" are in conversation with discourses of anatomy, medical diagnosis, optics, physics, natural history, and the mechanical arts, see Roychoudhury, *Phantasmatic Shakespeare*.

56. Edmund Spenser, *The Faerie Queene*, ed. A. C. Hamilton, text ed. Hiroshi Yamashita and Toshiyuki Suzuki, 2nd ed. (Harlow: Longman, 2007), 4.12.1.1. All citations are from this edition

and given by book, canto, and stanza number in the text. Where the whole stanza is not cited, I also provide line numbers.

57. Philippe Lacoue-Labarthe, *Typography: Mimesis, Philosophy, Politics*, ed. Christopher Fynsk (Cambridge, MA: Harvard University Press, 1989), 255–56.

58. John Harington borrows extensively from this passage in *A Brief Apology of Poetry* (1591), offering a glimpse into how writers were interacting with the *Defence* before its publication. See, *A Brief Apology of Poetry* (selected passages), in *Sidney's "The Defence of Poesy" and Selected Renaissance Literary Criticism*, ed. Gavin Alexander (London: Penguin, 2004), 266.

59. Stephen Halliwell, *The Aesthetics of Mimesis: Ancient Texts and Modern Problems* (Princeton, NJ: Princeton University Press, 2002), 5.

60. See Mann, *Trials of Orpheus*, which examines a particular aspect of the relation of language and *praxis*: "How do words produce action? And, more particularly, how do early modern English writers conceptualize the unseen 'force' of verbal eloquence?" (1).

61. Margreta De Grazia, "Lost Potential in Grammar and Nature: Sidney's *Astrophil and Stella*," *Studies in English Literature, 1500–1900* 21, no. 1 (1981): 21–35, 22. See also Lynne Magnusson, "'What May Be and Should Be': Grammar Moods and the Invention of History in *1 Henry VI*," in *Shakespeare's World of Words*, ed. Paul Yachnin (London: Arden Shakespeare, 2015), 147–70; Lynne Magnusson, "A Play of Modals: Grammar and Potential Action in Early Shakespeare," *Shakespeare Survey* 62 (2009): 69–80; Colleen Ruth Rosenfeld, "Poetry and the Potential Mood: The Counterfactual Form of Ben Jonson's 'To Fine Lady Would-Be,'" *Modern Philology* 112, no. 2 (2014): 336–57; and Colleen Ruth Rosenfeld, "In the Mood of Fiction," *English Literary Renaissance* 52, no. 3 (2022): 385–96.

62. William Lily and John Colet, *A Short Introduction of Grammar* (1549; facs. rpt., Menston, Bradford: Scholar Press, 1970), sig. B3v, quoted in De Grazia, "Lost Potential in Grammar and Nature," 22.

63. Thomas Blundeville, *The Arte of Logick* (London: Printed by William Stansby, 1617), 76, EEBO. The first edition was published in 1599.

64. Ibid., 79.

65. See, for instance, David Scott Kastan, *Shakespeare After Theory* (New York: Routledge, 1999), 49–50, which also argues that history deals with what was "thought." See also Barret, *Untold Futures*, 11. On discussions of evidence and early modern drama, see Quentin Skinner, *Forensic Shakespeare* (Oxford: Oxford University Press, 2014); Lorna Hutson, *The Invention of Suspicion: Law and Mimesis in Shakespeare and Renaissance Drama* (Oxford: Oxford University Press, 2007); and Lorna Hutson, *Circumstantial Shakespeare* (Oxford: Oxford University Press, 2015). See also Frances E. Dolan, *True Relations: Reading, Literature, and Evidence in Seventeenth-Century England* (Philadelphia: University of Pennsylvania Press, 2013).

66. See Catherine Gallagher, "When Did the Confederate States of America Free the Slaves?" *Representations* 98, no. 1 (2007): 53–61; Andrew H. Miller, "Lives Unled in Realist Fiction," *Representations* 98, no. 1 (2007): 118–34; and Stephanie Insley Hershinow, *Born Yesterday: Inexperience and the Early Realist Novel* (Baltimore: Johns Hopkins University Press, 2019), 8–14, 32, 37.

CHAPTER 1

1. Edmund Spenser, "Letter to Raleigh," in *The Faerie Queene*, 714–18, 716. For a reading of example as a positive middle ground between experience and maxim, see Jeffrey A. Dolven, *Scenes of Instruction in Renaissance Romance* (Chicago: University of Chicago Press, 2007), 137.

2. On how the "Letter" "crosses the familiar letter with a defence of poetry" and "provides all the cues necessary to understanding Spenser's didactic poetics, albeit not in a conveniently cut-and-dried memo," see Jane Grogan, *Exemplary Spenser: Visual and Poetic Pedagogy in The Faerie Queene* (Burlington, VT: Ashgate, 2009), 31–32, 17. For discrepancies between the "Letter" and the poem, see Joanne Craig, "The Image of Mortality: Myth and History in the *Faerie Queene*," *English Literary History* 39, no. 4 (1972): 520–44, esp. 521; Thomas P. Roche, *The Kindly Flame: A Study of the Third and Fourth Books of Spenser's Faerie Queene* (Princeton, NJ: Princeton University Press, 1964), 196–99; and Maureen Quilligan, *The Language of Allegory: Defining the Genre* (Ithaca, NY: Cornell University Press, 1979), 232.

3. Spenser, "Letter," 716.

4. Ibid., 714.

5. William Butler Yeats, *The Cutting of an Agate* (New York: Macmillan, 1912), 213–14.

6. For the mirror's promise of "perfect sight" as a "closure of the gap between signifier and signified" and its production of an "interpretive gap," see Elizabeth J. Bellamy, *Translations of Power: Narcissism and the Unconscious in Epic History* (Ithaca, NY: Cornell University Press, 1992), 203. The mirror functions as what Bruno Latour, highlighting the agential capacities of the nonhuman, classifies as a "mediator." A mediator "is an original event and creates what it translates as well as the entities between which it plays the mediating role." *We Have Never Been Modern*, trans. Catherine Porter (Cambridge, MA: Harvard University Press, 1993), 78. On posthumanist theories and Spenser's writings, see Ayesha Ramachandran and Melissa E. Sanchez, eds., "Spenser and 'the Human,'" special issue, *Spenser Studies* 30 (2015).

7. In an era when atlases, astrological charts, globes, and astronomical clocks served as explanatory models for a universe that seemed to be rapidly expanding in the European imaginary, Merlin's glass exemplifies the complex project of worldmaking, that is, the "methods by which early modern thinkers sought to imagine, shape, revise, control, and articulate the dimensions of the world." Ayesha Ramachandran, *The Worldmakers: Global Imagining in Early Modern Europe* (Chicago: University of Chicago Press, 2015), 6. On worlds and early modern fiction making, see Harry Berger Jr., *Second World and Green World: Studies in Renaissance Fiction-Making*, intro. John Patrick Lynch (Berkeley: University of California Press, 1988).

8. *OED Online*, s.v. "speculative, adj. and n."

9. Ibid., s.v. "speculation, n."

10. For instance, Stephen Greenblatt has shown how ideologies of control and conquest are intimately tied to Spenser's poetic ambitions, arguing that Spenser was "our originating and preeminent poet of empire." *Renaissance Self-Fashioning: From More to Shakespeare* (Chicago: University of Chicago Press, 1980), 174.

11. Patricia Parker, *Inescapable Romance: Studies in the Poetics of a Mode* (Princeton, NJ: Princeton University Press, 1979), 4; Jonathan Goldberg, *Endlesse Worke: Spenser and the Structures of Discourse* (Baltimore: Johns Hopkins University Press, 1981), 76n1.

12. The body of work on Spenser's philosophical sources is immense. For a classic study of Lucretian materialism in the poem, see Edwin Greenlaw, "Spenser and Lucretius," *Studies in Philology* 17, no. 4 (1920): 439–64. Greenlaw's argument had several critics; see, for example, Evelyn May Albright, "Spenser's Cosmic Philosophy and His Religion," *PMLA* 44, no. 3 (1929): 715–59; and Ronald B. Levinson, "Spenser and Bruno," *PMLA* 43, no. 3 (1928): 675–81. Yet scholarship today understands Spenser to be deeply engaged with Lucretian ideas. For representative works, see Jonathan Goldberg, *The Seeds of Things: Theorizing Sexuality and Materiality in Renaissance Representations* (New York: Fordham University Press, 2009); Gerard Passannante, *The Lucretian Renaissance: Philology and the Afterlife of Tradition* (Chicago: University of Chicago Press,

2011); and Ayesha Ramachandran, "Mutabilitie's Lucretian Metaphysics: Scepticism and Cosmic Process in Spenser's Cantos," in *Celebrating Mutabilitie: Essays on Edmund Spenser's Mutabilitie Cantos*, ed. Jane Grogan (Manchester: Manchester University Press, 2010), 220–45. For Spenser and Platonism, see Elizabeth Bieman, *Plato Baptized: Towards the Interpretation of Spenser's Mimetic Fictions* (Toronto: University of Toronto Press, 1988); Jon A. Quitslund, *Spenser's Supreme Fiction: Platonic Natural Philosophy and the Faerie Queene* (Toronto: University of Toronto Press, 2001); Kenneth Borris, Carol Kaske, and Jon Quitslund, eds., "Spenser and Platonism: An Expanded Special Volume," special issue, *Spenser Studies* 24 (2009); and Robert Ellrodt, *Neoplatonism in the Poetry of Spenser* (Geneva: Droz, 1960). For intersections between Neoplatonism and Lucretianism in Spenser, see Ayesha Ramachandran, "Edmund Spenser, Lucretian Neoplatonist: Cosmology in the *Fowre Hymnes*," in Borris, Kaske, and Quitslund, "Spenser and Platonism," 373–411. On Spenser and Aristotle, see William Fenn DeMoss, "Spenser's Twelve Moral Virtues 'According to Aristotle' (Continued)," *Modern Philology* 16, no. 1 (1918): 23–38; William Fenn DeMoss, "Spenser's Twelve Moral Virtues 'According to Aristotle' II," *Modern Philology* 16, no. 5 (1918): 245–70; and Crane, *Losing Touch with Nature*, 94–122. See also Spiller, *Science, Reading, and Renaissance Literature*, 59–100, on Neoplatonic metaphysics and Aristotelian biology in Spenser. On Spenser and Galen, see Michael Carl Schoenfeldt, *Bodies and Selves in Early Modern England: Physiology and Inwardness in Spenser, Shakespeare, Herbert, and Milton* (Cambridge: Cambridge University Press, 1999), 40–73.

13. These episodes have received significant attention in studies of Spenser's science and philosophy. See, for instance, Ramachandran, "Edmund Spenser, Lucretian Neoplatonist," esp. 380; Ramachandran, "Mutabilitie's Lucretian Metaphysics"; Wendy Beth Hyman, "Seizing Flowers in Spenser's Bower and Garden," *English Literary Renaissance* 37, no. 2 (2007): 193–214; and Angus Fletcher, "Complexity and the Spenserian Myth of Mutability," *Literary Imagination* 6, no. 1 (2004): 1–22.

14. Although I make references to the 1596 edition, my primary argument focuses on the 1590 edition of *The Faerie Queene*.

15. Gordon Teskey, *Allegory and Violence* (Ithaca, NY: Cornell University Press, 1996), 174.

16. Spenser, "Letter," 716–17.

17. On Spenser's use of "method" in relation to "discipline," see Spiller, *Science, Reading, and Renaissance Literature*, 63–65. The term "method" has a complex history, especially in dialectic and rhetoric. See Ong, *Ramus, Method, and the Decay of Dialogue*; and Jardine, *Francis Bacon*.

18. Spenser's own poem was interpreted as prophetic by later writers. For example, see the anonymous Royalist treatise *The Faerie Leveller* (London: Printed just levell anens the Saints Army: in the yeare of their Saintships ungodly Revelling for a godly Levelling, 1648), EEBO, which reads the argument between the Giant and Artegall in Book 5 as a prophecy.

19. Samuel Taylor Coleridge, *Coleridge's Miscellaneous Criticism*, ed. Thomas Middleton Raysor (London: Constable, 1936), 36.

20. Doležel, *Heterocosmica*, 13.

21. See Mary C. Fuller, *Voyages in Print: English Travel to America, 1576–1624* (New York: Cambridge University Press, 1995), esp. 1–15; and Mary Baine Campbell, *The Witness and the Other World: Exotic European Travel Writing, 400–1600* (Ithaca, NY: Cornell University Press, 1988), esp. 165–266, for changes in experiential travel writing in the early sixteenth century. On the "poetics of description" in travel literature, see Roychoudhury, *Phantasmatic Shakespeare*, 166.

22. Fuller, *Voyages in Print*, 11. On how the British deployed the language and idea of "discovery" and "gained a privileged epistemological position, whereby as 'discoverers' they could claim new knowledge which they could then process and circulate via the intractable colonial binarisms:

civilization and barbarism, tradition and modernity, and Christianity and heathenism, among others," see Jyotsna G. Singh, *Colonial Narratives / Cultural Dialogues: "Discoveries" of India in the Language of Colonialism* (London: Routledge, 1996), 2. For a recent study that recovers the imperial and racial ideologies underpinning discourses of—and the language of—"discovery," see Nedda Mehdizadeh, "Cosmography and/in the Academy: Authorizing the Ideological Pathways of Empire" (unpublished manuscript, August 9, 2022).

23. Fuller, *Voyages in Print*, 33.

24. Ibid., 13.

25. Campbell, *Witness and the Other World*, 211–12.

26. Humphrey Gilbert, *A Discovrse of a Discoverie for a New Passage to Cataia* (London: By Henry Middleton for Richarde Ihones, 1576), sig. ¶¶¶ iiir–¶¶¶ iiiv, EEBO. Fuller argues that Gilbert's account "opens up a middle category between the known and the impossible, which we might call the space of the credible" (*Voyages in Print*, 21).

27. Edward Phillips defines "*Utopia*" as "the feigned name of a Countrey described by Sir *Thomas More*, as the pattern of a well govern'd Common-wealth; hence it is taken by Metaphor for any imaginary, or feigned place." *The New World of English Words* (London: Printed by E. Tyler for Nath. Brooke, 1658), s.v. "Utopia," EEBO.

28. Campbell, *Wonder and Science*, 14–15. On utopia as a generative mode that is not other to history, see Marina Leslie, *Renaissance Utopias and the Problem of History* (Ithaca, NY: Cornell University Press, 1998), 14. On the relationship of fiction to empire and nation, mediated through understandings of the concept of "Nowhere," see Jeffrey Knapp, *An Empire Nowhere: England, America, and Literature from Utopia to The Tempest* (Berkeley: University of California Press, 1992).

29. Debapriya Sarkar, "The Utopian Hypothesis," *English Literary Renaissance* 52, no. 3 (2022): 371–84, 373.

30. In my use of the term "potentiality," I am influenced by Giorgio Agamben, "On Potentiality," in *Potentialities: Collected Essays in Philosophy*, ed. and trans. Daniel Heller-Roazen (Stanford, CA: Stanford University Press, 1999), 177–84.

31. For prevalent methods of empirical and practical knowledge operant in sixteenth-century England, see Crane, *Losing Touch with Nature*, esp. 1–18.

32. For a contrasting viewpoint, which marks the similarities between Spenser and Gilbert, see Shannon Miller, *Invested with Meaning: The Raleigh Circle in the New World* (Philadelphia: University of Pennsylvania Press, 1998), 34–39.

33. David Read, *Temperate Conquests: Spenser and the Spanish New World* (Detroit: Wayne State University Press, 2000), 15, 24. In the Proem to Book 2, Spenser "is writing the history of the future itself" (22).

34. Barret, *Untold Futures*, 83.

35. Campbell, *Wonder and Science*, 120. On Spenser and Bruno, see Levinson, "Spenser and Bruno"; and Steven J. Dick, *Plurality of Worlds: The Origins of the Extraterrestrial Life Debate from Democritus to Kant* (Cambridge: Cambridge University Press, 1982), 84. For shifts in ideas about cosmology and new findings in astronomy, see Peter Dear, *Revolutionizing the Sciences: European Knowledge and Its Ambitions, 1500–1700* (Princeton, NJ: Princeton University Press, 2001); and Shapin, *The Scientific Revolution*. For the "developing awareness" about "plurality of worlds and worldviews" in the sixteenth century, see Roland Greene, "A Primer of Spenser's Worldmaking: Alterity in the Bower of Bliss," in *Worldmaking Spenser: Explorations in the Early Modern Age*, ed. Patrick Cheney and Lauren Silberman (Lexington: University Press of Kentucky, 2000), 9–31, 9.

36. Ramachandran, *Worldmakers*, 8.

37. In "Pascal's Allegory of Persuasion," in *Allegory and Representation*, ed. Stephen Greenblatt (Baltimore: Johns Hopkins University Press, 1981), 1–25, Paul de Man argues that as a "referentially indirect mode" that attempts to deal with "furthest reaching truths," allegory cannot deal with representation of truth (2).

38. Jacqueline T. Miller, *Poetic License: Authority and Authorship in Medieval and Renaissance Contexts* (Oxford: Oxford University Press, 1986), 80. One way of mobilizing reference through allegory would be to displace it to a different topical domain. Puttenham highlights how, in "pastoral poesy, called eclogue," the poet uses allegory to comment on society without concern for the dangers of direct reference: "the poet devised the eclogue . . . not of purpose to counterfeit or represent the rustical manner of loves and communication, but under the veil of homely persons and in rude speeches to insinuate and glance at greater matters, and such as perchance had not been safe to have been disclosed in any other sort" (*The Art of English Poesy*, 127–28).

39. I borrow the phrase "temporal passage" from James J. Paxson, "The Allegory of Temporality and the Early Modern Calculus," *Configurations* 4, no. 1 (1996): 39–66, 54.

40. Spenser, "Letter," 714.

41. Angus Fletcher, *Allegory: The Theory of a Symbolic Mode* (Ithaca, NY: Cornell University Press, 1964), 176, 177. Scholars have discussed allegory's indeterminacies and the struggle between process and ideality in various ways. In Teskey's influential formulation, "the very word *allegory* evokes a schism in consciousness—between a life and a mystery, between the real and the ideal, between a literal tale and its moral—which is repaired, or at least concealed, by imagining a hierarchy on which we ascend toward truth. The opening of a schism, or, as I shall call it, a rift, and the subsequent effort to repair it by imaginative means lie behind many of our commonest uses of the term. On the whole, when we speak of *allegory* we refer to an enlightening or witty analogy between two things, both of some complexity, but one of less importance than the other" (*Allegory and Violence*, 2). Susanne L. Wofford underscores how in *The Faerie Queene* abstraction is opposed to narrative, arguing that the reader oscillates between the positions of the narrator, who imposes a hierarchical meaning, and the fictional characters, who remain oblivious of their allegorical statuses and functions within the text. *The Choice of Achilles: The Ideology of Figure in the Epic* (Stanford, CA: Stanford University Press, 1992).

42. Spenser, "Letter," 714 (emphasis mine).

43. Puttenham, *The Art of English Poesy*, 271.

44. See Quintilian, *Institutes of oratory*, ed. and trans. John Selby Watson, 2 vols. (London: G. Bell and Sons, 1895), 2:165.

45. Puttenham, *The Art of English Poesy*, 270.

46. Ibid., 271.

47. Ibid., 262–63.

48. Ibid., 262, 238.

49. Colin Burrow, "Spenser and Classical Traditions," in *The Cambridge Companion to Spenser*, ed. Andrew Hadfield (Cambridge: Cambridge University Press, 2001), 217–36, 229.

50. William A. Oram, "Spenserian Paralysis," *Studies in English Literature, 1500–1900* 41, no. 1 (2001): 49–70, 51; see also 59. See also Michael Slater, "Spenser's Poetics of 'Transfixion' in the Allegory of Chastity," *Studies in English Literature, 1500–1900* 54, no. 1 (2014): 41–58.

51. Louise Gilbert Freeman, "The Metamorphosis of Malbecco: Allegorical Violence and Ovidian Change," *Studies in Philology* 97, no. 3 (2000): 308–30, 324.

52. Kelly Lehtonen, "The Abjection of Malbecco: Forgotten Identity in Spenser's Legend of Chastity," *Spenser Studies* 29 (2014): 179–96, 190.

53. On his transformation into an abstraction, see Harold Skulsky, *Metamorphosis: The Mind in Exile* (Cambridge, MA: Harvard University Press, 1981), esp. 131–34; and Freeman, "The Metamorphosis of Malbecco," esp. 319–28. For Malbecco as a figure of two kinds of representation, "*exemplary* and *catalytic*," in which, respectively, "an allegorical figure directly bodies forth the psychic or material condition for which it is named," and "an allegorical figure functions as the precipitating cause or occasion of the condition for which it is named," see Linda Gregorson, "Protestant Erotics: Idolatry and Interpretation in Spenser's *Faerie Queene*," *English Literary History* 58, no. 1 (1991): 1–34, 5.

54. Teskey, *Allegory and Violence*, 29.

55. For how early modern writers used "deformity" to "describe departures from expected form," see Katherine Schaap Williams, *Unfixable Forms: Disability, Performance, and the Early Modern English Theater* (Ithaca, NY: Cornell University Press, 2021), 27. See also 10–18, 27–29. I primarily draw on the definition "Of irregular form; shapeless, formless" (3). Another definition, "Marred in shape, misshapen, distorted; unshapely, of an ill form. Now chiefly of persons: Misshapen in body or limbs" (2), can differently explain Malbecco's final state. *OED Online*, s.v. "deformed, adj."

56. On how the study of the "physics that underlie literary texts" can offer new understandings of form, see Liza Blake, "The Physics of Poetic Form in Arthur Golding's Translation of Ovid's *Metamorphoses*," *English Literary Renaissance* 51, no. 3 (2021): 331–55, 333. On the intertwining of early modern poetry and physics, see Liza Blake, "Hester Pulter's Particle Physics and the Poetics of Involution," *Journal of Early Modern Cultural Studies* 20, no. 2 (2020): 71–98, esp. 71–73.

57. See Aristotle, *Physics*, in *The Complete Works of Aristotle*, Book 4; and Titus Lucretius Carus, *The Nature of Things (De Rerum Natura)*, trans. A. E. Stallings (London: Penguin, 2007), Book 1.

58. See *OED Online*, s.v. "ousia, n.," and "substance, n." For a study of how Spenser's poem relies on a "dilated materiality" to curb the poem's tendency toward narrative expansion, see Debapriya Sarkar, "Dilated Materiality and Formal Restraint in *The Faerie Queene*," *Spenser Studies* 31–32 (2018): 137–66.

59. On Spenser as a thinker who engages with varied schools of thought, see Greenlaw, "Spenser and Lucretius," 464; and Ramachandran, "Edmund Spenser, Lucretian Neoplatonist," 376–77.

60. Wofford, *Choice of Achilles*, 299.

61. Of course, there are figures in the poem who correspond to what they seem. But as the Giant "*Disdayne*"—whom Guyon meets in Mammon's cave—suggests, even figures who seem to signify what they mean can exceed these imposed delineations (see 2.7.40–41). For a counterpoint to such multiplicity, see Chris Barrett, "Allegraphy and *The Faerie Queene*'s Significantly Unsignifying Ecology," *Studies in English Literature, 1500–1900* 56, no. 1 (2016): 1–21, who discusses "allegraphy," which "revels in the problem of incommensurability, cultivating a narrative world that routinely celebrates the failures of literature's enabling metaphoricity" (3).

62. Burrow, "Spenser and Classical Traditions," 229.

63. Jonathan Gil Harris, *Untimely Matter in the Time of Shakespeare* (Philadelphia: University of Pennsylvania Press, 2009), 11.

64. I read "Perfected" as "Made perfect or complete; faultless." *OED Online*, s.v. "perfected, adj."

65. Spenser, "Letter," 715.

66. Spenser, Ibid. 716.

67. Ibid.

68. Magnificence has a complicated history. As Hugh MacLachlan and Philip B. Rollinson note in their entry "magnanimity, magnificence," "Spenser had available to him an extended tradition of the virtues in which magnificence was frequently defined as the doing of great deeds for the sake of glory, and he may have seen magnificence as an expansion of Aristotelian magnanimity. Thus when he says that magnificence is the 'perfection' of the other virtues, he could mean that it is their 'completion,' their being brought into action—action of the highest order available to man—for the sake of honor or glory." In *The Spenser Encyclopedia*, gen. ed. A. C. Hamilton et al. (Toronto: University of Toronto Press, 1990), 1175–77, 1176. Aristotle defines "magnanimity" and not "magnificence" as "the perfection of the virtues" (1175), and both occupy complex positions in Scholastic ethics, which also influences Spenser here. On how "magnificence" in *The Faerie Queene* relates to Aristotle's writings, see DeMoss, "Spenser's Twelve Moral Virtues 'According to Aristotle' (Continued)"; and DeMoss, "Spenser's Twelve Moral Virtues 'According to Aristotle' II."

69. MacLachlan and Rollinson, "magnanimity, magnificence," 1176.

70. Spenser, "Letter," 715.

71. For names as "rigid designators" that function across worlds, see Doležel, *Heterocosmica*, 18.

72. On how in this moment "it is as though the poet were simultaneously inviting us to overlook Arthur's death and refusing to let us do so," see Judith H. Anderson, "Arthur, Argante, and the Ideal Vision: An Exercise in Speculation and Parody," in *The Passing of Arthur: New Essays in Arthurian Tradition*, ed., Christopher Baswell and William Sharpe (New York: Garland, 1988), 193–206, 199.

73. Quilligan, *The Language of Allegory*, 24.

74. Barret, *Untold Futures*, 64.

75. Bellamy, *Translations of Power*, 228.

76. Sarah Wall-Randell locates pleasure and wonder in this instance of limited knowledge, arguing that Arthur's actions demonstrate "pleasurable *misreading*." *The Immaterial Book: Reading and Romance in Early Modern England* (Ann Arbor: University of Michigan Press, 2013), esp. 34. See also Barret, *Untold Futures*, 66–81, for how the poem reinvents history through Arthur reading his own past.

77. Harris, *Untimely Matter in the Time of Shakespeare*, 11.

78. Comparing Arthur and Aeneas as epic heroes who both experience the "uncanny moment when they confront *themselves* in the very representations of a history they are presumably outside of" (*Translations of Power*, 225), Bellamy argues that Spenser's protagonist undergoes "no . . . formation of historical unconscious" (228). Wall-Randell notes that "Arthur apparently has found his 'self,' suddenly, at the very center of things, at the paused wellspring of the historical record" (*The Immaterial Book*, 25), but "cannot truly recognize himself as the heir of Uther Pendragon" (29).

79. See Edmund Spenser, *The Shepheardes Calendar*, in *The Yale Edition of the Shorter Poems of Edmund Spenser*, ed. William A. Oram, et al. (New Haven, CT: Yale University Press, 1989).

80. Puttenham, *The Art of English Poesy*, 94, 97. For Sidney's poetry as prophetic, see Roger E. Moore, "Sir Philip Sidney's Defense of Prophesying," *Studies in English Literature, 1500–1900* 50, no. 1 (2010): 35–62.

81. For examples of the passive role of the prophet, see Jerrod Rosenbaum, "Spenser's Merlin Rehabilitated," *Spenser Studies* 29 (2014): 149–78, esp. 160, 161, 174. For an argument about how

E. K. in *The Shepheardes Calender* "attempts to make the *vates* not the victim of mortal or divine coercion, but a powerful, masculine figure whose verse ravishes his male listeners," see Brian Pietras, "Erasing Evander's Mother: Spenser, Virgil, and the Dangers of Vatic Authorship," *Spenser Studies* 31–32 (2018): 43–69, 46. Burrow argues that the "vision of the future which Merlin gives Britomart is the first real burst of Virgilian prophecy in *The Faerie Queene*" ("Spenser and Classical Traditions," 221).

82. On Merlin as a "figure for the poet," see William Blackburn, "Spenser's Merlin," *Renaissance and Reformation/Renaissance et Reforme* 4, no. 2 (1980): 179–98, 179.

83. Rosenbaum, "Spenser's Merlin Rehabilitated," 150.

84. Merlin's actions reflect the following meanings of the verb "divine": "To have supernatural or magical insight into (things to come); to have presentiment of; hence *gen.* to predict or prophesy by some kind of special inspiration or intuition" (3), and "To foretell by divine or superhuman power; to prophesy" (8). *OED Online*, s.v. "divine, v."

85. See Michael Witmore, *Culture of Accidents: Unexpected Knowledges in Early Modern England* (Stanford, CA: Stanford University Press, 2001).

86. Spenser, "Letter," 718.

87. Sidney states: "For who will be taught, if he be not moved with desire to be taught? And what so much good doth that teaching bring forth (I speak still of moral doctrine) as that it moveth one to do that which it doth teach?" (*A Defence of Poetry*, 39).

88. For a reading of how one of the poem's aims is to "persuade" Elizabeth, see Colin Burrow, *Epic Romance: Homer to Milton* (New York: Oxford University Press, 1993), 102.

89. On how Spenser's poetics "depends on a consistently anti-apocalyptic poetics," see Susanne L. Wofford, "Britomart's Petrarchan Lament: Allegory and Narrative in *The Faerie Queene* III, iv," *Comparative Literature* 39, no. 1 (1987): 28–57, 54.

90. The etymology of hypothesis is "foundation, base; hence, basis of an argument, supposition, also, subject matter, etc." *OED Online*, s.v. "hypothesis, n."

CHAPTER 2

1. Mary Thomas Crane, "What Was Performance?" *Criticism* 43, no. 2 (2001): 169–87, 172.

2. Scholars trace the topicality of the play primarily by focusing on politics, sovereignty, and witchcraft. See, for example, Kastan, *Shakespeare After Theory*, 165–82; Stephen Greenblatt, "Shakespeare Bewitched," in *New Historical Literary Study: Essays on Reproducing Texts, Representing History*, ed. Jeffrey N. Cox and Larry J. Reynolds (Princeton, NJ: Princeton University Press, 1993), 108–35; Peter Stallybrass, "Macbeth and Witchcraft," in *Focus on Macbeth*, ed. John Russell Brown (Boston: Routledge, 1982), 189–209; and Alan Sinfield, *Faultlines: Cultural Materialism and the Politics of Dissident Reading* (Berkeley: University of California Press, 1992). *Macbeth*'s use of equivocal language, most explicitly dramatized in the porter's claim, "here's an equivocator" (2.3.7), is often discussed in relation to Henry Garnet's trial in 1606 and *A Treatise of Equivocation*. See, for instance, Frank L. Huntley, "*Macbeth* and the Background of Jesuitical Equivocation," *PMLA* 79, no. 4 (1964): 390–400.

3. David Scott Kastan, *Shakespeare and the Shapes of Time* (Hanover, NH: University Press of New England, 1982), 100, 91. Kastan argues that *Macbeth* is an "exception" to the "shape of Shakespeare's tragedies" (100) because it does not "close"; the "end of *Macbeth* allows us to look confidently beyond Macbeth's death" (91). *Macbeth* scholarship has frequently returned to issues of temporality, history, and genealogy to explore the play's ruminations on time. For representative

works, see Harris, *Untimely Matter in the Time of Shakespeare*, 119–39; Janet Adelman, "Born of Woman, Fantasies of Maternal Power in *Macbeth*," in *Cannibals, Witches, and Divorce: Estranging the Renaissance*, ed. Marjorie Garber (Baltimore: Johns Hopkins University Press, 1987), 90–121; Donald W. Foster, "*Macbeth's* War on Time," *English Literary Renaissance* 16, no. 2 (1986): 319–42; Luisa Guj, "Macbeth and the Seeds of Time," *Shakespeare Studies* 18 (1986): 175–88; and Julia MacDonald, "Demonic Time in Macbeth," *Ben Jonson Journal* 17, no. 1 (2010): 76–96. On how the play is a dramatization of "concurrent and occasionally competing temporal models," see Rhodri Lewis, "Polychronic Macbeth," *Modern Philology* 117, no. 3 (2020): 323–46, 325. On the temporality of tragedy and tragic time more broadly, see Rebecca Bushnell, *Tragic Time, in Drama, Film, and Videogames: The Future in an Instant* (London: Palgrave Macmillan, 2016). On how ideas related to time—such as anachronism, periods, and chronology—shape both Shakespeare's plays and Shakespeare criticism, see Margreta De Grazia, *Four Shakespearean Period Pieces* (Chicago: University of Chicago Press, 2021).

4. On witchcraft in the period, see Stallybrass, "Macbeth and Witchcraft"; Greenblatt, "Shakespeare Bewitched"; Adelman, "Born of Woman"; and Keith Thomas, *Religion and the Decline of Magic* (New York: Scribner, 1971), 435–586. On how discourses of witchcraft intersected with discussions in science, philosophy, religion, and law, see Shapiro, *Probability and Certainty in Seventeenth-Century England*, esp. 194–226. For an example of how the witches serve as signifiers of ethnography and environment, see Mary Floyd-Wilson, "English Epicures and Scottish Witches," *Shakespeare Quarterly* 57, no. 2 (2006): 131–61.

5. On "maker's knowledge," see Pérez-Ramos, *Francis Bacon's Idea of Science*, 48–62. On how Renaissance literature and science are "forms of 'making,'" see Spiller, *Science, Reading, and Renaissance Literature*, 3.

6. For example, see Sharon L. Jansen Jaech, "Political Prophecy and Macbeth's 'Sweet Bodements,'" *Shakespeare Quarterly* 34, no. 3 (1983): 290–97.

7. Marjorie Garber, "'What's Past Is Prologue': Temporality and Prophecy in Shakespeare's History Plays," in *Renaissance Genres: Essays on Theory, History, and Interpretation*, ed. Barbara Kiefer Lewalski (Cambridge, MA: Harvard University Press, 1986), 301–31, 308, 318. Garber marks the audience's powerlessness, which "is paradoxically a function of our very belief in the truth of history" (305). Various aspects of her analysis of prophecies resonate, though not perfectly, with *Macbeth*: "The history play as such is thus lodged in the paradoxical temporality of what the French call the *futur antérieur*, the prior future, the tense of what 'will have occurred'" (306–7). This relationship to history, and the emphasis on the past, affects "audience-response" (306): "Hearing these anachronistic prophecies, we know their truth and are powerless to alter the course of a history that has already taken place. Only Shakespeare can do that, and he does, modifying his sources as he sees fit" (330–31).

8. See Eamon, *Science and the Secrets of Nature*, esp. 267–360.

9. Ibid., 131.

10. Although my argument does not take this approach, there exist connections between Sidney's emphasis on non-affirmation and *Macbeth's* focus on equivocation.

11. Eamon, *Science and the Secrets of Nature*, 4.

12. John Harvey, *A Discoursiue Probleme concerning Prophesies* . . . (London: By Iohn Iackson, for Richard Watkins, 1588), 37, EEBO.

13. See, for instance, Patrick Curry, *Prophecy and Power: Astrology in Early Modern England* (Cambridge: Polity Press, 1989); and Howard Dobin, *Merlin's Disciples: Prophecy, Poetry, and Power in Renaissance England* (Stanford, CA: Stanford University Press, 1990). Thomas (*Religion and the Decline of Magic*) outlines the influential (and subsequently critiqued) thesis that dimin-

ishing faith in prophecy led to the decline of magic. This modernizing narrative has been challenged by works like Tim Thornton's *Prophecy, Politics and the People in Early Modern England* (Rochester: Boydell, 2006).

14. Moore, "Sir Philip Sidney's Defense of Prophesying," 42.

15. See Ibid.," 54–57, for the perception that prophecy was supposed to be edifying and useful. Moore connects this idea to the figure of the vatic poet who can claim access to divine truth.

16. In Merlin's revelations to Britomart, we glimpse a successful, even optimistic, depiction of how prophecy can maintain and solidify political lineage.

17. Francis Bacon, *Essays or Counsels, Civil and Moral*, in *Francis Bacon: The Major Works*, 414.

18. Turner, *English Renaissance Stage*, 47. For an example of the breakdown between *praxis* and *poiesis* in Shakespearean drama, see Richard Halpern, "Eclipse of Action: *Hamlet* and the Political Economy of Playing," *Shakespeare Quarterly* 59, no. 4 (2008): 450–82. According to Halpern, drama "crosses the categories of making and doing" because it "is a product of work, but one that imitates, in order to preserve, the realm of action" (458).

19. Eamon, *Science and the Secrets of Nature*, 54.

20. Harvey, *A Discoursiue Probleme*, 16.

21. Henry Howard, *A defensatiue against the poyson of supposed Prophesies . . .* (London: Printed by Iohn Charlewood, printer to the right Honourable Earle of Arundell, 1583), I1r, EEBO.

22. Ibid., E3r.

23. Ibid., G3r.

24. Ibid., G3v.

25. Thomas, *Religion and the Decline of Magic*, 139.

26. Eamon, *Science and the Secrets of Nature*, 10.

27. Ibid., 131, 7.

28. Ibid., 131. See also Wendy Wall, *Recipes for Thought: Knowledge and Taste in the Early Modern English Kitchen* (Philadelphia: University of Pennsylvania Press, 2015), 3, 13–14.

29. Julia Reinhard Lupton, "Thinking with Things: Hannah Woolley to Hannah Arendt," *Postmedieval* 3, no. 1 (2012): 63–79, 66.

30. On the terminology, see also Wall, *Recipes for Thought*, esp. 3–4.

31. Eamon, *Science and the Secrets of Nature*, 131.

32. See Michelle DiMeo, "Authorship and Medical Networks: Reading Attributions in Early Modern Manuscript Recipe Books," in *Reading and Writing Recipe Books, 1550–1800*, ed. Michelle DiMeo and Sara Pennell (Manchester: Manchester University Press, 2013), 25–46.

33. See, for instance, the work of the Early Modern Recipes Online Collective, https://emroc.hypotheses.org; and The Recipes Project, https://recipes.hypotheses.org. For a more practically oriented project, see Cooking in the Archives, http://rarecooking.com/. See also Elizabeth Spiller, *Seventeenth-Century English Recipe Books: Cooking, Physic and Chirurgery in the Works of Elizabeth Grey and Aletheia Talbot* (Burlington, VT: Ashgate, 2008); Rebecca Laroche, Elaine Leong, Jennifer Munroe, Hillary M. Nunn, Lisa Smith, and Amy L. Tigner, "Becoming Visible: Recipes in the Making," *Early Modern women* 13, no. 1 (2018): 133–43; Rebecca Laroche, *Medical Authority and Englishwomen's Herbal Texts, 1550–1650* (Burlington, VT: Ashgate, 2009); Jennifer Park, "Artisans of the Skin: Recipe Culture, Surface Thinking, and 17th-Century Racial Embodiment" (paper presented at the MLA Annual Convention, Seattle, WA, January 9, 2020); and Elaine Leong, *Recipes and Everyday Knowledge: Medicine, Science, and the Household in Early Modern England* (Chicago: University of Chicago Press, 2018).

34. Susanna Packe, Cookbook of Susanna Packe [manuscript], 1674, p. 221, Folger V.a.215, luna.folger.edu. Thanks to Jennifer Munroe and Wendy Wall for their guidance about specific recipe collections.

35. Hugh Plat, *Delightes for Ladies, to adorne their Persons, Tables, closets and distillatories. With Beauties, banquets, perfumes and Waters* (London: Printed by Peter Short, 1602), F4r, EEBO.

36. The Granville family, Cookery and medicinal recipes of the Granville family [manuscript], ca. 1640–ca. 1750, p. 39, Folger V.a.430, luna.folger.edu. For the attribution of the recipe books, see Kristine Kowalchuk, ed., *Preserving on Paper: Seventeenth-Century Englishwomen's Receipt Books* (Toronto: University of Toronto Press, 2017), 61–65.

37. Anne Carr, "Choyce receits collected out of the book of receits, of the Lady Vere Wilkinson [manuscript]/begun to be written by the Right Honble the Lady Anne Carr," 1673/4, p. 63, Folger V.a.612, luna.folger.edu.

38. Wall, *Recipes for Thought*, 3.

39. For textual history, and the possible spuriousness of 4.1, see William Shakespeare, *The Arden Shakespeare Complete Works*, ed. Richard Proudfoot, Ann Thompson, and David Scott Kastan (London: Methuen Drama, 2011), 773. See also Shakespeare, *The Norton Shakespeare*, 2713, 2719; and William Shakespeare, *The Riverside Shakespeare*, ed. Blakemore Evans, with J. J. M. Tobin, 2nd ed. (Boston: Houghton Mifflin, 1997), 1388. Martin Wiggins, with Catherine Richardson, *British Drama, 1533–1642: A Catalogue*, 8 vols. (Oxford: Oxford University Press, 2015), notes, "The extant text may have been adapted by Thomas Middleton; his hand has been traced in 3.5 and the Hecate parts of 4.1" (5:1606).

40. Gitanjali G. Shahani identifies the references to "Turk," "Jew," and "Tartar" as one of the "earliest experiences of tasting the other." *Tasting Difference: Food, Race, and Cultural Encounters in Early Modern Literature* (Ithaca, NY: Cornell University Press, 2020), 9. See also 8–10.

41. Eamon, *Science and the Secrets of Nature*, 144. Eamon studies "Alessio's 'secrets,'" a third of which were medical recipes that claimed to provide an alternative to the "official drugs" composed "according to the principles of classical pharmacology," whose "ingredients [were] authorized by Theophrastus, Galen, and Dioscorides." He claims to draw on a practical medical tradition, "the tried-and-true secrets of surgeons, empirics, gentlemen, housewives, monks, and ordinary peasants" (ibid.).

42. For the various echoes, doublings, and comparisons among characters and events in the play, see Kastan, *Shakespeare After Theory*, 165–82. In mobilizing the oppositions of "foul" and "fair," this scene also taps into how "fairness" was, in Kim F. Hall's words, a "site of crucial delineations of cultural difference"; in *Macbeth*'s opening scenes, the foul/fair binary also seems to become a marker of interconnected disorderliness of the physical and political realms. *Things of Darkness: Economies of Race and Gender in Early Modern England* (Ithaca, NY: Cornell University Press, 1995), 177. Francesca Royster argues that "the witch's equivocating refrain, 'Fair is foul and foul is fair,'" "would seem to put into question the reliability of fairness, or whiteness, as a marker of goodness in the play, even as it frequently uses the language of blackness to convey evil and the desire to create more evil. . . . [I]t is also clear that the inversion of fairness and foulness in the person of the fair-skinned (and often quakingly pale) Macbeth and Lady Macbeth is powerful enough to set nature off its course, loosing a whole cycle of unnatural acts." "Riddling Whiteness, Riddling Certainty: Roman Polanski's *Macbeth*," in *Weyward Macbeth: Intersections of Race and Performance*, ed. Scott L. Newstok and Ayanna Thompson (New York: Palgrave Macmillan, 2010), 173–81, 178. See also B. K. Adams, "Fair/Foul," in *Shakespeare/Text: Contemporary Readings in Textual Studies, Editing and Performance*, ed. Claire M. L. Bourne (London: Bloomsbury Arden Shakespeare, 2021), 29–49.

43. Puttenham, *The Art of English Poesy*, 272. See also Steven Mullaney, "Lying Like Truth: Riddle, Representation and Treason in Renaissance England," *English Literary History* 47, no. 1 (1980): 32–47, for how "ambiguitas" or "amphibology" functions in *Macbeth*. In *Genre* (New York: Routledge, 2006), John Frow claims that "the prophetic riddles in *Macbeth* have an intertextual force: that is, they refer to the genres of prophecy and riddle, and actualise something of the semantic potential of each. From the prophecy they take the sense of an inevitable fate; from the riddle, the structure of an apparent paradox which is resolved in an unexpected way. . . . By welding these two forms together, the play fuses the 'non-time' of the riddle (Dorst 1983: 423) with the prophecy's drive towards the future" (40).

44. Puttenham, *The Art of English Poesy*, 273.

45. I borrow the term "implicative" from A. H. Miller, "Implicative Criticism." See esp. 347.

46. On the varied significations of wonders, and how they were related to and even became objects of natural inquiry, see Daston and Park, *Wonders and the Order of Nature*.

47. On the King's power to "confer new names that transform the *public* identity of the individual," see David Lucking, "Imperfect Speakers: Macbeth and the Name of King," *English Studies* 87, no. 4 (2006): 415–25, 418.

48. Burrow, *Epic Romance*, 110.

49. On how analogy works in producing knowledge of the physical world, see Mary B. Hesse, *Models and Analogies in Science* (Notre Dame, IN: University of Notre Dame Press, 1966). On early modern natural philosophy and analogy, see Katharine Park, "Bacon's 'Enchanted Glass,'" *Isis* 75, no. 2 (1984): 290–302.

50. I draw on the meaning "To affect by invocation or incantation; to charm, bewitch. (By the Protestant Reformers applied opprobriously to consecration.)" *OED Online*, s.v. "conjure, v.," 7.

51. Emma Smith, *The Cambridge Introduction to Shakespeare* (Cambridge: Cambridge University Press, 2007), 71.

52. Stallybrass, "Macbeth and Witchcraft," 200.

53. For multiple connotations of the "two-fold balls" and "treble scepters" in relation to James's rule, see E. B. Lyle, "The 'Twofold Balls and Treble Scepters' in *Macbeth*," *Shakespeare Quarterly* 28, no. 4 (1977): 516–19.

54. The English king becomes a repository of various kinds of knowledge. For his "royal touch" as a mechanism of healing, see Katherine Schaap Williams, "'Strange Virtue': Staging Acts of Cure," in *Disability, Health, and Happiness in the Shakespearean Body*, ed. Sujata Iyengar (New York: Routledge, 2014), 93–108.

55. Some other instances where this is used: "tyrant" (4.3.12, 4.3.45, 5.2.11); "tyranny" (4.3.32); "confident tyrant" (5.4.8); "untitled tyrant" (4.3.104). See Marjorie Garber, *Shakespeare's Ghost Writers: Literature as Uncanny Causality* (New York: Methuen, 1987), for Macbeth's transformation into "an object lesson, a spectacle, a warning against tyranny, a figure for theater and for art" (114).

56. For an exploration of reproductive futurity in Shakespeare, see Lee Edelman, "Against Survival: Queerness in a Time That's Out of Joint," *Shakespeare Quarterly* 62, no. 2 (2011): 148–69. More broadly on politics and reproductive futurity, see Lee Edelman, *No Future: Queer Theory and the Death Drive* (Durham, NC: Duke University Press, 2004).

57. Brian Walsh, *Shakespeare, the Queen's Men, and the Elizabethan Performance of History* (Cambridge: Cambridge University Press, 2009), 25. According to Walsh, "[the] temporality of drama . . . was then considered a defining element of its ontology" (27). In "'Deep Prescience': Succession and the Politics of Prophecy in *Friar Bacon and Friar Bungay*," *Medieval and Renaissance Drama in England* 23 (2010): 63–85, Walsh expands on the present tense of such eventness,

studying how theater "continually alerts playgoers to the fact of the *present tense* they occupy and in which they are witnessing, and participating in, a real-time event of theater" (75).

58. I draw on Jacques Lezra's explication of the duality of events in Deleuze: "Deleuze refers lucidly to the 'double structure of the event,' split always between 'the present moment of its actualization' and 'the future and the past of the event considered in itself, sidestepping each present, being free of the limitations of a particular state of affairs, impersonal and pre-individual, neutral, neither general nor particular.'" *Unspeakable Subjects: The Genealogy of the Event in Early Modern Europe* (Stanford, CA: Stanford University Press, 1997), 39.

59. Walsh, *Shakespeare*, 1.

60. Drawing on Peter Brooks's argument that the distinction between plot and story breaks down from the perspective of the reader, Turner argues that "a successful narrative arguably produces not simply a memory of events but also the anticipation of events to come—and the operation of both in tandem results in the reader's total apprehension of the narrative 'form'" (*English Renaissance Stage*, 24).

CHAPTER 3

1. Abraham Cowley, "To the Royal Society," in Sprat, *The History of the Royal-Society of London*, lines 93–98.

2. This work is wide-ranging. See, for instance, Pérez-Ramos, *Francis Bacon's Idea of Science*; Jardine, *Francis Bacon*; Julian Martin, *Francis Bacon, the State, and the Reform of Natural Philosophy* (Cambridge: Cambridge University Press, 1992); Eamon, *Science and the Secrets of Nature*; Daston and Park, *Wonders and the Order of Nature*; P. H. Smith, *The Body of the Artisan*; and Deborah E. Harkness, *The Jewel House: Elizabethan London and the Scientific Revolution* (New Haven, CT: Yale University Press, 2007).

3. I borrow the term "romancical" from Margaret Cavendish, *The Blazing World and Other Writings*, ed. Kate Lilley (New York: Penguin, 1994), 124. All citations of this text are from this edition and hereafter cited parenthetically by page number.

4. The commonplace appears, among other places, in Seneca, *Ad Lucilium epistulae morales*. See Thomas M. Greene, *The Light in Troy: Imitation and Discovery in Renaissance Poetry* (New Haven, CT: Yale University Press, 1982), esp. 73–74, 199.

5. See Introduction, n45.

6. See Mann and Sarkar, "Introduction: Capturing Proteus," 10. For the classic study in feminist historiography, see Carolyn Merchant, *The Death of Nature: Women, Ecology, and the Scientific Revolution* (San Francisco: Harper and Row, 1980). For a debate on this topic, see Brian Vickers, "Francis Bacon, Feminist Historiography, and the Dominion of Nature," *Journal of the History of Ideas* 69, no. 1 (2008): 117–41; and Katharine Park, "Response to Brian Vickers, 'Francis Bacon, Feminist Historiography, and the Dominion of Nature,'" *Journal of the History of Ideas* 69, no. 1 (2008): 143–46.

7. For an argument that Bacon did not advocate the "torture" of nature, see Peter Pesic, "Proteus Unbound: Francis Bacon's Successors and the Defense of Experiment," *Studies in Philology* 98, no. 4 (2001): 428–56.

8. Parker, *Inescapable Romance*, 76.

9. Henry S. Turner, *The Corporate Commonwealth: Pluralism and Political Fictions in England, 1516–1651* (Chicago: University of Chicago Press, 2016), argues that "Bacon describes [nature] as a pluralist universe made up of independent bodies" (191).

10. The definition of "form" is complicated in Bacon's works. For the classic study, see Mary B. Hesse, "Francis Bacon," in *A Critical History of Western Philosophy*, ed. D. J. O'Connor (New York: Free Press of Glencoe, 1964), 141–52. See also Witmore, *Culture of Accidents*, 116–17; and Pérez-Ramos, *Francis Bacon's Idea of Science*, 83–132.

11. For definitions and classifications, see Amelia Zurcher, "Serious Extravagance: Romance Writing in Seventeenth-Century England," *Literature Compass* 8, no. 6 (2011): 376–89, esp. 376–79. Steve Mentz complicates the commonplace that English romances are failed novels in *Romance for Sale in Early Modern England: The Rise of Prose Fiction* (Burlington, VT: Ashgate, 2006). See esp. 8–11.

12. For romance as a "mode or tendency," a "liminal space," and "the dilation of a threshold," see Parker, *Inescapable Romance*, 4–5. For romance as "meme," that is, "an idea that behaves like a gene in its ability to replicate faithfully and abundantly, but also on occasion to adapt, mutate, and therefore survive in different forms and cultures," see Helen Cooper, *The English Romance in Time: Transforming Motifs from Geoffrey of Monmouth to the Death of Shakespeare* (Oxford: Oxford University Press, 2004), 3.

13. Barbara Fuchs, *Romance* (New York: Routledge, 2004), 2.

14. Ibid., 9.

15. I draw on the following definitions of "strategy": "A plan, a scheme, and related uses" (II), esp. "The art or practice of planning the future direction or outcome of something; the formulation or implementation of a plan, scheme, or course of action, esp. of a long-term or ambitious nature" (4b). *OED Online*, s.v. "strategy, n."

16. See Bernard Weinberg, *A History of Literary Criticism in the Italian Renaissance* (Chicago: University of Chicago Press, 1961).

17. Ibid., 958.

18. Ibid., 955.

19. Ibid., 968.

20. Ibid., 1044.

21. Ibid.

22. Ibid., 981–82.

23. David Quint, *Epic and Empire: Politics and Generic Form from Virgil to Milton* (Princeton, NJ: Princeton University Press, 1993), 9. Thanks to Katherine Eggert for suggesting I think further about this issue.

24. See Hesse, "Francis Bacon"; and Levao, "Francis Bacon."

25. Harkness, *The Jewel House*, 246. Crane argues that while Bacon "did not make much substantive contribution to real knowledge about the world," his "boosterism for his inductive method" and "polemics" had a more lasting impact on natural philosophy than the work of other practitioners. *Losing Touch with Nature*, 175.

26. Francis Bacon, *Preparative Toward Natural and Experimental History*, in *The New Organon and Related Writings*, 271. All quotations of *Preparative* are from this edition and hereafter cited parenthetically by page number.

27. Witmore, *Culture of Accidents*, 129.

28. Wolfram Schmidgen, *Exquisite Mixture: The Virtues of Impurity in Early Modern England* (Philadelphia: University of Pennsylvania Press, 2013).

29. See, for instance, Kuhn, "Mathematical Versus Experimental Traditions."

30. Jeffrey Todd Knight, *Bound to Read: Compilations, Collections, and the Making of Renaissance Literature* (Philadelphia: University of Pennsylvania Press, 2013), 8. See also Megan Heffernan, *Making the Miscellany: Poetry, Print, and the History of the Book in Early Modern England* (Philadelphia: University of Pennsylvania Press, 2021).

31. On inductive method as a form of commonplacing, see Crane, *Losing Touch with Nature*, 171. See also Adam Smyth, "Commonplace Book Culture: A List of Sixteen Traits," in *Women and Writing, c. 1340–c. 1650: The Domestication of Print Culture*, ed. Anne Lawrence-Mathers and Phillipa Hardman (Woodbridge: York Medieval, with Boydell, 2010), 90–110; Mary Thomas Crane, *Framing Authority: Sayings, Self, and Society in Sixteenth-Century England* (Princeton, NJ: Princeton University Press, 1993); and Ann Moss, *Printed Commonplace-Books and the Structuring of Renaissance Thought* (New York: Oxford University Press, 1996).

32. Pérez-Ramos terms Bacon's *inductio* as "inductive," "deductive," and "intuitive" (*Francis Bacon's Idea of Science*, 241).

33. See Eamon, *Science and the Secrets of Nature*, 269.

34. Stephen Clucas, "'A Knowledge Broken': Francis Bacon's Aphoristic Style and the Crisis of Scholastic and Humanistic Knowledge Systems," in *English Renaissance Prose: History, Language and Politics*, ed. Neil Rhodes (Tempe, AZ: Medieval and Renaissance Texts and Studies, 1997), 147–72, 167.

35. James Stephens, "Science and the Aphorism: Bacon's Theory of the Philosophical Style," *Speech Monographs* 37, no. 3 (1970): 157–71, 161. Alvin Snider ("Francis Bacon and the Authority of Aphorism," *Prose Studies* 11, no. 2 [1988]: 60–71) argues that the aphorism "by nature is anticanonical" (64).

36. See Clucas, "'Knowledge Broken.'"

37. Lorraine Daston provides a prehistory of objectivity and demonstrates Bacon's foundational role in ensuring that "facts came to be detached from the context of theory." "Baconian Facts, Academic Civility, and the Prehistory of Objectivity," *Annals of Scholarship* 8 (1991): 337–63, 338.

38. "Interpretation" is a key term for Bacon. See Michael C. Clody, "Deciphering the Language of Nature: Cryptography, Secrecy, and Alterity in Francis Bacon," *Configurations* 19, no. 1 (2011): 117–42, on how the interpretation of nature's alphabet occurs in Bacon.

39. Snider, "Francis Bacon and the Authority of Aphorism," 60.

40. Stephens, "Science and the Aphorism," 157.

41. Bacon, *De Augmentis*, 4:449.

42. Bacon, *The Advancement of Learning*, 642n233.

43. Bacon, *The Great Instauration*, 16.

44. The etymological connection is explicit: *errāre* means "to wander." *OED Online*, s.v. "error, n."

45. See Daston and Park, *Wonders and the Order of Nature*, 220–45.

46. Clody, "Deciphering the Language of Nature," 117.

47. For Bacon's technocratic view of science, see Pérez-Ramos, *Francis Bacon's Idea of Science*, 3–6. For the argument that "*The New Atlantis*, in its unprecedented intention a technological utopia, fails as such, but succeeds in unadvertised ways, featuring both more and less social engineering than is at first apparent," see Christopher Kendrick, *Utopia, Carnival, and Commonwealth in Renaissance England* (Toronto: University of Toronto Press, 2004), 330.

48. In "Mining Tacitus: Secrets of Empire, Nature and Art in the Reason of State," *British Journal for the History of Science* 45, no. 2 (2012): 189–212, Vera Keller demonstrates how Bacon provides a "collaborative and secretive view of research" in *New Atlantis*; since "Bacon's ultimate end was not the state, but knowledge, . . . he imagined an arena of research free from both the public and the state" (199).

49. Denise Albanese, "The *New Atlantis* and the Uses of Utopia," *English Literary History* 57, no. 3 (1990): 503–28, 504.

50. William Poole, "General Introduction," in Francis Lodwick, *A Country Not Named (MS. Sloane 913, Fols. 1r–33r): An Edition with an Annotated Primary Bibliography and an Introductory Essay on Lodwick and His Intellectual Context* (Tempe: Arizona Center for Medieval and Renaissance Studies, 2007), 3–67, 23. See also Sarkar, "Utopian Hypothesis."

51. Tobin Siebers, "Introduction: What Does Postmodernism Want? Utopia," in *Heterotopia: Postmodern Utopia and the Body Politic*, ed. Tobin Siebers (Ann Arbor: University of Michigan Press, 1995), 1–38, 3.

52. H. Vaihinger, *The Philosophy of "As If"* (London: K. Paul, Trench, Trubner, 1924), 85.

53. Albanese, "*New Atlantis*," 505.

54. Poole, "General Introduction," outlines different modes of utopianism: "the philosophic account of the ideal state, the fantastic voyage, and the travel narrative" (25). On the tensions between utopia, ideal commonwealth, and romance in *New Atlantis*, see Christopher Kendrick, "The Imperial Laboratory: Discovering Forms in 'The New Atlantis,'" *English Literary History* 70, no. 4 (2003): 1021–42.

55. Francis Bacon, *New Atlantis*, in *Francis Bacon: The Major Works*, 457. All citations of the text are from this edition and hereafter cited parenthetically by page number.

56. See Latour, *Science in Action*, esp. 247–57, for center-peripheries and "metrologies."

57. For the argument that Bacon aspires to a "possessive corporatism," see Kendrick, *Utopia, Carnival, and Commonwealth*, 302.

58. On how Bacon's "*desiderata* list was an open invitation for others to continue the work of advancement," see Vera Keller, *Knowledge and the Public Interest, 1575–1725* (New York: Cambridge University Press, 2015), 130; see also 127–66. On the imaginative aspects of wish lists, see Debapriya Sarkar, "Imagining Early Modern Wish-Lists and Their Environs," in *Object-Oriented Environs*, ed. Jeffrey Jerome Cohen and Julian Yates (New York: Punctum Books, 2016), 123–33. On the status of desiderata, and its relation to other forms such as query lists, see Keller, *Knowledge and the Public Interest*, esp. 1–31. For the relation of *desiderata* and *optative*, see Vera Keller, "Accounting for Invention: Guido Pancirolli's Lost and Found Things and the Development of *Desiderata*," *Journal of the History of Ideas* 73, no. 2 (2012): 223–45. As Keller points out, "The *optativa* were clearly operative. However, if properly directed, *optativa* might lead man's enquiry upwards towards speculative knowledge" (237); she argues, "Bacon's *desiderata* were epistemic, and not material objects" or *optativa* (238). On questionnaires, *interrogatoria*, and query list, see Justin Stagl, *A History of Curiosity: The Theory of Travel, 1500–1800* (Chur, Switzerland: Harwood Academic, 1995). See also Daniel Carey, "Inquiries, Heads, and Directions: Orienting Early Modern Travel," in *Travel Narratives, the New Science, and Literary Discourse, 1569–1750*, ed. Judy A. Hayden (Burlington, VT: Ashgate, 2012), 25–52; and Daniel Carey, "Hakluyt's Instructions: The Principal Navigations and Sixteenth-Century Travel Advice," *Studies in Travel Writing* 13, no. 2 (2009): 167–85.

59. For lists in early modern knowledge production, see James Delbourgo and Staffan Müller-Wille, "Introduction to Focus: 'Listmania,'" *Isis* 103, no. 4 (2012): 710–15, 711.

60. Keller, *Knowledge and the Public Interest*, 166; see also 128–30.

61. Ibid., 16.

62. See Charles Webster, *The Great Instauration: Science, Medicine, and Reform, 1626–1660*, 2nd ed. (New York: Peter Lang, 2002); on Bacon's influence on the Hartlib circle, see Keller, *Knowledge and the Public Interest*, esp. 167–68.

63. On the variety of the immediate afterlife of Baconianism, see Webster, *The Great Instauration*. On influences of Bacon's thought and the practices and theories of the early Royal Society, see

William T. Lynch, *Solomon's Child: Method in the Early Royal Society of London* (Stanford, CA: Stanford University Press, 2001), esp. 1–33.

64. Bronwen Price, introduction to *Francis Bacon's New Atlantis: New Interdisciplinary Essays*, ed. Bronwen Price (Manchester: Manchester University Press, 2002), 1–27, argues it was "most likely Robert Hooke" (15).

65. R. H., *New Atlantis. Begun by the Lord Verulam, Viscount St. Albans: and continued by R.H. Esquire. Wherein is set forth a platform of monarchical government. With a pleasant intermixture of divers rare inventions, and wholsom customs, fit to be introduced into all kingdoms, states, and common-wealths* (London: Printed for John Crooke at the signe of the Ship in St. Pauls Church-Yard, 1660), sig. a5v, EEBO.

66. Ibid., sig. b6v.

67. Ibid., sig. b7v.

68. For Bacon's influence on Hartlibian reforms, see Todd Butler, "Revitalizing Nation and Mind: The Failed Promise of Seventeenth-Century Educational Reform," in *Political Turmoil: Early Modern British Literature in Transition, 1623–1660*, ed. Stephen B. Dobranski (Cambridge: Cambridge University Press, 2019), 237–52. See also Chloë Houston, "Utopia and Education in the Seventeenth Century: Bacon's Salomon's House and Its Influence," in *New Worlds Reflected: Travel and Utopia in the Early Modern Period*, ed. Chloë Houston (Farnham: Ashgate, 2010), 161–78.

69. [Gabriel Plattes], *A Description of the Famous Kingdome of Macaria* (London: Printed for Francis Constable, 1641), 2, 3, EEBO. (NB: Although we now know this text was written by Plattes, EEBO currently lists Hartlib as its author.) All citations of this text are from this edition and hereafter cited parenthetically by page number.

70. [William Petty], *The advice of W.P. to Mr. Samuel Hartlib. For the Advancement of Some Particular Parts of Learning* (London, 1648), 10, EEBO.

71. For representative works that mark the discrepancies between Bacon's methods and the practices of the Royal Society, see Daston and Park, *Wonders and the Order of Nature*, 230; Hesse, "Francis Bacon," 152; and Dear, *Revolutionizing the Sciences*, 139.

72. Sprat, *The History of the Royal-Society of London*, 61.

73. Ibid., 113.

CHAPTER 4

1. Cavendish routinely, though not uniformly, personifies the concept of "Nature" as she elaborates its various properties and workings across her writing. This deliberate construction of the concept—and her careful constructedness of its capacities in different texts—is central to my arguments about her poetics. In order to mark that Cavendish's concept of "Nature" is a created entity that she deploys for specific creative and intellectual purposes, rather than an extant object, I have capitalized the first letter of the term (except in direct quotations) throughout this chapter.

2. Virginia Woolf, *A Room of One's Own*, in *The Norton Anthology of English Literature*, gen. ed. M. H. Abrams, 7th ed., 2 vols. (New York: W. W. Norton, 2000), 2:2186.

3. This scholarship on the scope of her literary writing is vast. For representative works on these topics, see Catherine Gallagher, "Embracing the Absolute: The Politics of the Female Subject in Seventeenth-Century England," *Genders* 1 (1988): 24–39; Sandra Sherman, "Trembling Texts: Margaret Cavendish and the Dialectic of Authorship," *English Literary Renaissance* 24, no. 1

(1994): 184–210; Sylvia Bowerbank, "The Spider's Delight: Margaret Cavendish and the 'Female' Imagination," *English Literary Renaissance* 14, no. 3 (1984): 392–408; Sujata Iyengar, "Royalist, Romancist, Racialist: Rank, Gender, and Race in the Science and Fiction of Margaret Cavendish," *English Literary History* 69, no. 3 (2002): 649–72. See also Lara Dodds, *The Literary Invention of Margaret Cavendish* (Pittsburgh: Duquesne University Press, 2013); and Lisa Walters, *Margaret Cavendish: Gender, Science and Politics* (Cambridge: Cambridge University Press, 2014). A common strategy has been to trace Cavendish's philosophical influences and present her work in the context of contemporary philosophy and thought. For representative works, see Jacqueline Broad, "Margaret Cavendish," in *Women Philosophers of the Seventeenth Century* (Cambridge: Cambridge University Press, 2003), 35–64; Goldberg, *Seeds of Things*, 122–53; Rogers, *The Matter of Revolution*, 177–211; David Norbrook, "Margaret Cavendish and Lucy Hutchinson: Identity, Ideology and Politics," *In-between: Essays and Studies in Literary Criticism* 9, nos. 1–2 (2000): 179–203; Sarah Hutton, "Margaret Cavendish and Henry More," in *A Princely Brave Woman: Essays on Margaret Cavendish, Duchess of Newcastle*, ed. Stephen Clucas (Burlington, VT: Ashgate, 2003), 185–98; and Stephen Clucas, "Margaret Cavendish's Materialist Critique of Van Helmontian Chymistry," *Ambix: Journal of the Society for the History of Alchemy and Chemistry* 58, no. 1 (2011): 1–12. See also the following biographies: Emma L. E. Rees, *Margaret Cavendish: Gender, Genre, Exile* (New York: Manchester University Press, 2003); and Anna Battigelli, *Margaret Cavendish and the Exiles of the Mind* (Lexington: University Press of Kentucky, 1998).

4. On "the competing demands of the Duchess's radical feminism and social conservatism," see Rachel Trubowitz, "The Reenchantment of Utopia and the Female Monarchical Self: Margaret Cavendish's *Blazing World*," *Tulsa Studies in Women's Literature* 11, no. 2 (1992): 229–45, 229. See also Peter Dear, "A Philosophical Duchess: Understanding Margaret Cavendish and the Royal Society," in Cummins Burchell *Science, Literature and Rhetoric in Early Modern England*, 125–42; and Aït-Touati, *Fictions of the Cosmos*, 178–85.

5. Margaret Cavendish, *Poems, and Fancies* (London: Printed by T. R. for J. Martin, and J. Allestrye, 1653), sig. A6r, EEBO. All in-text citations are from this edition and given by signature for prose and title and line number for poetry. I use the original title from the 1653 ed. to distinguish it from later editions (both modern and early modern).

6. *Poems, and Fancies* was revised and reprinted in 1664 and 1668. For the textual history of these editions, see Liza Blake, "Textual and Editorial Introduction," in *Margaret Cavendish's Poems and Fancies: A Digital Critical Edition*, ed. Liza Blake (website published May 2019), http://library2.utm.utoronto.ca/poemsandfancies/textual-and-editorial-introduction/.

7. The term appears in "The Poetress's Hasty Resolution" and "The Poetress's Petition," in *Poems, and Fancies*. See sig. A8r–A8v.

8. Margaret Cavendish, *The Philosophical and Physical Opinions* (London: Printed for J. Martin and J. Allestrye at the Bell in St. Pauls Church-Yard, 1655), sig. B2r, EEBO. All future citations are in-text and from this edition.

9. The exact details of Cavendish's shift from atomism to vitalism have long been a subject of scholarly debate. For instance, Rogers (*The Matter of Revolution*, esp. 188–89) identifies it with her 1663 works, but Eileen O'Neill makes the case for an "animistic materialism" as early as 1655 in her introduction to *Observations upon Experimental Philosophy*, by Margaret Cavendish, ed. Eileen O'Neill (Cambridge: Cambridge University Press, 2001), x–xxxvi, xx. Spiller (*Science, Reading, and Renaissance Literature*, 161) argues that Cavendish advocated "a strong and hierarchical form of materialism" from *Philosophical and Physical Opinions* onward. In "Margaret Cavendish's Nonfeminist Natural Philosophy," *Configurations* 12, no. 2 (2004): 195–227, Deborah Boyle locates a vitalist materialism (including a "vitalist atomism" [200]) throughout her corpus.

10. Margaret Cavendish, *Natures pictures drawn by fancies pencil to the life* (London: Printed for J. Martin and J. Allestrye at the Bell in St. Pauls Church-Yard, 1656), sig. c3v, EEBO. All future citations are in-text and from this edition.

11. Scholars often address questions of genre in Cavendish's works by examining how *The Blazing World* revises utopian and romance conventions. See, for instance, Trubowitz, "The Reenchantment of Utopia and the Female Monarchical Self"; Kate Lilley, "Blazing Worlds: Seventeenth-Century Women's Utopian Writing," in *Women, Texts and Histories, 1575–1760*, ed. Clare Brant and Diane Purkiss (New York: Routledge, 1992), 102–33; Marina Leslie, "Gender, Genre and the Utopian Body in Margaret Cavendish's *Blazing World*," *Utopian Studies* 7, no. 1 (1996): 6–24; and Geraldine Wagner, "Romancing Multiplicity: Female Subjectivity and the Body Divisible in Margaret Cavendish's *Blazing World*," *Early Modern Literary Studies* 9, no. 1 (2003), http://purl.oclc.org/emls/09-1/wagnblaz.htm.

12. See, for instance, Gallagher, "Embracing the Absolute"; Sherman, "Trembling Texts"; Campbell, *Wonder and Science*, esp. 206–13; Aït-Touati, *Fictions of the Cosmos*, 188–89; Spiller, *Science, Reading, and Renaissance Literature*, 174; and Harris, *Untimely Matter in the Time of Shakespeare*, 151–61.

13. I am referring to the moment when the Empress "command[s]" her experimental philosophers (the bear-men) to "break [their telescopes]," which are "false informers" that "delude [their] senses." Challenging the bear-men, who "take more delight in artificial delusions, than in natural truths," she states that they should instead trust only "their natural eyes" and their "sense and reason," since "nature has made [their] sense and reason more regular than art has [their] glasses" (141–42). Here, Cavendish satirizes the efficacy of apparatuses such as Robert Hooke's microscopes and offers one of her most pointed criticisms of the experimental methods of the Royal Society. This scene and Cavendish's response to Hooke have garnered significant scholarly attention. See, for instance, Campbell, *Wonder and Science*, 182, 213–18; Aït-Touati, *Fictions of the Cosmos*, 176; Eve Keller, "Producing Petty Gods: Margaret Cavendish's Critique of Experimental Science," *English Literary History* 64, no. 2 (1997): 447–71; and Shannon Miller, *Engendering the Fall: John Milton and Seventeenth-Century Women Writers* (Philadelphia: University of Pennsylvania Press, 2008), 144–45.

14. Margaret Cavendish, *Observations upon Experimental Philosophy*, ed. Eileen O'Neill (Cambridge: Cambridge University Press, 2001), 85. Further citations are from this edition and given parenthetically in the text.

15. Margaret Cavendish, *Philosophicall Fancies* (London: Printed by Tho: Roycroft, for J. Martin, and J. Allestrye, at the Bell in St. Pauls Church-Yard, 1653), 5, 14–15, EEBO. All future in-text citations are from this edition and given by page number for prose and title and line number for poetry.

16. Cavendish echoes versions of this argument in her works. For instance, in *Observations* she states: "nature is neither absolutely necessitated, nor has an absolute free will: for, she is so much necessitated, that she depends upon the all-powerful God, and cannot work beyond herself, or beyond her own nature; and yet hath so much liberty, that in her particulars she works as she pleaseth, and as God has given her power; but she being wise, acts according to her infinite natural wisdom, which is the cause of her orderly government in all particular productions, changes and dissolutions" (109).

17. See for instance, *Observations*, which marks artificial experiments as a version of art—"nature's sporting or playing actions" (105)—and describes art as "the emulating ape of nature, [which] makes often vain and useless things" (59). In *The Blazing World*, Cavendish declares art "nature's changeling" (157).

18. *Observations* echoes these claims, stating that Nature is "a self-moving, and consequently a self-living and self-knowing infinite body" (125).

19. Sarah Hutton, "In Dialogue with Thomas Hobbes: Margaret Cavendish's Natural Philosophy," *Women's Writing* 4, no. 3 (1997): 421–32, 423. See also Rogers (*The Matter of Revolution*, esp. 178, 186–90, 199–208), who opposes a masculinist mechanism to Cavendish's vitalist feminism. Cf. D. Boyle, "Margaret Cavendish's Nonfeminist Natural Philosophy," who questions the premise that Cavendish's philosophy is feminist.

20. Spiller, *Science, Reading, and Renaissance Literature*, differentiates Hooke's passive, mechanist philosophy of reading from Cavendish's "vitalist theory of reading that allows for a more active reader who is able to destroy as well as create knowledge" (23).

21. Others replicate the conceit of originality. E. Toppe's letter in *Poems, and Fancies* declares, "You are not onely the first English Poet of your Sex, but the first that ever wrote this way: therefore whosoever that writes afterwards, must own you for their Pattern. . . . For whatsoever is written afterwards, it will be but a Copy of your Originall" (sig. A5v).

22. See Lisa T. Sarasohn, *The Natural Philosophy of Margaret Cavendish: Reason and Fancy During the Scientific Revolution* (Baltimore: Johns Hopkins University Press, 2010), esp. 39, 156–57, 180, for instances of Cavendish's self-presentation as an uneducated author whose ideas were evolving. Sarasohn argues that Cavendish's writerly self is constructed by the motto "I am, therefore I think (193)." Other instances of privileging naturalness can be seen in what Brandie R. Siegfried identifies as Cavendish's emphasis on the "ways in which ontology precedes epistemology." "Anecdotal and Cabalistic Forms in *Observations upon Experimental Philosophy*," in *Authorial Conquests: Essays on Genre in the Writings of Margaret Cavendish*, ed. Line Cottegnies and Nancy Weitz (Madison, NJ: Fairleigh Dickinson University Press, 2003), 59–79, 61. See also Spiller, *Science, Reading, and Renaissance Literature*, 143; and Battigelli, *Margaret Cavendish and the Exiles of the Mind*, 41. The latter points out that Cavendish did not know or learn French.

23. Margaret Cavendish, *The Worlds Olio* (London: Printed for J. Martin and J. Allestrye at the Bell in St. Pauls Church-Yard, 1655), 26, EEBO. All citations of this text are from this edition and hereafter cited parenthetically by page number. For the primacy of creation over imitation in Cavendish's writing, see Battigelli, *Margaret Cavendish and the Exiles of the Mind*, 68; and Campbell, *Wonder and Science*, 203.

24. Wroth is on Cavendish's mind at various times. She refers to Wroth as "the *Lady* that wrote the *Romancy*" in "To All Noble, and Worthy Ladies" (*Poems, and Fancies*), before quoting lines from Edward Denny's poem criticizing Wroth (sig. A3v).

25. Lara Dodds describes this cluster as "an important attempt to create a suitable poetic representation of a world reordered by mechanical speculation, astronomical observation, and microscopy." "'Poore Donne Was Out': Reading and Writing Donne in the Works of Margaret Cavendish," *John Donne Journal* 29 (2010): 133–74, 155.

26. Jessie Hock, "Fanciful Poetics and Skeptical Epistemology in Margaret Cavendish's *Poems and Fancies*," *Studies in Philology* 115, no. 4 (2018): 766–802, 786, 789. See 786n66 for a summary of the various ways in which Cavendish could have had access to Lucretian ideas. On Cavendish and Lucretius, see also Rees, *Margaret Cavendish: Gender, Genre, Exile*, esp. 54–79; Battigelli, *Margaret Cavendish and the Exiles of the Mind*, 45–61; and Goldberg, *Seeds of Things*, 122–53. See Robert Hugh Kargon, *Atomism in England from Hariot to Newton* (Oxford: Clarendon Press, 1966), 63–76, who situates Cavendish's atomism within discussions of the Newcastle circle. For how she deploys this atomism poetically, see Sarasohn, *Natural Philosophy of Margaret Cavendish*, 34–53.

27. *Poems, and Fancies* contains poems on topics as varied as "Of Vacuum," "The Attraction of the Earth," "The Power of Fire," "Of Stars," and "What Atoms Cause Sicknesse."

28. For another vantage from which to approach Cavendish's investment in the whole, see Blake, "Textual and Editorial Introduction," which marks the author's role in the rearrangement of the poems as "an attempt to think through concepts of arrangement, atomism, mutual interrelation, and the logic of collection." See also Liza Blake, "Margaret Cavendish's Forms: Literary Formalism and the Figures of Cavendish's Atom Poems," in *Feminist Formalism and Early Modern Women's Writing: Readings, Conversations, Pedagogies"*, ed. Lara Dodds and Michelle M. Dowd (Lincoln: University of Nebraska Press, 2022), 38–55; and Liza Blake, "Choose Your Own Poems and Fancies" (fellow lecture, Folger Shakespeare Library, Washington, DC, April 2019).

29. On Cavendish's skepticism, see Lisa T. Sarasohn, "A Science Turned Upside Down: Feminism and the Natural Philosophy of Margaret Cavendish," *Huntington Library Quarterly* 47, no. 4 (1984): 289–307, esp. 292–93; D. Boyle, "Margaret Cavendish's Nonfeminist Natural Philosophy"; Stephen Clucas, "Variation, Irregularity, and Probabilism: Margaret Cavendish and Natural Philosophy as Rhetoric," in Clucas, *Princely Brave Woman*, 199–209; and Hock, "Fanciful Poetics and Skeptical Epistemology," 776.

30. See, for instance, "The Sun doth set the Aire on a light, as Some Opinions hold," "Of the Center," and "The Infinites of Matter."

31. Dodds, "'Poore Donne Was Out,'" reads this poem as a "response to or rewriting of Donne's *The First Anniversary*," arguing that unlike Donne, Cavendish offers a "celebration of the fundamental epistemological and ontological transformations of the new" (157).

32. Ibid., 160.

33. See Karen Detlefsen, "Atomism, Monism, and Causation in the Natural Philosophy of Margaret Cavendish," *Oxford Studies in Early Modern Philosophy* 3 (2006): 199–240. See also Rogers, *The Matter of Revolution*.

34. Dodds, "'Poore Donne Was Out,'" 163.

35. For its hybrid genre, including romance and science fiction, see Nicole Pohl, "'Of Mixt Natures': Questions of Genre in Margaret Cavendish's *The Blazing World*," in Clucas, *Princely Brave Woman*, 51–68.

36. William Poole, "Francis Godwin, Henry Neville, Margaret Cavendish, H. G. Wells: Some Utopian Debts," ANQ: A *Quarterly Journal of Short Articles, Notes, and Reviews* 16, no. 3 (2003): 12–18, 17. Iyengar argues England is the "tacit parallel and counterexample" ("Royalist, Romancist, Racialist," 661) but marks absolute differences between fictional and real worlds (667–68).

37. The Empress's signification as author, ruler, and goddess has been variously examined. See, for instance, Oddvar Holmesland, "Margaret Cavendish's *The Blazing World*: Natural Art and the Body Politic," *Studies in Philology* 96, no. 4 (1999): 457–79, for a study of the "imperialist vision" that defines her as a deity and that uses the "image of the expanded self" to mark "a widening artistic and epistemological space" (478).

38. On the significations of the beast-men, see Carol Thomas Neely, "Women/Utopia/Fetish: Disavowal and Satisfied Desire in Margaret Cavendish's *New Blazing World* and Gloria Anzaldúa's *Borderlands/La Frontera*," in Siebers, *Heterotopia*, 58–95, esp. 79; Bronwen Price, "Journeys Beyond Frontiers: Knowledge, Subjectivity and Outer Space in Margaret Cavendish's *The Blazing World* (1666)," in *The Arts of 17th-Century Science: Representations of the Natural World in European and North American Culture*, ed. Claire Jowitt and Diane Watt (Aldershot: Ashgate, 2002) 127–45, esp. 132; and Carrie Hintz, "'But One Opinion': Fear of Dissent in Cavendish's *New Blazing World*," *Utopian Studies* 7, no. 1 (1996): 25–37.

39. Cf. Rogers's argument that Cavendish's philosophy was dominated by resignation and a recognition that "a utopia of female volition . . . might never materialize" (*The Matter of Revolution*, 211).

40. Angus Fletcher, in "The Irregular Aesthetic of *The Blazing World*," *Studies in English Literature, 1500–1900* 47, no. 1 (2007): 123–41, reads Cavendish's version of Nature as a "response to misogynist stereotypes" (129) and argues that "Cavendish's distinctive view of irregularity should be traced not to a desire to reject hierarchy but rather to her place in a feminist tradition that sought to redefine authority by associating it with women's inconstant dispositions" (125).

41. Marjorie Hope Nicolson, *Voyages to the Moon* (New York: Macmillan, 1948), 224.

42. Cavendish's intellectual responses to Hobbes, as well as Hobbes's role in the Newcastle circle, have been a consistent focus of scholars. See, for instance, Spiller, *Science, Reading, and Renaissance Literature*; Rogers, *The Matter of Revolution*; Hutton, "In Dialogue with Thomas Hobbes"; Battigelli, *Margaret Cavendish and the Exiles of the Mind*, 64–84; Sarasohn, *Natural Philosophy of Margaret Cavendish*, Detlefsen, "Atomism, Monism, and Causation"; and Neil Ankers, "Paradigms and Politics: Hobbes and Cavendish Contrasted," in Clucas, *Princely Brave Woman*, 242–54.

43. William Cavendish, "To the Duchess of Newcastle, on Her New Blazing World," in M. Cavendish, *Blazing World and Other Writings*, lines 9–10.

44. Andrew Marvell, "On Mr. Milton's *Paradise Lost*" (1674), in Milton, *The Complete Poetry and Essential Prose of John Milton*, ed. William Kerrigan, John Rumrich, and Stephen M. Fallon (New York: Modern Library, 2007), line 23.

45. Ibid., line 42.

46. Ibid., line 2.

CHAPTER 5

1. John Milton, *Paradise Lost*, in *The Complete Poetry and Essential Prose of John Milton*, 1.1–4. All future in-text quotations from Milton's works are from this edition. *Paradise Lost* is cited by book and line number, and the prose is cited by page number.

2. Robert Hooke, *Micrographia* (London: Printed by Jo. Martyn, and Ja. Allestry, printers to the Royal Society, 1665), Preface, sig. b2r–b2v, EEBO.

3. For how these issues are interlinked, see Mann, *Outlaw Rhetoric*, 205–8. For how Adam functioned as a key figure in discussions of natural knowledge, see Harrison, *The Fall of Man and the Foundations of Science*, esp. 139–85; and Picciotto, *Labors of Innocence*, 1–28.

4. This stands in contrast to Milton's references and allusions to Galileo, a topic that has received sustained attention from scholars. For classic work on the topic, see Marjorie Hope Nicolson, "Milton and the Telescope," *English Literary History* 2, no. 1 (1935): 1–32. See also Annabel Patterson, "Imagining New Worlds: Milton, Galileo, and the 'Good Old Cause,'" in *The Witness of Times: Manifestations of Ideology in Seventeenth Century England*, ed. Katherine Z. Keller and Gerald J. Schiffhorst (Pittsburgh: Duquesne University Press, 1993), 238–60; John C. Ulreich, "Two Great World Systems: Galileo, Milton, and the Problem of Truth," *Cithara: Essays in the Judaeo-Christian Tradition* 43, no. 1 (2003): 25–36; Harinder Singh Marjara, *Contemplation of Created Things: Science in Paradise Lost* (Toronto: University of Toronto Press, 1992); Amy Boesky, "Milton, Galileo, and Sunspots: Optics and Certainty in *Paradise Lost*," *Milton Studies* 34 (1997): 23–43; S. Miller, *Engendering the Fall*; and Jacqueline L. Cowan, "Milton, the Royal Society, and the

Galileo Problem," *Studies in English Literature, 1500–1900* 58, no. 3 (2018): 569–89, esp. 577 and 585n1.

5. *OED Online*, s.v. "event, n," II.3.a, 1.1.

6. Lezra, *Unspeakable Subjects*, 39.

7. My understanding of Miltonic events is influenced by scholarship on Miltonic moments, especially that of Parker, who marks the transformational moments in the epic through notions of betweenness and pendency, the "threshold of choice" (*Inescapable Romance*, 6) that lies at the heart of the poem, and J. Martin Evans, who, considering Milton's early poems, defines the Miltonic moment as a "'betweenness'" that "resides in a temporal interstice between two events, or two sequences of events, on a chronological continuum." *The Miltonic Moment* (Lexington: University Press of Kentucky, 1998), 6.

8. Witmore, *Culture of Accidents*, 8.

9. Stanley Fish, *Surprised by Sin: The Reader in Paradise Lost*, 2nd ed. (Cambridge, MA: Harvard University Press, 1998), 250.

10. Karen L. Edwards, *Milton and the Natural World: Science and Poetry in Paradise Lost* (Cambridge: Cambridge University Press, 1999), 33. The study of Eve as an experimenter often leads scholars to an inevitable Fall and an always/already fallen Eve, a premise that has been revised by critics including John Leonard, in *Naming in Paradise: Milton and the Language of Adam and Eve* (Oxford: Clarendon Press, 1990); and Diane Kelsey McColley, in *Milton's Eve* (Urbana: University of Illinois Press, 1983). The broader connection of the scene of the Fall and early modern science is not surprising. As Harrison demonstrates in *The Fall of Man and the Foundations of Science*, the idea that the Fall can be reversed and mankind restored through knowledge is common in the period. Picciotto has also established the importance of the innocent figure of Adam to the ideals of early modern natural philosophers. *Labors of Innocence*, esp. 129–87. See also Joanna Picciotto, "Reforming the Garden: The Experimentalist Eden and *Paradise Lost*," *English Literary History* 72, no. 1 (2005): 23–78.

11. Edwards, *Milton and the Natural World*, 18.

12. S. Miller, *Engendering the Fall*, 136–68. For an ecofeminist reading of this scene, see Jennifer Munroe, "First 'Mother of Science': Milton's Eve, Knowledge, and Nature," in *Ecofeminist Approaches to Early Modernity*, ed. Jennifer Munroe and Rebecca Laroche (New York: Palgrave Macmillan, 2011), 37–54.

13. Picciotto, *Labors of Innocence*, 475. For exposition of this difference between Adam and Eve, see esp. 474–76.

14. Ibid., 26, 16, 14.

15. Ibid., 439.

16. Shapin and Schaffer, *Leviathan and the Air-Pump*, 60.

17. Picciotto, *Labors of Innocence*, 4.

18. Kester Svendsen, *Milton and Science* (Cambridge, MA: Harvard University Press, 1956).

19. For various strands of this scholarship, see Edwards, *Milton and the Natural World*; Stephen M. Fallon, *Milton Among the Philosophers: Poetry and Materialism in Seventeenth-Century England* (Ithaca, NY: Cornell University Press, 1991); Marjara, *Contemplation of Created Things*; Rogers, *The Matter of Revolution*, 103–43; Catherine Gimelli Martin, "'What If the Sun Be Centre to the World': Milton's Epistemology, Cosmology, and Paradise of Fools Reconsidered," *Modern Philology* 99, no. 2 (2001): 231–65; Lara Dodds, "Milton's Other Worlds," in *Uncircumscribed Mind: Reading Milton Deeply*, ed. Charles W. Durham and Kristin A. Pruitt (Selinsgrove, PA: Susquehanna University Press, 2008), 164–82; and S. Miller, *Engendering the Fall*, 136–68. For works that focus on mathematics, see Erin Webster, "Milton's Pandæmonium and the Infinitesimal Calculus,"

English Literary Renaissance 45, no. 3 (2015): 425–58; Shankar Raman, "Milton, Leibniz, and the Measure of Motion," in Marchitello and Tribble, *Palgrave Handbook of Early Modern Literature and Science*, 277–93; and Travis D. Williams, "Unspeakable Creation: Writing in *Paradise Lost* and Early Modern Mathematics," *Philological Quarterly* 98, nos. 1–2 (2019): 181–200. For Milton's vitalist materialism and the history of reading, see Elizabeth Spiller, "Milton, the Poetics of Matter, and the Sciences of Reading," in Marchitello and Tribble, *Palgrave Handbook of Early Modern Literature and Science*, 159–77. For classic works, see Nicolson, "Milton and the Telescope"; and Grant McColley, "The Astronomy of *Paradise Lost*," *Studies in Philology* 34, no. 2 (1937): 209–47.

20. William Poole, "Milton and Science: A Caveat," *Milton Quarterly* 38, no. 1 (2004): 18–34, 18. Poole is responding to connections proposed in works such as Nicholas von Maltzahn, "The Royal Society and the Provenance of Milton's *History of Britain* (1670)," *Milton Quarterly* 32, no. 3 (1998): 90–95.

21. Picciotto, *Labors of Innocence*, 8.

22. *OED Online*, s.v. "experience, n." and "experiment, n."

23. Spiller, *Science, Reading, and Renaissance Literature*, 27. See esp. 1–31.

24. See Dear, *Revolutionizing the Sciences*, 131–48; and Kuhn, "Mathematical Versus Experimental Traditions."

25. See Steven Shapin, "The House of Experiment in Seventeenth-Century England," *Isis* 79, no. 3 (1988): 373–404, for an account of the different kinds of settings, including private spaces, in which experiments were performed.

26. Dear, *Discipline and Experience*, 4.

27. Phillips, *The New World of English Words*, s.v. "experience."

28. Dear, *Discipline and Experience*, 22.

29. Dear, *Revolutionizing the Sciences*, 6.

30. Aristotle, *Posterior Analytics*, trans. Jonathan Barnes, 2nd ed. (New York: Oxford University Press, 1994), 2.19. See also N. K. Sugimura, *"Matter of Glorious Trial:" Spiritual and Material Substance in Paradise Lost* (New Haven, CT: Yale University Press, 2009), 58–59. Sugimura marks the complex relations between the particular and the universal.

31. Aristotle, *Posterior Analytics*, 2.19.

32. Peter Dear, "The Meanings of Experience," in *The Cambridge History of Science*, Volume 3. *Early Modern Science* ed. Katharine Park and Lorraine Daston (Cambridge: Cambridge University Press, 2006), 3:106–31, 107.

33. On experience and probable knowledge, see Dear, *Discipline and Experience*, 3, 23; and Shapiro, *Probability and Certainty in Seventeenth-Century England*, 15.

34. Aristotle, *Topica*, in *Basic Works*, ed. Richard McKeon (New York: Modern Library, 2001), 1.100.1.

35. John of Salisbury, *The Metalogicon of John of Salisbury: A Twelfth-Century Defense of the Verbal and Logical Arts of the Trivium*, trans. Daniel D. McGarry (Gloucester, MA: Peter Smith, 1955), 2.3.

36. Blundeville, *The Arte of Logick*, 169.

37. Thomas Spencer, *The art of logick delivered in the precepts of aristotle and ramvs* (London: Printed by Iohn Dawson for Nicholas Bourne, at the south entrance of the Royall Exchange, 1628), 288, EEBO.

38. In this era of uncertainty, Shapiro argues, "larger and larger portions of intellectual endeavor that once had been expected to attain the status of science by achieving demonstrative truth were now relegated to the domain of one level or another of probability" (*Probability and Certainty in Seventeenth-Century England*, 5). As a result, "experience, conjecture, and opinion, which

once had little or no role in philosophy or physics, and probability, belief, and credibility, once the possession of the rhetorician, theologian, and lawyer, now became relevant and even crucial categories for natural scientists and philosophers" (4).

39. Dear, *Discipline and Experience*, 23.

40. Dear, *Revolutionizing the Sciences*, 3.

41. See Shapin and Schaffer, *Leviathan and the—Air Pump*. For a study that focuses on the ideals of representation rather than on social history, see Picciotto, *Labors of Innocence*, esp. 26, 129–87.

42. Dear, *Revolutionizing the Sciences*, 7.

43. Dear, *Discipline and Experience*, 25.

44. On the relation of "disinterestedness" and "male gentility," see Michael McKeon, *The Secret History of Domesticity: Public, Private, and the Division of Knowledge* (Baltimore: Johns Hopkins University Press, 2005), 347. See also Dear, "From Truth to Disinterestedness," 625–29. For an example of construction of disinterestedness in the seventeenth century, see Robert Boyle, "A Letter of the Honorable Robert Boyle Concerning the Vegetable Nature of Amber Greece, according to an Extract Taken Out of a Dutch East Indian Journal," *Philosophical Transactions* 8 (1673): 6113–15.

45. Steven Shapin, *A Social History of Truth: Civility and Science in Seventeenth-Century England* (Chicago: University of Chicago Press, 1994), xxvi.

46. Sprat, *The History of the Royal-Society of London*, 72.

47. Ibid.

48. McKeon, *Secret History of Domesticity*, 347.

49. Sprat, *The History of the Royal-Society of London*, 66, 67.

50. For how Aristotelian notions of probable knowledge persisted in modified forms in experimental philosophy, see Debapriya Sarkar, "Shakespeare and the Social History of Truth," *Shakespeare Studies* 49 (2021): 94–106.

51. The narrator associates Satan with the experiential early on, describing his "reiterated crimes" (1.214) and marking how he "darts his experienced eye" (1.568).

52. Edwards, *Milton and the Natural World*, 17.

53. Dear, *Discipline and Experience*, 25.

54. On innocent curiosity in *Paradise Lost*, see Picciotto, *Labors of Innocence*, esp. 466–69.

55. Fish, *Surprised by Sin*, 248.

56. For Eve's affinity to the place of Eden, see D. K. McColley, *Milton's Eve*; and Ken Hiltner, *Milton and Ecology* (Cambridge: Cambridge University Press, 2003), 39–54. For the value of place, see also John Gillies, "Space and Place in *Paradise Lost*," *English Literary History* 74, no. 1 (2007): 27–57; Maura Brady, "Space and the Persistence of Place in *Paradise Lost*," *Milton Quarterly* 41, no. 3 (2007): 167–82; and Andrew Mattison, *Milton's Uncertain Eden: Understanding Place in Paradise Lost* (New York: Routledge, 2007).

57. Edwards, *Milton and the Natural World*, 21.

58. Ibid., 36.

59. Ibid., 18.

60. The "monist materialist" view is comprehensively established by Fallon, *Milton Among the Philosophers*. See also William Kerrigan, *The Sacred Complex: On the Psychogenesis of Paradise Lost* (Cambridge, MA: Harvard University Press, 1983), 193–262; and William Walker, "Milton's Dualistic Theory of Religious Toleration in A *Treatise of Civil Power*, *Of Christian Doctrine*, and *Paradise Lost*," *Modern Philology* 99, no. 2 (2001): 201–30. For works that complicate this position, see Rachel J. Trubowitz, "Body Politics in *Paradise Lost*," *PMLA* 121, no. 2 (2006): 388–404; and Sugimura, *"Matter of Glorious Trial,"* which offers a full-scale revision of materialism in Milton's oeuvre.

61. I draw on works such as Marsilio Ficino, *Three Books on Life* (1489), and Heinrich Cornelius Agrippa von Nettesheim, *Three Books of Occult Philosophy . . . Translated Out of the Latin into the English Tongue by J.F* (1651), which understand "spirit" as an intermediary between the body and soul, or as a medium between matter and soul.

62. Edwards, *Milton and the Natural World*, 37.

63. For representative works on Milton's cosmology and the place of "new worlds" in early modern science, see Nicolson, "Milton and the Telescope"; Patterson, "Imagining New Worlds"; C. G. Martin, "'What If the Sun Be Centre to the World'"; and Dodds, "Milton's Other Worlds." The "new worlds" also refer to colonial enterprises, travel, and speculations on cosmology.

64. See Regina M. Schwartz, *Remembering and Repeating: Biblical Creation in Paradise Lost* (Cambridge: Cambridge University Press, 1988), for the iterative and ritually repetitive nature of Milton's exploration of origins. The concept of the "new-created world" recurs throughout the epic, including in Raphael's narration of creation (7.617–22); and when Raphael warns Adam not to seek to know too much (8.172–78).

65. On Eve's naming, see Leonard, *Naming in Paradise*, esp. 33–51; and James Nohrnberg, "Naming Milton's Eve," *Milton Studies* 60, nos. 1–2 (2018): 1–28. On naming and concepts, see "Sugimura, *'Matter of Glorious Trial*,'" 40–80.

66. On analogy in *Paradise Lost*, see Marjara, *Contemplation of Created Things*; and Lara Dodds, "'Great Things to Small May Be Compared': Rhetorical Microscopy in *Paradise Lost*," *Milton Studies* 47 (2008): 96–117. On the main issues in analogy, see Joshua P. Hochschild, *The Semantics of Analogy: Rereading Cajetan's "De Nominum Analogia"* (Notre Dame, IN: University of Notre Dame Press, 2010). For how analogy generates knowledge of the physical world, see Hesse, *Models and Analogies in Science*.

67. Fallon, *Milton Among the Philosophers*, 194–222.

68. On Milton and utopia, see Amy Boesky, "Milton's Heaven and the Model of the English Utopia," *Studies in English Literature, 1500–1900* 36, no. 1 (1996): 91–110.

69. The scholarship on Michael as a teacher is extensive; for representative works, see Michael Allen, "Divine Instruction: *Of Education* and the Pedagogy of Raphael, Michael, and the Father," *Milton Quarterly* 26, no. 4 (1992): 113–21; Ann Baynes Coiro, "'To Repair the Ruins of Our First Parents': *Of Education* and Fallen Adam," *Studies in English Literature, 1500–1900* 28, no. 1 (1988): 133–47; and Angelica Duran, *The Age of Milton and the Scientific Revolution* (Pittsburgh: Duquesne University Press, 2007). On the genealogical aspects of the tutorial in relation to politics, patriarchy, and monarchy in Books 11 and 12, see Erin Murphy, *Familial Forms: Politics and Genealogy in Seventeenth-Century English Literature* (Newark: University of Delaware Press, 2011), 116–39.

70. For visual and aural revelations in the epic, see Joseph Anthony Wittreich Jr., *Visionary Poetics: Milton's Tradition and His Legacy* (San Marino, CA: Huntington Library, 1979). See also Peter E. Medine, John T. Shawcross, and David V. Urban, eds., *Visionary Milton: Essays on Prophecy and Violence* (Pittsburgh: Duquesne University Press, 2010). For the argument that "Milton believed himself a prophet," see William Kerrigan, *The Prophetic Milton* (Charlottesville: University Press of Virginia, 1974), 11.

71. Adam's education is composed of such moments of disjunction, what Alastair Fowler marks as "Adam's newfangled, postlapsarian tendency to local devotions." See John Milton, *Paradise Lost*, ed. Alastair Fowler, 2nd ed. (Harlow: Longman, 1998), 614, note to lines 335–54.

72. For the classic study of typology in Milton's works, see William G. Madsen, *From Shadowy Types to Truth: Studies in Milton's Symbolism* (New Haven, CT: Yale University Press, 1968).

See also William Walker, "Typology and *Paradise Lost*, Books XI and XII," *Milton Studies* 25 (1989): 245–64; Regina M. Schwartz, "From Shadowy Types to Shadowy Types: The Unendings of *Paradise Lost*," *Milton Studies* 24 (1988): 123–39; David Loewenstein, *Milton and the Drama of History: Historical Vision, Iconoclasm, and the Literary Imagination* (Cambridge: Cambridge University Press, 1990), 92–125; and Murphy, *Familial Forms*, 138. For the ways in which typology was understood in the seventeenth century, see Barbara Kiefer Lewalski, *Protestant Poetics and the Seventeenth-Century Religious Lyric* (Princeton, NJ: Princeton University Press, 1979), 111–44.

73. See Loewenstein, *Milton and the Drama of History*, for a comprehensive treatment of the differences between Milton's polemical prose and his later poems. In comparing Michael's tutorial with *Areopagitica*, he writes, "Michael's final account has become a long series of successive falls and repetitions ensuing from that original fall when the world first groaned under her own weight; as a process, it lacks the emphasis on energetic social confrontation and conflict that so distinctly characterized Milton's dynamic sense of historical renovation in a text like *Areopagitica*" (118–19). For similarities between the two works, and their deferral of ends, see Schwartz, "From Shadowy Types to Shadowy Types."

74. Loewenstein situates this duality within what he terms "the drama of history" (*Milton and the Drama of History*, 2): "What is so striking about this passage [4.1–8] is the narrator's highly personal response—his difficulty in disengaging himself from the tragic drama of human history he writes about" (102).

75. Ann Baynes Coiro, "Drama in the Epic Style: Narrator, Muse, and Audience in *Paradise Lost*," *Milton Studies* 51 (2010): 63–100, 71.

76. On the uncertain position and role of Milton's muse, see ibid., esp. 85–93.

CODA

1. Scholarship and public discourse on these topics are vast. For recent work in early modern literary studies that marks the early modern origins of this "crisis," see Eggert, *Disknowledge*, 245; and McEleney, *Futile Pleasures*, 1–5.

2. In using the language of "cultivation," I draw on Patricia Akhimie, *Shakespeare and the Cultivation of Difference: Race and Conduct in the Early Modern World* (New York: Routledge, 2018).

3. For the specific problem of distinguishing fact from fiction in Sidney, see J. T. Miller, *Poetic License*, esp. 77–78.

4. Campbell, *Wonder and Science*, 1–2.

5. For representative works, see Hall, *Things of Darkness*; Ian Smith, *Race and Rhetoric in the Renaissance: Barbarian Errors* (New York: Palgrave Macmillan, 2009); Ania Loomba, *Shakespeare, Race, and Colonialism* (Oxford: Oxford University Press, 2002); Margo Hendricks and Patricia Parker, eds., *Women, "Race," and Writing in the Early Modern Period* (New York: Routledge, 1994); Ayanna Thompson, *Performing Race and Torture on the Early Modern Stage* (New York: Routledge, 2008); Jyotsna G. Singh, ed., *A Companion to the Global Renaissance: English Literature and Culture in the Era of Expansion* (Malden, MA: Wiley-Blackwell, 2009); and Peter Erickson and Kim F. Hall, eds., "Shakespeare and Race," special issue, *Shakespeare Quarterly* 67, no. 1 (2016). For a discussion of the state of the field, see Urvashi Chakravarty, "The Renaissance of Race and the Future of Early Modern Race Studies," *English Literary Renaissance* 50, no. 1 (2020): 17–24. The scholarship on these topics is extensive and spans several decades, and my limited

bibliography here can only begin to gesture toward the connections between studies of race and colonialism and the questions raised in this book.

6. Debapriya Sarkar, "Literary Justice: The Participatory Ethics of Early Modern Possible Worlds," in *Teaching Social Justice Through Shakespeare: Why Renaissance Literature Matters Now*, ed. Hillary Eklund and Wendy Beth Hyman (Edinburgh: Edinburgh University Press, 2019), 174–84, 174–75.

7. See Akhimie, *Shakespeare and the Cultivation of Difference*, 1–48.

8. Silvio Zavala, *Sir Thomas More in New Spain: A Utopian Adventure of the Renaissance* (London: Hispanic and Luso-Brazilian Councils, 1955).

9. My thinking on patterns of creation and endings has been shaped by Kathryn Yusoff, *A Billion Black Anthropocenes or None* (Minneapolis: University of Minnesota Press, 2019), who argues: "The Anthropocene might seem to offer a dystopic future that laments the end of the world, but imperialism and ongoing (settler) colonialisms have been ending worlds for as long as they have been in existence" (xiii).

10. See *The Colbert Report*, "The Word—Truthiness," Comedy Central, October 17, 2005, https://www.cc.com/video/63ite2/the-colbert-report-the-word-truthiness; and Kellyanne Conway, "Conway: Press Secretary Gave 'Alternative Facts,'" interview by Chuck Todd, NBC News, January 22, 2017, https://www.nbcnews.com/meet-the-press/video/conway-press-secretary-gave-alternative-facts-860142147643.

11. See, for instance, Bruno Latour, "Why Has Critique Run Out of Steam? From Matters of Fact to Matters of Concern," *Critical Inquiry* 30, no. 2 (2004): 225–48.

12. Jan Golinski, *Making Natural Knowledge: Constructivism and the History of Science* (New York: Cambridge University Press, 1998), describes the "constructivist" outlook as that "which regards scientific knowledge primarily as a human product, made with locally situated cultural and material resources, rather than as simply the revelation of a pre-given order of nature" (ix).

13. A. H. Miller, "Implicative Criticism," 348.

14. Ibid., 347.

BIBLIOGRAPHY

PRIMARY SOURCES

Aristotle. *Basic Works*. Ed. Richard McKeon. New York: Modern Library, 2001.

———. *Physics*. In *The Complete Works of Aristotle*, ed. Jonathan Barnes. Princeton, NJ: Princeton University Press, 1984.

———. *Poetics*. In *The Complete Works of Aristotle*.

———. *Posterior Analytics*. Trans. Jonathan Barnes. 2nd ed. New York: Oxford University Press, 1994.

———. Topica. In Basic Works.

Bacon, Francis. *The Advancement of Learning*. In *Francis Bacon: The Major Works*.

———. *De Augmentis*. In *The Works of Francis Bacon*, ed. James Spedding, Robert Leslie Ellis, and Douglas Denon Heath. 15 vols. London: Longman, 1857–64.

———. *Essays or Counsels, Civil and Moral*. In *Francis Bacon: The Major Works*.

———. *Francis Bacon: The Major Works*. Ed. Brian Vickers. Oxford: Oxford World's Classics, 2008.

———. *The Great Instauration*. In *The New Organon and Related Writings*.

———. *New Atlantis*. In *Francis Bacon: The Major Works*.

———. *The New Organon*. In *The New Organon and Related Writings*.

———. *The New Organon and Related Writings*. Ed. Fulton H. Anderson. New York: Liberal Arts Press, 1960.

———. *Preparative Toward Natural and Experimental History*. In *The New Organon and Related Writings*.

———. *Sylva Sylvarum*. In *Francis Bacon: The Major Works*.

Blundeville, Thomas. *The Arte of Logick*. London: Printed by William Stansby, 1617. EEBO.

Boyle, Robert. *A Free Enquiry into the Vulgarly Receiv'd Notion of Nature made in an Essay Address'd to a Friend*. London: Printed by H. Clark for John Taylor, 1686.

———. "A Letter of the Honorable Robert Boyle Concerning the Vegetable Nature of Amber Greece, according to an Extract Taken Out of a Dutch East Indian Journal." *Philosophical Transactions* 8 (1673): 6113–15.

Carr, Anne. "Choyce receits collected out of the book of receits, of the Lady Vere Wilkinson [manuscript]/begun to be written by the Right Honble the Lady Anne Carr," 1673/4. Folger V.a.612. Luna.folger.edu.

Cavendish, Margaret. *The Blazing World and Other Writings*. Ed. Kate Lilley. New York: Penguin, 1994.

———. *Grounds of Natural Philosophy*. London: Printed by A. Maxwell, 1668. EEBO.

———. *Natures pictures drawn by fancies pencil to the life*. London: Printed for J. Martin and J. Allestrye at the Bell in St. Pauls Church-Yard, 1656. EEBO.
———. *Observations upon Experimental Philosophy*. Ed. Eileen O'Neill. Cambridge: Cambridge University Press, 2001.
———. *The Philosophical and Physical Opinions*. London: Printed for J. Martin and J. Allestrye at the Bell in St. Pauls Church-Yard, 1655. EEBO.
———. *Philosophicall Fancies*. London: Printed by Tho: Roycroft, for J. Martin, and J. Allestrye, at the Bell in St. Pauls Church-Yard, 1653. EEBO.
———. *Poems, and Fancies*. London: Printed by T. R. for J. Martin, and J. Allestrye, 1653. EEBO.
———. *The Worlds Olio*. London: Printed for J. Martin and J. Allestrye at the Bell in St. Pauls Church-Yard, 1655. EEBO.
Cavendish, William. "To the Duchess of Newcastle, on Her New Blazing World." In M. Cavendish, *Blazing World and Other Writings*, 121.
Cowley, Abraham. "To the Royal Society." In Sprat, *The istory of the Royal-Society of London*, sig. B1r–B3v.
The Faerie Leveller. London: Printed just levell anens the Saints Army: in the yeare of their Saintships ungodly Revelling for a godly Levelling, 1648. EEBO.
Florio, John. *A VVorlde of Wordes, Or most Copious, and Exact Dictionarie in Italian and English*. London: By Arnold Hatfield for Edw. Blount, 1598. EEBO.
Galileo. *The Assayer* (excerpts from). In *Discoveries and Opinions of Galileo*.
———. *Discoveries and Opinions of Galileo*. Ed. and trans. Stillman Drake. Garden City, NY: Doubleday, 1957.
———. "Letters on Sunspots" (excerpts from). In *Discoveries and Opinions of Galileo*.
Gilbert, Humphrey. A *Discourse of a Discoverie for a New Passage to Cataia*. London: By Henry Middleton for Richarde Ihones, 1576. EEBO.
Gilbert, William. *De Magnete*. Trans. P. Fleury Mottelay. New York: Dover Publications, 1958.
Gosson, Stephen. *The Schoole of Abuse*. London: By [Thomas Dawson for] Thomas VVoodcocke, 1579.
The Granville Family. Cookery and medicinal recipes of the Granville family [manuscript], ca. 1640–ca. 1750. Folger V.a.430. Luna.folger.edu.
H., R. *New Atlantis. Begun by the Lord Verulam, Viscount St. Albans: and continued by R.H. Esquire. Wherein is set forth a platform of monarchical government. With a pleasant intermixture of divers rare inventions, and wholsom customs, fit to be introduced into all kingdoms, states, and common-wealths*. London: Printed for John Crooke at the signe of the Ship in St. Pauls Church-Yard, 1660. EEBO.
Harington, John. *A Brief Apology of Poetry* (selected passages). In *Sidney's "The Defence of Poesy" and Selected Renaissance Literary Criticism*, ed. Gavin Alexander. London: Penguin, 2004.
Harvey, John. *A Discoursiue Probleme concerning Prophesies*. . . . London: By Iohn Iackson, for Richard Watkins, 1588. EEBO.
Hollyband, Claudius. *A Dictionarie French and English: Published for the Benefite of the Studious in that Language*. London: Imprinted at London by T. O. for Thomas Woodcock, 1593. EEBO.
Hooke, Robert. *Micrographia*. London: Printed by Jo. Martyn, and Ja. Allestry, printers to the Royal Society, 1665. EEBO.
Howard, Henry. *A defensatiue against the poyson of supposed Prophesies*. . . . London: Printed by Iohn Charlewood, printer to the right Honourable Earle of Arundell, 1583. EEBO.
John of Salisbury. *The Metalogicon of John of Salisbury: A Twelfth-Century Defense of the Verbal and Logical Arts of the Trivium*. Trans. Daniel D. McGarry. Gloucester, MA: Peter Smith, 1955.

Lily, William, and John Colet. *A Short Introduction of Grammar.* 1549. Facs. rpt. Menston, Bradford: Scholar Press, 1970.

Lucretius Carus, Titus. *The Nature of Things (De Rerum Natura).* Trans. A. E. Stallings. London: Penguin, 2007.

Marvell, Andrew. "On Mr. Milton's *Paradise Lost*" (1674). In Milton, *The Complete Poetry and Essential Prose of John Milton.*

Milton, John. *Areopagitica.* In *The Complete Poetry and Essential Prose of John Milton.*

———. *The Complete Poetry and Essential Prose of John Milton.* Ed. William Kerrigan, John Rumrich, and Stephen M. Fallon. New York: Modern Library, 2007.

———. *Of Education.* In *The Complete Poetry and Essential Prose of John Milton.*

———. *Paradise Lost.* In *The Complete Poetry and Essential Prose of John Milton.*

———. *Paradise Lost.* Ed. Alastair Fowler. 2nd ed. Harlow: Longman, 1998.

———. *The Reason of Church-Government Urg'd Against Prelaty.* In *The Complete Poetry and Essential Prose of John Milton.*

More, Thomas. *Utopia.* Ed. George M. Logan and Robert M. Adams. Cambridge: Cambridge University Press, 1989.

Packe, Susanna. Cookbook of Susanna Packe [manuscript], 1674. Folger V.a.215. Luna.folger.edu.

[Petty, William]. *The advice of W.P. to Mr. Samuel Hartlib. For the Advancement of Some Particular Parts of Learning.* London, 1648. EEBO.

Phillips, Edward. *The New World of English Words.* London: Printed by E. Tyler for Nath. Brooke, 1658. EEBO.

Piemontese, Alessio. *The secretes of the reuerende Maister Alexis of Piemount. . . . Translated out of Frenche into Englishe, by Wyllyam Warde.* London: By Iohn Kingstone for Nicolas Inglande, dwellinge in Poules churchyard, 1558.

Plat, Hugh. *Delightes for Ladies, to adorne their Persons, Tables, closets and distillatories. With Beauties, banquets, perfumes and Waters.* London: Printed by Peter Short, 1602. EEBO.

[Plattes, Gabriel]. *A Description of the Famous Kingdome of Macaria.* London: Printed for Francis Constable, 1641. EEBO.

Puttenham, George. *The Art of English Poesy.* Ed. Frank Whigham and Wayne A. Rebhorn. Ithaca, NY: Cornell University Press, 2007.

Quintilian. *Institutes of oratory.* Ed. and trans. John Selby Watson. 2 vols. London: G. Bell and Sons, 1895.

Shakespeare, William. *The Arden Shakespeare Complete Works.* Ed. Richard Proudfoot, Ann Thompson, and David Scott Kastan. London: Methuen Drama, 2011.

———. *A Midsummer Night's Dream.* In *The Norton Shakespeare.*

———. *Macbeth.* In *The Norton Shakespeare.*

———. *The Norton Shakespeare.* Gen. ed. Stephen Greenblatt. Ed. Walter Cohen, Suzanne Gossett, Jean E. Howard, Katharine Eisaman Maus, and Gordon McMullan. 3rd ed. New York: W. W. Norton, 2016.

———. *The Riverside Shakespeare.* Ed. Blakemore Evans, with J. J. M. Tobin. 2nd ed. Boston: Houghton Mifflin, 1997.

Sidney, Philip. *A Defence of Poetry.* Ed. Jan Van Dorsten. Oxford: Oxford University Press, 1966.

Spencer, Thomas. *The art of logick delivered in the precepts of aristotle and ramvs.* London: Printed by Iohn Dawson for Nicholas Bourne, at the south entrance of the Royall Exchange, 1628. EEBO.

Spenser, Edmund. *The Faerie Queene.* Ed. A. C. Hamilton, text ed. Hiroshi Yamashita and Toshiyuki Suzuki. 2nd ed. Harlow: Longman, 2007.

———. "Letter to Raleigh." In *The Faerie Queene*, 714–18.
———. *The Shepheardes Calendar*. In *The Yale Edition of the Shorter Poems of Edmund Spenser*, ed. William A. Oram, Einar Bjorvand, Ronald Bond, Thomas H. Cain, Alexander Dunlop, and Richard Schell. New Haven, CT: Yale University Press, 1989.
Sprat, Thomas. *The History of the Royal-Society of London: For the Improving of Natural Knowledge*. London: Printed by T. R. for J. Martyn and J. Allestry, 1667. EEBO.
Weamys, Anna. *A continuation of Sir Philip Sydney's Arcadia: wherein is handled the loves of Amphialus and Helena Queen of Corinth, Prince Plangus and Erona. With the historie of the loves of old Claius and young Strephon to Vrania*. London: Printed by William Bentley, and are to be sold by Thomas Heath, near the Pyazza of the Coven-Garden, 1651.
Wroth, Mary, *The First Part of the Countess of Mongtomery's Urania*. Ed. Josephine A. Roberts. Medieval & Renaissance Texts & Studies. Binghamton: State University of New York at Binghamton, 1995.

SECONDARY SOURCES

Adams, B. K. "Fair/Foul." In *Shakespeare/Text: Contemporary Readings in Textual Studies, Editing and Performance*, ed. Claire M. L. Bourne, 29–49. London: Bloomsbury Arden Shakespeare, 2021.
Adelman, Janet. "Born of Woman, Fantasies of Maternal Power in *Macbeth*." In *Cannibals, Witches, and Divorce: Estranging the Renaissance*, ed. Marjorie Garber, 90–121. Baltimore: Johns Hopkins University Press, 1987.
Agamben, Giorgio. "On Potentiality." In *Potentialities: Collected Essays in Philosophy*, 177–84. Ed. and trans. Daniel Heller-Roazen. Stanford, CA: Stanford University Press, 1999.
Aït-Touati, Frédérique. *Fictions of the Cosmos: Science and Literature in the Seventeenth Century*. Trans. Susan Emanuel. Chicago: University of Chicago Press, 2011.
Akhimie, Patricia. *Shakespeare and the Cultivation of Difference: Race and Conduct in the Early Modern World*. New York: Routledge, 2018.
Albanese, Denise. "The *New Atlantis* and the Uses of Utopia." *English Literary History* 57, no. 3 (1990): 503–28.
———. *New Science, New World*. Durham, NC: Duke University Press, 1996.
Albright, Evelyn May. "Spenser's Cosmic Philosophy and His Religion." *PMLA* 44, no. 3 (1929): 715–59.
Allen, Michael. "Divine Instruction: *Of Education* and the Pedagogy of Raphael, Michael, and the Father." *Milton Quarterly* 26, no. 4 (1992): 113–21.
Anderson, Judith H. "Arthur, Argante, and the Ideal Vision: An Exercise in Speculation and Parody." In *The Passing of Arthur: New Essays in Arthurian Tradition*, ed. Christopher Baswell and William Sharpe, 193–206. New York: Garland, 1988.
Ankers, Neil. "Paradigms and Politics: Hobbes and Cavendish Contrasted." In Clucas, *Princely Brave Woman*, 242–54.
Barret, J. K. *Untold Futures: Time and Literary Culture in Renaissance England*. Ithaca, NY: Cornell University Press, 2016.
Barrett, Chris. "Allegraphy and *The Faerie Queene*'s Significantly Unsignifying Ecology." *Studies in English Literature, 1500–1900* 56, no. 1 (2016): 1–21.
Battigelli, Anna. *Margaret Cavendish and the Exiles of the Mind*. Lexington: University Press of Kentucky, 1998.

Bellamy, Elizabeth J. *Translations of Power: Narcissism and the Unconscious in Epic History*. Ithaca, NY: Cornell University Press, 1992.

Berger, Harry, Jr. *Second World and Green World: Studies in Renaissance Fiction-Making*. Intro. John Patrick Lynch. Berkeley: University of California Press, 1988.

Bieman, Elizabeth. *Plato Baptized: Towards the Interpretation of Spenser's Mimetic Fictions*. Toronto: University of Toronto Press, 1988.

Blackburn, William. "Spenser's Merlin." *Renaissance and Reformation/Renaissance et Reforme* 4, no. 2 (1980): 179–98.

Blake, Liza. "Choose Your Own Poems and Fancies." Fellow lecture, Folger Shakespeare Library, Washington, DC, April 2019.

———. "Hester Pulter's Particle Physics and the Poetics of Involution." *Journal of Early Modern Cultural Studies* 20, no. 2 (2020): 71–98.

———. "Margaret Cavendish's Forms: Literary Formalism and the Figures of Cavendish's Atom Poems." In *Feminist Formalism and Early Modern Women's Writing: Readings, Conversations, Pedagogies*, ed. Lara Dodds and Michelle M. Dowd, 38–55. Lincoln: University of Nebraska Press, 2022.

———. "The Physics of Poetic Form in Arthur Golding's Translation of Ovid's *Metamorphoses*." *English Literary Renaissance* 51, no. 3 (2021): 331–55.

———. "Textual and Editorial Introduction." In *Margaret Cavendish's Poems and Fancies: A Digital Critical Edition*, ed. Liza Blake. Website published May 2019. http://library2.utm.utoronto.ca/poemsandfancies/textual-and-editorial-introduction/.

Boesky, Amy. "Milton, Galileo, and Sunspots: Optics and Certainty in *Paradise Lost*." *Milton Studies* 34 (1997): 23–43.

———. "Milton's Heaven and the Model of the English Utopia." *Studies in English Literature, 1500–1900* 36, no. 1 (1996): 91–110.

Borris, Kenneth, Carol Kaske, and Jon Quitslund, eds. "Spenser and Platonism: An Expanded Special Volume." Special issue, *Spenser Studies* 24 (2009).

Bowerbank, Sylvia. "The Spider's Delight: Margaret Cavendish and the 'Female' Imagination." *English Literary Renaissance* 14, no. 3 (1984): 392–408.

Boyle, Deborah. "Margaret Cavendish's Nonfeminist Natural Philosophy." *Configurations* 12, no. 2 (2004): 195–227.

Brady, Maura. "Space and the Persistence of Place in *Paradise Lost*." *Milton Quarterly* 41, no. 3 (2007): 167–82.

Broad, Jacqueline. "Margaret Cavendish." In *Women Philosophers of the Seventeenth Century*, 35–64. Cambridge: Cambridge University Press, 2003.

Burrow, Colin. *Epic Romance: Homer to Milton*. New York: Oxford University Press, 1993.

———. "Spenser and Classical Traditions." In *The Cambridge Companion to Spenser*, ed. Andrew Hadfield, 217–36. Cambridge: Cambridge University Press, 2001.

Bushnell, Rebecca. *Tragic Time in Drama, Film, and Videogames: The Future in an Instant*. London: Palgrave Macmillan, 2016.

Butler, Todd. "Revitalizing Nation and Mind: The Failed Promise of Seventeenth-Century Educational Reform." In *Political Turmoil: Early Modern British Literature in Transition, 1623–1660*, ed. Stephen B. Dobranski, 237–52. Cambridge: Cambridge University Press, 2019.

Campbell, Mary Baine. *The Witness and the Other World: Exotic European Travel Writing, 400–1600*. Ithaca, NY: Cornell University Press, 1988.

———. *Wonder and Science: Imagining Worlds in Early Modern Europe*. Ithaca, NY: Cornell University Press, 1999.

Carey, Daniel. "Hakluyt's Instructions: The Principal Navigations and Sixteenth-Century Travel Advice." *Studies in Travel Writing* 13, no. 2 (2009): 167–85.

———. "Inquiries, Heads, and Directions: Orienting Early Modern Travel." In *Travel Narratives, the New Science, and Literary Discourse, 1569–1750*, ed. Judy A. Hayden, 25–52. Burlington, VT: Ashgate, 2012.

Chakravarty, Urvashi. "The Renaissance of Race and the Future of Early Modern Race Studies." *English Literary Renaissance* 50, no. 1 (2020): 17–24.

Chico, Tita. *The Experimental Imagination: Literary Knowledge and Science in the British Enlightenment.* Stanford, CA: Stanford University Press, 2018.

Clody, Michael C. "Deciphering the Language of Nature: Cryptography, Secrecy, and Alterity in Francis Bacon." *Configurations* 19, no. 1 (2011): 117–42.

Clucas, Stephen. "'A Knowledge Broken': Francis Bacon's Aphoristic Style and the Crisis of Scholastic and Humanistic Knowledge Systems." In *English Renaissance Prose: History, Language and Politics*, ed. Neil Rhodes, 147–72. Tempe, AZ: Medieval and Renaissance Texts and Studies, 1997.

———. "Margaret Cavendish's Materialist Critique of Van Helmontian Chymistry." *Ambix: Journal of the Society for the History of Alchemy and Chemistry* 58, no. 1 (2011): 1–12.

———, ed. *A Princely Brave Woman: Essays on Margaret Cavendish, Duchess of Newcastle.* Burlington, VT: Ashgate, 2003.

———. "Variation, Irregularity, and Probabilism: Margaret Cavendish and Natural Philosophy as Rhetoric." In Clucas, *Princely Brave Woman*, 199–209.

Coiro, Ann Baynes. "Drama in the Epic Style: Narrator, Muse, and Audience in *Paradise Lost*." *Milton Studies* 51 (2010): 63–100.

———. "'To Repair the Ruins of Our First Parents': *Of Education* and Fallen Adam." *Studies in English Literature, 1500–1900* 28, no. 1 (1988): 133–47.

The Colbert Report. "The Word—Truthiness." Comedy Central, October 17, 2005. https://www.cc.com/video/63ite2/the-colbert-report-the-word-truthiness.

Coleridge, Samuel Taylor. *Coleridge's Miscellaneous Criticism.* Ed. Thomas Middleton Raysor. London: Constable, 1936.

Conway, Kellyanne. "Conway: Press Secretary Gave 'Alternative Facts.'" Interview by Chuck Todd. NBC News, January 22, 2017. https://www.nbcnews.com/meet-the-press/video/conway-press-secretary-gave-alternative-facts-860142147643.

Cooper, Helen. *The English Romance in Time: Transforming Motifs from Geoffrey of Monmouth to the Death of Shakespeare.* Oxford: Oxford University Press, 2004.

Cowan, Jacqueline L. "Milton, the Royal Society, and the Galileo Problem." *Studies in English Literature, 1500–1900* 58, no. 3 (2018): 569–89.

Craig, Joanne. "The Image of Mortality: Myth and History in the *Faerie Queene*." *English Literary History* 39, no. 4 (1972): 520–44.

Crane, Mary Thomas. *Framing Authority: Sayings, Self, and Society in Sixteenth-Century England.* Princeton, NJ: Princeton University Press, 1993.

———. *Losing Touch with Nature: Literature and the New Science in Sixteenth-Century England.* Baltimore: Johns Hopkins University Press, 2014.

———. *Shakespeare's Brain: Reading with Cognitive Theory.* Princeton, NJ: Princeton University Press, 2001.

———. "What Was Performance?" *Criticism* 43, no. 2 (2001): 169–87.

Culler, Jonathan. "The Closeness of Close Reading." *ADFL Bulletin* 41, no. 3 (2011): 8–13.

Cummins, Juliet, and David Burchell, eds. *Science, Literature and Rhetoric in Early Modern England*. Burlington, VT: Ashgate, 2007.

Curry, Patrick. *Prophecy and Power: Astrology in Early Modern England*. Cambridge: Polity Press, 1989.

Daston, Lorraine. "Baconian Facts, Academic Civility, and the Prehistory of Objectivity." *Annals of Scholarship* 8 (1991): 337–63.

———. *Classical Probability in the Enlightenment*. Princeton, NJ: Princeton University Press, 1988.

Daston, Lorraine, and Katharine Park. *Wonders and the Order of Nature, 1150–1750*. New York: Zone Books; Cambridge, MA: Distributed by the MIT Press, 1998.

Dear, Peter. *Discipline and Experience: The Mathematical Way in the Scientific Revolution*. Chicago: University of Chicago Press, 1995.

———. "From Truth to Disinterestedness in the Seventeenth Century." *Social Studies of Science* 22, no. 4 (1992): 619–31.

———. "The Meanings of Experience." In *The Cambridge History of Science*, Volume 3. Early Modern Science ed. Katharine Park and Lorraine Daston, 3:106–31. Cambridge: Cambridge University Press, 2006.

———. "A Philosophical Duchess: Understanding Margaret Cavendish and the Royal Society." In Cummins and Burchell, *Science, Literature and Rhetoric in Early Modern England*, 125–42.

———. *Revolutionizing the Sciences: European Knowledge and Its Ambitions, 1500–1700*. Princeton, NJ: Princeton University Press, 2001.

De Grazia, Margreta. *Four Shakespearean Period Pieces*. Chicago: University of Chicago Press, 2021.

———. "Lost Potential in Grammar and Nature: Sidney's *Astrophil and Stella*." *Studies in English Literature, 1500–1900* 21, no. 1 (1981): 21–35.

Delbourgo, James, and Staffan Müller-Wille. "Introduction to Focus: 'Listmania.'" *Isis* 103, no. 4 (2012): 710–15.

de Man, Paul. "Pascal's Allegory of Persuasion." In *Allegory and Representation*, ed. Stephen Greenblatt, 1–25. Baltimore: Johns Hopkins University Press, 1981.

DeMoss, William Fenn. "Spenser's Twelve Moral Virtues 'According to Aristotle' (Continued)." *Modern Philology* 16, no. 1 (1918): 23–38.

———. "Spenser's Twelve Moral Virtues 'According to Aristotle' II." *Modern Philology* 16, no. 5 (1918): 245–70.

Detlefsen, Karen. "Atomism, Monism, and Causation in the Natural Philosophy of Margaret Cavendish." *Oxford Studies in Early Modern Philosophy* 3 (2006): 199–240.

Dick, Steven J. *Plurality of Worlds: The Origins of the Extraterrestrial Life Debate from Democritus to Kant*. Cambridge: Cambridge University Press, 1982.

DiMeo, Michelle. "Authorship and Medical Networks: Reading Attributions in Early Modern Manuscript Recipe Books." In *Reading and Writing Recipe Books, 1550–1800*, ed. Michelle DiMeo and Sara Pennell, 25–46. Manchester: Manchester University Press, 2013.

Dobin, Howard. *Merlin's Disciples: Prophecy, Poetry, and Power in Renaissance England*. Stanford, CA: Stanford University Press, 1990.

Dodds, Lara. "'Great Things to Small May Be Compared': Rhetorical Microscopy in *Paradise Lost*." *Milton Studies* 47 (2008): 96–117.

———. *The Literary Invention of Margaret Cavendish*. Pittsburgh: Duquesne University Press, 2013.

———. "Milton's Other Worlds." In *Uncircumscribed Mind: Reading Milton Deeply*, ed. Charles W. Durham and Kristin A. Pruitt, 164–82. Selinsgrove, PA: Susquehanna University Press, 2008.

———. "'Poore Donne Was Out': Reading and Writing Donne in the Works of Margaret Cavendish." *John Donne Journal* 29 (2010): 133–74.

Dolan, Frances E. *True Relations: Reading, Literature, and Evidence in Seventeenth-Century England*. Philadelphia: University of Pennsylvania Press, 2013.

Doležel, Lubomír. *Heterocosmica: Fiction and Possible Worlds*. Baltimore: Johns Hopkins University Press, 1998.

Dolven, Jeffrey A. *Scenes of Instruction in Renaissance Romance*. Chicago: University of Chicago Press, 2007.

Dubrow, Heather. *Echoes of Desire: English Petrarchism and Its Counterdiscourses*. Ithaca, NY: Cornell University Press, 1995.

Duran, Angelica. *The Age of Milton and the Scientific Revolution*. Pittsburgh: Duquesne University Press, 2007.

Eamon, William. *Science and the Secrets of Nature: Books of Secrets in Medieval and Early Modern Culture*. Princeton, NJ: Princeton University Press, 1994.

Edelman, Lee. "Against Survival: Queerness in a Time That's Out of Joint." *Shakespeare Quarterly* 62, no. 2 (2011): 148–69.

———. *No Future: Queer Theory and the Death Drive*. Durham, NC: Duke University Press, 2004.

Edwards, Karen L. *Milton and the Natural World: Science and Poetry in Paradise Lost*. Cambridge: Cambridge University Press, 1999.

Eggert, Katherine. *Disknowledge: Literature, Alchemy, and the End of Humanism in Renaissance England*. Philadelphia: University of Pennsylvania Press, 2015.

Eisenstein, Elizabeth L. *The Printing Press as an Agent of Change: Communications and Cultural Transformations in Early Modern Europe*. Cambridge: Cambridge University Press, 1979.

Elam, Keir. *The Semiotics of Theatre and Drama*. 2nd ed. New York: Routledge, 2002.

Ellrodt, Robert. *Neoplatonism in the Poetry of Spenser*. Geneva: Droz, 1960.

Erickson, Peter, and Kim F. Hall, eds. "Shakespeare and Race." Special issue, *Shakespeare Quarterly* 67, no. 1 (2016).

Evans, J. Martin. *The Miltonic Moment*. Lexington: University Press of Kentucky, 1998.

Fallon, Stephen M. *Milton Among the Philosophers: Poetry and Materialism in Seventeenth-Century England*. Ithaca, NY: Cornell University Press, 1991.

Findlen, Paula. *Possessing Nature: Museums, Collecting, and Scientific Culture in Early Modern Italy*. Berkeley: University of California Press, 1994.

Fish, Stanley. *Surprised by Sin: The Reader in Paradise Lost*. 2nd ed. Cambridge, MA: Harvard University Press, 1998.

Fletcher, Angus. *Allegory: The Theory of a Symbolic Mode*. Ithaca, NY: Cornell University Press, 1964.

Fletcher, Angus. "Complexity and the Spenserian Myth of Mutability." *Literary Imagination* 6, no. 1 (2004): 1–22.

Fletcher, Angus. "The Irregular Aesthetic of *The Blazing World*." *Studies in English Literature, 1500–1900* 47, no. 1 (2007): 123–41.

Floyd-Wilson, Mary. "English Epicures and Scottish Witches." *Shakespeare Quarterly* 57, no. 2 (2006): 131–61.

Foster, Donald W. "*Macbeth*'s War on Time." *English Literary Renaissance* 16, no. 2 (1986): 319–42.

Freeman, Louise Gilbert. "The Metamorphosis of Malbecco: Allegorical Violence and Ovidian Change." *Studies in Philology* 97, no. 3 (2000): 308–30.

Frow, John. *Genre*. New York: Routledge, 2006.

Fuchs, Barbara. *Romance*. New York: Routledge, 2004.

Fuller, Mary C. *Voyages in Print: English Travel to America, 1576–1624*. New York: Cambridge University Press, 1995.

Gadberry, Andrea. *Cartesian Poetics: The Art of Thinking*. Chicago: University of Chicago Press, 2020.

Gallagher, Catherine. "Embracing the Absolute: The Politics of the Female Subject in Seventeenth-Century England." *Genders* 1 (1988): 24–39.

———. "When Did the Confederate States of America Free the Slaves?" *Representations* 98, no. 1 (2007): 53–61.

Gallop, Jane. "The Ethics of Reading: Close Encounters." *Journal of Curriculum Theorizing* 16, no. 3 (2000): 7–17.

Garber, Marjorie. *Shakespeare's Ghost Writers: Literature as Uncanny Causality*. New York: Methuen, 1987.

———. "'What's Past Is Prologue': Temporality and Prophecy in Shakespeare's History Plays." In *Renaissance Genres: Essays on Theory, History, and Interpretation*, ed. Barbara Kiefer Lewalski, 301–31. Cambridge, MA: Harvard University Press, 1986.

Gillies, John. "Space and Place in *Paradise Lost*." *English Literary History* 74, no. 1 (2007): 27–57.

Glimp, David, and Michelle R. Warren, eds. *Arts of Calculation: Quantifying Thought in Early Modern Europe*. New York: Palgrave Macmillan, 2004.

Goldberg, Jonathan. *Endlesse Worke: Spenser and the Structures of Discourse*. Baltimore: Johns Hopkins University Press, 1981.

———. *The Seeds of Things: Theorizing Sexuality and Materiality in Renaissance Representations*. New York: Fordham University Press, 2009.

Golinski, Jan. *Making Natural Knowledge: Constructivism and the History of Science*. New York: Cambridge University Press, 1998.

Goodman, Nelson. *Ways of Worldmaking*. Indianapolis: Hackett, 1978.

Grafton, Anthony. *Defenders of the Text: The Traditions of Scholarship in an Age of Science, 1450–1800*. Cambridge, MA: Harvard University Press, 1991.

Grayling, A. C. *An Introduction to Philosophical Logic*. 3rd ed. Oxford: Blackwell, 1997.

Greenblatt, Stephen. *Renaissance Self-Fashioning: From More to Shakespeare*. Chicago: University of Chicago Press, 1980.

———. "Shakespeare Bewitched." In *New Historical Literary Study: Essays on Reproducing Texts, Representing History*, ed. Jeffrey N. Cox and Larry J. Reynolds, 108–35. Princeton, NJ: Princeton University Press, 1993.

Greene, Roland. *Five Words: Critical Semantics in the Age of Shakespeare and Cervantes*. Chicago: University of Chicago Press, 2013.

———. "A Primer of Spenser's Worldmaking: Alterity in the Bower of Bliss." In *Worldmaking Spenser: Explorations in the Early Modern Age*, ed. Patrick Cheney and Lauren Silberman, 9–31. Lexington: University Press of Kentucky, 2000.

Greene, Thomas M. *The Light in Troy: Imitation and Discovery in Renaissance Poetry*. New Haven, CT: Yale University Press, 1982.

Greenlaw, Edwin. "Spenser and Lucretius." *Studies in Philology* 17, no. 4 (1920): 439–64.

Gregorson, Linda. "Protestant Erotics: Idolatry and Interpretation in Spenser's *Faerie Queene*." *English Literary History* 58, no. 1 (1991): 1–34.

Grogan, Jane. *Exemplary Spenser: Visual and Poetic Pedagogy in The Faerie Queene*. Burlington, VT: Ashgate, 2009.

Guj, Luisa. "Macbeth and the Seeds of Time." *Shakespeare Studies* 18 (1986): 175–88.

Hacking, Ian. *The Emergence of Probability: A Philosophical Study of Early Ideas About Probability, Induction and Statistical Inference*. Cambridge: Cambridge University Press, 1975.

Hall, Kim F. *Things of Darkness: Economies of Race and Gender in Early Modern England*. Ithaca, NY: Cornell University Press, 1995.

Halliwell, Stephen. *The Aesthetics of Mimesis: Ancient Texts and Modern Problems*. Princeton, NJ: Princeton University Press, 2002.

Halpern, Richard. "Eclipse of Action: *Hamlet* and the Political Economy of Playing." *Shakespeare Quarterly* 59, no. 4 (2008): 450–82.

Hamilton, A. C. "Sidney's Idea of the 'Right Poet.'" *Comparative Literature* 9, no. 1 (1957): 51–59.

Harkness, Deborah E. *The Jewel House: Elizabethan London and the Scientific Revolution*. New Haven, CT: Yale University Press, 2007.

Harris, Jonathan Gil. *Untimely Matter in the Time of Shakespeare*. Philadelphia: University of Pennsylvania Press, 2009.

Harrison, Peter. *The Fall of Man and the Foundations of Science*. Cambridge: Cambridge University Press, 2007.

Heffernan, Megan. *Making the Miscellany: Poetry, Print, and the History of the Book in Early Modern England*. Philadelphia: University of Pennsylvania Press, 2021.

Hendricks, Margo, and Patricia Parker, eds. *Women, "Race," and Writing in the Early Modern Period*. New York: Routledge, 1994.

Heninger, S. K., Jr. *Sidney and Spenser: The Poet as Maker*. University Park: Pennsylvania State University Press, 1989.

Hershinow, Stephanie Insley. *Born Yesterday: Inexperience and the Early Realist Novel*. Baltimore: Johns Hopkins University Press, 2019.

Hesse, Mary B. "Francis Bacon" In *A Critical History of Western Philosophy*, ed. D. J. O'Connor, 141–52. New York: Free Press of Glencoe, 1964.

———. *Models and Analogies in Science*. Notre Dame, IN: University of Notre Dame Press, 1966.

Hiltner, Ken. *Milton and Ecology*. Cambridge: Cambridge University Press, 2003.

Hintz, Carrie. "'But One Opinion': Fear of Dissent in Cavendish's *New Blazing World*." *Utopian Studies* 7, no. 1 (1996): 25–37.

Hochschild, Joshua P. *The Semantics of Analogy: Rereading Cajetan's "De Nominum Analogia."* Notre Dame, IN: University of Notre Dame Press, 2010.

Hock, Jessie. "Fanciful Poetics and Skeptical Epistemology in Margaret Cavendish's *Poems and Fancies*." *Studies in Philology* 115, no. 4 (2018): 766–802.

Holmesland, Oddvar. "Margaret Cavendish's *The Blazing World*: Natural Art and the Body Politic." *Studies in Philology* 96, no. 4 (1999): 457–79.

Houston, Chloë. "Utopia and Education in the Seventeenth Century: Bacon's Salomon's House and Its Influence." In *New Worlds Reflected: Travel and Utopia in the Early Modern Period*, ed. Chloë Houston, 161–78. Farnham: Ashgate, 2010.

Howell, Wilbur Samuel. *Logic and Rhetoric in England, 1500–1700*. Princeton, NJ: Princeton University Press, 1956.

Huntley, Frank L. "*Macbeth* and the Background of Jesuitical Equivocation." *PMLA* 79, no. 4 (1964): 390–400.

Hutson, Lorna. *Circumstantial Shakespeare*. Oxford: Oxford University Press, 2015.

———. *The Invention of Suspicion: Law and Mimesis in Shakespeare and Renaissance Drama*. Oxford: Oxford University Press, 2007.
Hutton, Sarah. "In Dialogue with Thomas Hobbes: Margaret Cavendish's Natural Philosophy." *Women's Writing* 4, no. 3 (1997): 421–32.
———. "Margaret Cavendish and Henry More." In Clucas, *Princely Brave Woman*, 185–98.
Hyman, Wendy Beth. *Impossible Desire and the Limits of Knowledge in Renaissance Poetry*. Oxford: Oxford University Press, 2019.
———. "Seizing Flowers in Spenser's Bower and Garden." *English Literary Renaissance* 37, no. 2 (2007): 193–214.
Iyengar, Sujata. "Royalist, Romancist, Racialist: Rank, Gender, and Race in the Science and Fiction of Margaret Cavendish." *English Literary History* 69, no. 3 (2002): 649–72.
Jansen Jaech, Sharon L. "Political Prophecy and Macbeth's 'Sweet Bodements.'" *Shakespeare Quarterly* 34, no. 3 (1983): 290–97.
Jardine, Lisa. *Francis Bacon: Discovery and the Art of Discourse*. Cambridge: Cambridge University Press, 1974.
———. "Lorenzo Valla and the Intellectual Origins of Humanist Dialectic." *Journal of the History of Philosophy* 15, no. 2 (1977): 143–64.
Johns, Adrian. *The Nature of the Book: Print and Knowledge in the Making*. Chicago: University of Chicago Press, 1998.
Johnson, Laurie, John Sutton, and Evelyn Tribble, eds. *Embodied Cognition and Shakespeare's Theatre: The Early Modern Body-Mind*. New York: Routledge, 2014.
Kargon, Robert Hugh. *Atomism in England from Hariot to Newton*. Oxford: Clarendon Press, 1966.
Kastan, David Scott. *Shakespeare After Theory*. New York: Routledge, 1999.
———. *Shakespeare and the Shapes of Time*. Hanover, NH: University Press of New England, 1982.
Keller, Eve. "Producing Petty Gods: Margaret Cavendish's Critique of Experimental Science." *English Literary History* 64, no. 2 (1997): 447–71.
Keller, Vera. "Accounting for Invention: Guido Pancirolli's Lost and Found Things and the Development of *Desiderata*." *Journal of the History of Ideas* 73, no. 2 (2012): 223–45.
———. *Knowledge and the Public Interest, 1575–1725*. New York: Cambridge University Press, 2015.
———. "Mining Tacitus: Secrets of Empire, Nature and Art in the Reason of State." *British Journal for the History of Science* 45, no. 2 (2012): 189–212.
Kendrick, Christopher. "The Imperial Laboratory: Discovering Forms in 'The New Atlantis.'" *English Literary History* 70, no. 4 (2003): 1021–42.
———. *Utopia, Carnival, and Commonwealth in Renaissance England*. Toronto: University of Toronto Press, 2004.
Kerrigan, William. *The Prophetic Milton*. Charlottesville: University Press of Virginia, 1974.
———. *The Sacred Complex: On the Psychogenesis of Paradise Lost*. Cambridge, MA: Harvard University Press, 1983.
Knapp, Jeffrey. *An Empire Nowhere: England, America, and Literature from Utopia to The Tempest*. Berkeley: University of California Press, 1992.
Knight, Jeffrey Todd. *Bound to Read: Compilations, Collections, and the Making of Renaissance Literature*. Philadelphia: University of Pennsylvania Press, 2013.
Kowalchuk, Kristine, ed. *Preserving on Paper: Seventeenth-Century Englishwomen's Receipt Books*. Toronto: University of Toronto Press, 2017.

Kripke, Saul A. *Naming and Necessity*. Cambridge, MA: Harvard University Press, 1980.
Kuhn, Thomas. "Mathematical Versus Experimental Traditions in the Development of Physical Science." In *The Essential Tension: Selected Studies in Scientific Tradition and Change*, 31–65. Chicago: University of Chicago Press, 1977.
Lacoue-Labarthe, Philippe. *Typography: Mimesis, Philosophy, Politics*. Ed. Christopher Fynsk. Cambridge, MA: Harvard University Press, 1989.
Laroche, Rebecca. *Medical Authority and Englishwomen's Herbal Texts, 1550–1650*. Burlington, VT: Ashgate, 2009.
Laroche, Rebecca, Elaine Leong, Jennifer Munroe, Hillary M. Nunn, Lisa Smith, and Amy L. Tigner. "Becoming Visible: Recipes in the Making." *Early Modern Women* 13, no. 1 (2018): 133–43.
Latour, Bruno. *Science in Action: How to Follow Scientists and Engineers Through Society*. Cambridge, MA: Harvard University Press, 1987.
———. *We Have Never Been Modern*. Trans. Catherine Porter. Cambridge, MA: Harvard University Press, 1993.
———. "Why Has Critique Run Out of Steam? From Matters of Fact to Matters of Concern." *Critical Inquiry* 30, no. 2 (2004): 225–48.
Lehtonen, Kelly. "The Abjection of Malbecco: Forgotten Identity in Spenser's Legend of Chastity." *Spenser Studies* 29 (2014): 179–96.
Lentricchia, Frank, and Andrew Dubois, eds. *Close Reading: The Reader*. Durham, NC: Duke University Press, 2003.
Leonard, John. *Naming in Paradise: Milton and the Language of Adam and Eve*. Oxford: Clarendon Press, 1990.
Leong, Elaine. *Recipes and Everyday Knowledge: Medicine, Science, and the Household in Early Modern England*. Chicago: University of Chicago Press, 2018.
Leslie, Marina. "Gender, Genre and the Utopian Body in Margaret Cavendish's *Blazing World*." *Utopian Studies* 7, no. 1 (1996): 6–24.
———. *Renaissance Utopias and the Problem of History*. Ithaca, NY: Cornell University Press, 1998.
Levao, Ronald. "Francis Bacon and the Mobility of Science." In "Seeing Science." Special issue, *Representations* 40 (1992): 1–32.
———. "Sidney's Feigned Apology." *PMLA* 94, no. 2 (1979): 223–33.
Levine, Caroline. *Forms: Whole, Rhythm, Hierarchy, Network*. Princeton, NJ: Princeton University Press, 2015.
Levinson, Ronald B. "Spenser and Bruno." *PMLA* 43, no. 3 (1928): 675–81.
Lewalski, Barbara Kiefer. *Protestant Poetics and the Seventeenth-Century Religious Lyric*. Princeton, NJ: Princeton University Press, 1979.
Lewis, David. *Counterfactuals*. Cambridge, MA: Harvard University Press, 1973. Rev. printing, Oxford: Blackwell, 1986.
———. *On the Plurality of Worlds*. New York: B. Blackwell, 1986.
Lewis, Rhodri. "Francis Bacon, Allegory and the Uses of Myth." *Review of English Studies* 61, no. 250 (2010): 360–89.
———. "Polychronic Macbeth." *Modern Philology* 117, no. 3 (2020): 323–46.
Lezra, Jacques. *Unspeakable Subjects: The Genealogy of the Event in Early Modern Europe*. Stanford, CA: Stanford University Press, 1997.
Lilley, Kate. "Blazing Worlds: Seventeenth-Century Women's Utopian Writing." In *Women, Texts and Histories, 1575–1760*, ed. Clare Brant and Diane Purkiss, 102–33. New York: Routledge, 1992.

Lipking, Lawrence. *What Galileo Saw: Imagining the Scientific Revolution*. Ithaca, NY: Cornell University Press, 2014.

Loewenstein, David. *Milton and the Drama of History: Historical Vision, Iconoclasm, and the Literary Imagination*. Cambridge: Cambridge University Press, 1990.

Long, Pamela O. *Openness, Secrecy, Authorship: Technical Arts and the Culture of Knowledge from Antiquity to the Renaissance*. Baltimore: Johns Hopkins University Press, 2001.

Loomba, Ania. *Shakespeare, Race, and Colonialism*. Oxford: Oxford University Press, 2002.

Lucking, David. "Imperfect Speakers: Macbeth and the Name of King." *English Studies* 87, no. 4 (2006): 415–25.

Lupton, Julia Reinhard. "Thinking with Things: Hannah Woolley to Hannah Arendt." *Postmedieval* 3, no. 1 (2012): 63–79.

Lyle, E. B. "The 'Twofold Balls and Treble Scepters' in *Macbeth*." *Shakespeare Quarterly* 28, no. 4 (1977): 516–19.

Lynch, William T. *Solomon's Child: Method in the Early Royal Society of London*. Stanford, CA: Stanford University Press, 2001.

MacDonald, Julia. "Demonic Time in Macbeth." *Ben Jonson Journal* 17, no. 1 (2010): 76–96.

Mack, Michael. *Sidney's Poetics: Imitating Creation*. Washington, DC: Catholic University of America Press, 2005.

MacLachlan, Hugh, and Philip B. Rollinson. "Magnanimity, Magnificence." In *The Spenser Encyclopedia*, gen. ed. A. C. Hamilton, 1175–77. Toronto: University of Toronto Press, 1990.

Madsen, William G. *From Shadowy Types to Truth: Studies in Milton's Symbolism*. New Haven, CT: Yale University Press, 1968.

Magnusson, Lynne. "A Play of Modals: Grammar and Potential Action in Early Shakespeare." *Shakespeare Survey* 62 (2009): 69–80.

———. "'What May Be and Should Be': Grammar Moods and the Invention of History in *1 Henry VI*." In *Shakespeare's World of Words*, ed. Paul Yachnin, 147–70. London: Arden Shakespeare, 2015.

Maltzahn, Nicholas von. "The Royal Society and the Provenance of Milton's *History of Britain* (1670)." *Milton Quarterly* 32, no. 3 (1998): 90–95.

Mann, Jenny C. *Outlaw Rhetoric: Figuring Vernacular Eloquence in Shakespeare's England*. Ithaca, NY: Cornell University Press, 2012.

———. *The Trials of Orpheus: Poetry, Science, and the Early Modern Sublime*. Princeton, NJ: Princeton University Press, 2021.

Mann, Jenny C., and Debapriya Sarkar, eds. "Imagining Early Modern Scientific Forms." Special issue, *Philological Quarterly* 98, nos. 1–2 (2019).

———. "Introduction: Capturing Proteus." In "Imagining Early Modern Scientific Forms," 1–22.

Marchitello, Howard. *The Machine in the Text: Science and Literature in the Age of Shakespeare and Galileo*. Oxford: Oxford University Press, 2011.

Marchitello, Howard, and Evelyn Tribble, eds. *The Palgrave Handbook of Early Modern Literature and Science*. London: Palgrave Macmillan, 2017.

Marjara, Harinder Singh. *Contemplation of Created Things: Science in Paradise Lost*. Toronto: University of Toronto Press, 1992.

Martin, Catherine Gimelli. "'What If the Sun Be Centre to the World': Milton's Epistemology, Cosmology, and Paradise of Fools Reconsidered." *Modern Philology* 99, no. 2 (2001): 231–65.

Martin, Julian. *Francis Bacon, the State, and the Reform of Natural Philosophy*. Cambridge: Cambridge University Press, 1992.

Mattison, Andrew. *Milton's Uncertain Eden: Understanding Place in Paradise Lost*. New York: Routledge, 2007.

Mazzio, Carla, ed. "Shakespeare and Science." Special issue, *South Central Review* 26, nos. 1–2 (2009).

McColley, Diane Kelsey. *Milton's Eve*. Urbana: University of Illinois Press, 1983.

McColley, Grant. "The Astronomy of *Paradise Lost*." *Studies in Philology* 34, no. 2 (1937): 209–47.

McEleney, Corey. *Futile Pleasures: Early Modern Literature and the Limits of Utility*. New York: Fordham University Press, 2017.

McIntyre, John P. "Sidney's 'Golden World.'" *Comparative Literature* 14, no. 4 (1962): 356–65.

McKeon, Michael. *The Secret History of Domesticity: Public, Private, and the Division of Knowledge*. Baltimore: Johns Hopkins University Press, 2005.

Medine, Peter E., John T. Shawcross, and David V. Urban, eds. *Visionary Milton: Essays on Prophecy and Violence*. Pittsburgh: Duquesne University Press, 2010.

Mehdizadeh, Nedda. "Cosmography and/in the Academy: Authorizing the Ideological Pathways of Empire." Unpublished manuscript, August 9, 2022.

Mentz, Steve. *Romance for Sale in Early Modern England: The Rise of Prose Fiction*. Burlington, VT: Ashgate, 2006.

Merchant, Carolyn. *The Death of Nature: Women, Ecology, and the Scientific Revolution*. San Francisco: Harper and Row, 1980.

Miller, Andrew H. "Implicative Criticism or the Display of Thinking." *New Literary History* 44, no. 3 (2013): 345–60.

———. "Lives Unled in Realist Fiction." *Representations* 98, no. 1 (2007): 118–34.

Miller, Jacqueline T. *Poetic License: Authority and Authorship in Medieval and Renaissance Contexts*. Oxford: Oxford University Press, 1986.

Miller, Shannon. *Engendering the Fall: John Milton and Seventeenth-Century Women Writers*. Philadelphia: University of Pennsylvania Press, 2008.

———. *Invested with Meaning: The Raleigh Circle in the New World*. Philadelphia: University of Pennsylvania Press, 1998.

Moore, Roger E. "Sir Philip Sidney's Defense of Prophesying." *Studies in English Literature, 1500–1900* 50, no. 1 (2010): 35–62.

Moss, Ann. *Printed Commonplace-Books and the Structuring of Renaissance Thought*. New York: Oxford University Press, 1996.

Mullaney, Steven. "Lying Like Truth: Riddle, Representation and Treason in Renaissance England." *English Literary History* 47, no. 1 (1980): 32–47.

Munroe, Jennifer. "First 'Mother of Science': Milton's Eve, Knowledge, and Nature." In *Ecofeminist Approaches to Early Modernity*, ed. Jennifer Munroe and Rebecca Laroche, 37–54. New York: Palgrave Macmillan, 2011.

Murphy, Erin. *Familial Forms: Politics and Genealogy in Seventeenth-Century English Literature*. Newark: University of Delaware Press, 2011.

Nardizzi, Vin. "Daphne Described: Ovidian Poetry and Speculative Natural History in Gerard's *Herball*." *Philological Quarterly* 98, nos. 1–2 (2019): 137–56.

Neely, Carol Thomas. "Women/Utopia/Fetish: Disavowal and Satisfied Desire in Margaret Cavendish's *New Blazing World* and Gloria Anzaldúa's *Borderlands/La Frontera*." In Siebers, *Heterotopia*, 58–95.

Newman, William R. *Promethean Ambitions: Alchemy and the Quest to Perfect Nature*. Chicago: University of Chicago Press, 2004.

Nicolson, Marjorie Hope. "Milton and the Telescope." *English Literary History* 2, no. 1 (1935): 1–32.

———. *Science and Imagination*. Ithaca, NY: Cornell University Press, 1956.
———. *Voyages to the Moon*. New York: Macmillan, 1948.
Nohrnberg, James. "Naming Milton's Eve." *Milton Studies* 60, nos. 1–2 (2018): 1–28.
Norbrook, David. "Margaret Cavendish and Lucy Hutchinson: Identity, Ideology and Politics." *In-between: Essays and Studies in Literary Criticism* 9, nos. 1–2 (2000): 179–203.
O'Neill, Eileen. Introduction to M. Cavendish, *Observations upon Experimental Philosophy*, x–xxxvi.
Ong, Walter J. *Ramus, Method, and the Decay of Dialogue: From the Art of Discourse to the Art of Reason*. Cambridge, MA: Harvard University Press, 1958.
Oram, William A. "Spenserian Paralysis." *Studies in English Literature, 1500–1900* 41, no. 1 (2001): 49–70.
Palfrey, Simon. *Shakespeare's Possible Worlds*. Cambridge: Cambridge University Press, 2014.
Park, Jennifer. "Artisans of the Skin: Recipe Culture, Surface Thinking, and 17th-Century Racial Embodiment." Paper presented at the MLA Annual Convention, Seattle, WA, January 9, 2020.
Park, Katharine. "Bacon's 'Enchanted Glass.'" *Isis* 75, no. 2 (1984): 290–302.
———. "Response to Brian Vickers, 'Francis Bacon, Feminist Historiography, and the Dominion of Nature.'" *Journal of the History of Ideas* 69, no. 1 (2008): 143–46.
Parker, Patricia. *Inescapable Romance: Studies in the Poetics of a Mode*. Princeton, NJ: Princeton University Press, 1979.
Partee, Morriss Henry. "Sir Philip Sidney and the Renaissance Knowledge of Plato." *English Studies* 51, no. 5 (1970): 411–24.
Passannante, Gerard. *The Lucretian Renaissance: Philology and the Afterlife of Tradition*. Chicago: University of Chicago Press, 2011.
Paster, Gail Kern. *The Body Embarrassed: Drama and the Disciplines of Shame in Early Modern England*. Ithaca, NY: Cornell University Press, 1993.
Patey, Douglas Lane. *Probability and Literary Form: Philosophic Theory and Literary Practice in the Augustan Age*. Cambridge: Cambridge University Press, 1984.
Patterson, Annabel. "Imagining New Worlds: Milton, Galileo, and the 'Good Old Cause.'" In *The Witness of Times: Manifestations of Ideology in Seventeenth Century England*, ed. Katherine Z. Keller and Gerald J. Schiffhorst, 238–60. Pittsburgh: Duquesne University Press, 1993.
Pavel, Thomas G. *Fictional Worlds*. Cambridge, MA: Harvard University Press, 1986.
Paxson, James J. "The Allegory of Temporality and the Early Modern Calculus." *Configurations* 4, no. 1 (1996): 39–66.
Pérez-Ramos, Antonio. *Francis Bacon's Idea of Science and the Maker's Knowledge Tradition*. New York: Oxford University Press, 1988.
Pesic, Peter. "Proteus Unbound: Francis Bacon's Successors and the Defense of Experiment." *Studies in Philology* 98, no. 4 (2001): 428–56.
Picciotto, Joanna. *Labors of Innocence in Early Modern England*. Cambridge, MA: Harvard University Press, 2010.
———. "Reforming the Garden: The Experimentalist Eden and *Paradise Lost*." *English Literary History* 72, no. 1 (2005): 23–78.
Pietras, Brian. "Erasing Evander's Mother: Spenser, Virgil, and the Dangers of Vatic Authorship." *Spenser Studies* 31–32 (2018): 43–69.
Plantinga, Alvin. "Actualism and Possible Worlds." *Theoria* 42, nos. 1–3 (1976): 139–60.
Pohl, Nicole. "'Of Mixt Natures': Questions of Genre in Margaret Cavendish's *The Blazing World*." In Clucas, *Princely Brave Woman*, 51–68.

Poole, William. "Francis Godwin, Henry Neville, Margaret Cavendish, H. G. Wells: Some Utopian Debts." *ANQ: A Quarterly Journal of Short Articles, Notes, and Reviews* 16, no. 3 (2003): 12–18.

———. "General Introduction." In Francis Lodwick, *A Country Not Named (MS. Sloane 913, Fols. 1r–33r): An Edition with an Annotated Primary Bibliography and an Introductory Essay on Lodwick and His Intellectual Context*, 3–67. Tempe: Arizona Center for Medieval and Renaissance Studies, 2007.

———. "Milton and Science: A Caveat." *Milton Quarterly* 38, no. 1 (2004): 18–34.

Preston, Claire. *The Poetics of Scientific Investigation in Seventeenth-Century England*. Oxford: Oxford University Press, 2015.

Price, Bronwen. Introduction to *Francis Bacon's New Atlantis: New Interdisciplinary Essays*, ed. Bronwen Price, 1–27. Manchester: Manchester University Press, 2002.

———. "Journeys Beyond Frontiers: Knowledge, Subjectivity and Outer Space in Margaret Cavendish's *The Blazing World* (1666)." In *The Arts of 17th-Century Science: Representations of the Natural World in European and North American Culture*, ed. Claire Jowitt and Diane Watt, 127–45. Aldershot: Ashgate, 2002.

Quilligan, Maureen. *The Language of Allegory: Defining the Genre*. Ithaca, NY: Cornell University Press, 1979.

Quint, David. *Epic and Empire: Politics and Generic Form from Virgil to Milton*. Princeton, NJ: Princeton University Press, 1993.

Quitslund, Jon A. *Spenser's Supreme Fiction: Platonic Natural Philosophy and the Faerie Queene*. Toronto: University of Toronto Press, 2001.

Ramachandran, Ayesha. "Edmund Spenser, Lucretian Neoplatonist: Cosmology in the *Fowre Hymnes*." In Borris, Kaske, and Quitslund, "Spenser and Platonism," 373–411.

———. "Mutabilitie's Lucretian Metaphysics: Scepticism and Cosmic Process in Spenser's Cantos." In *Celebrating Mutabilitie: Essays on Edmund Spenser's Mutabilitie Cantos*, ed. Jane Grogan, 220–45. Manchester: Manchester University Press, 2010.

———. *The Worldmakers: Global Imagining in Early Modern Europe*. Chicago: University of Chicago Press, 2015.

Ramachandran, Ayesha, and Melissa E. Sanchez, eds. "Spenser and 'the Human.'" Special issue, *Spenser Studies* 30 (2015).

Raman, Shankar. "Milton, Leibniz, and the Measure of Motion." In Marchitello and Tribble, *Palgrave Handbook of Early Modern Literature and Science*, 277–93.

Read, David. *Temperate Conquests: Spenser and the Spanish New World*. Detroit: Wayne State University Press, 2000.

Rees, Emma L. E. *Margaret Cavendish: Gender, Genre, Exile*. New York: Manchester University Press, 2003.

Roche, Thomas P. *The Kindly Flame: A Study of the Third and Fourth Books of Spenser's Faerie Queene*. Princeton, NJ: Princeton University Press, 1964.

Rogers, John. *The Matter of Revolution: Science, Poetry, and Politics in the Age of Milton*. Ithaca, NY: Cornell University Press, 1996.

Rosenbaum, Jerrod. "Spenser's Merlin Rehabilitated." *Spenser Studies* 29 (2014): 149–78.

Rosenfeld, Colleen Ruth. *Indecorous Thinking: Figures of Speech in Early Modern Poetics*. New York: Fordham University Press, 2018.

———. "In the Mood of Fiction." *English Literary Renaissance* 52, no. 3 (2022): 385–96.

———. "Opening Remarks." Paper presented at the Renaissance Project Symposium [online], Pomona College, Claremont, CA, June 1, 2021.

———. "Poetry and the Potential Mood: The Counterfactual Form of Ben Jonson's 'To Fine Lady Would-Be'." *Modern Philology* 112, no. 2 (2014): 336–57.

Roychoudhury, Suparna. *Phantasmatic Shakespeare: Imagination in the Age of Early Modern Science.* Ithaca, NY: Cornell University Press, 2018.

Royster, Francesca. "Riddling Whiteness, Riddling Certainty: Roman Polanski's *Macbeth*." In *Weyward Macbeth: Intersections of Race and Performance*, ed. Scott L. Newstok and Ayanna Thompson, 173–81. New York: Palgrave Macmillan, 2010.

Ryan, Marie-Laure. *Possible Worlds, Artificial Intelligence, and Narrative Theory.* Bloomington: Indiana University Press, 1991.

Sarasohn, Lisa T. *The Natural Philosophy of Margaret Cavendish: Reason and Fancy During the Scientific Revolution.* Baltimore: Johns Hopkins University Press, 2010.

———. "A Science Turned Upside Down: Feminism and the Natural Philosophy of Margaret Cavendish." *Huntington Library Quarterly* 47, no. 4 (1984): 289–307.

Sarkar, Debapriya. "Dilated Materiality and Formal Restraint in *The Faerie Queene*." *Spenser Studies* 31 (2018): 137–66.

———. "Imagining Early Modern Wish-Lists and Their Environs." In *Object-Oriented Environs*, ed. Jeffrey Jerome Cohen and Julian Yates, 123–33. New York: Punctum Books, 2016.

———. "Literary Justice: The Participatory Ethics of Early Modern Possible Worlds." In *Teaching Social Justice Through Shakespeare: Why Renaissance Literature Matters Now*, ed. Hillary Eklund and Wendy Beth Hyman, 174–84. Edinburgh: Edinburgh University Press, 2019.

———. "Shakespeare and the Social History of Truth." *Shakespeare Studies* 49 (2021): 94–106.

———. "The Utopian Hypothesis." *English Literary Renaissance* 52, no. 3 (2022): 371–84.

Sawday, Jonathan. *The Body Emblazoned: Dissection and the Human Body in Renaissance Culture.* New York: Routledge, 1995.

Schmidgen, Wolfram. *Exquisite Mixture: The Virtues of Impurity in Early Modern England.* Philadelphia: University of Pennsylvania Press, 2013.

Schoenfeldt, Michael Carl. *Bodies and Selves in Early Modern England: Physiology and Inwardness in Spenser, Shakespeare, Herbert, and Milton.* Cambridge: Cambridge University Press, 1999.

Schwartz, Regina M. "From Shadowy Types to Shadowy Types: The Unendings of *Paradise Lost*." *Milton Studies* 24 (1988): 123–39.

———. *Remembering and Repeating: Biblical Creation in Paradise Lost.* Cambridge: Cambridge University Press, 1988.

Shahani, Gitanjali G. *Tasting Difference: Food, Race, and Cultural Encounters in Early Modern Literature.* Ithaca, NY: Cornell University Press, 2020.

Shapin, Steven. "The House of Experiment in Seventeenth-Century England." *Isis* 79, no. 3 (1988): 373–404.

———. "The Invisible Technician." *American Scientist* 77, no. 6 (1989): 554–63.

———. *The Scientific Revolution.* Chicago: University of Chicago Press, 1996.

———. *A Social History of Truth: Civility and Science in Seventeenth-Century England.* Chicago: University of Chicago Press, 1994.

Shapin, Steven, and Simon Schaffer. *Leviathan and the Air–Pump: Hobbes, Boyle, and the Experimental Life.* Princeton, NJ: Princeton University Press, 1985.

Shapiro, Barbara J. *Probability and Certainty in Seventeenth-Century England: A Study of the Relationships Between Natural Science, Religion, History, Law, and Literature.* Princeton, NJ: Princeton University Press, 1983.

Sherman, Sandra. "Trembling Texts: Margaret Cavendish and the Dialectic of Authorship." *English Literary Renaissance* 24, no. 1 (1994): 184–210.

Siebers, Tobin, ed. *Heterotopia: Postmodern Utopia and the Body Politic*. Ann Arbor: University of Michigan Press, 1995.

———. "Introduction: What Does Postmodernism Want? Utopia." In Siebers, *Heterotopia*, 1–38.

Siegfried, Brandie R. "Anecdotal and Cabalistic Forms in *Observations upon Experimental Philosophy*." In *Authorial Conquests: Essays on Genre in the Writings of Margaret Cavendish*, ed. Line Cottegnies and Nancy Weitz, 59–79. Madison, NJ: Fairleigh Dickinson University Press, 2003.

Simon, David Carroll. *Light Without Heat: The Observational Mood from Bacon to Milton*. Ithaca, NY: Cornell University Press, 2018.

Sinfield, Alan. *Faultlines: Cultural Materialism and the Politics of Dissident Reading*. Berkeley: University of California Press, 1992.

———. *Colonial Narratives / Cultural Dialogues: "Discoveries" of India in the Language of Colonialism*. London: Routledge, 1996.

Singh, Jyotsna G., ed. *A Companion to the Global Renaissance: English Literature and Culture in the Era of Expansion*. Malden, MA: Wiley-Blackwell, 2009.

Skinner, Quentin. *Forensic Shakespeare*. Oxford: Oxford University Press, 2014.

Skulsky, Harold. *Metamorphosis: The Mind in Exile*. Cambridge, MA: Harvard University Press, 1981.

Slater, Michael. "Spenser's Poetics of 'Transfixion' in the Allegory of Chastity." *Studies in English Literature, 1500–1900* 54, no. 1 (2014): 41–58.

Smith, Emma. *The Cambridge Introduction to Shakespeare*. Cambridge: Cambridge University Press, 2007.

Smith, Ian. *Race and Rhetoric in the Renaissance: Barbarian Errors*. New York: Palgrave Macmillan, 2009.

Smith, Pamela H. *The Body of the Artisan: Art and Experience in the Scientific Revolution*. Chicago: University of Chicago Press, 2004.

Smyth, Adam. "Commonplace Book Culture: A List of Sixteen Traits." In *Women and Writing, c. 1340–c. 1650: The Domestication of Print Culture*, ed. Anne Lawrence-Mathers and Phillipa Hardman, 90–110. Woodbridge: York Medieval, with Boydell, 2010.

Snider, Alvin. "Francis Bacon and the Authority of Aphorism." *Prose Studies* 11, no. 2 (1988): 60–71.

Solomon, Julie Robin. "'To Know, To Fly, To Conjure': Situating Baconian Science at the Juncture of Early Modern Modes of Reading." *Renaissance Quarterly* 44, no. 3 (1991): 513–58.

Spiller, Elizabeth. "Milton, the Poetics of Matter, and the Sciences of Reading." In Marchitello and Tribble, *Palgrave Handbook of Early Modern Literature and Science*, 159–77.

———. *Science, Reading, and Renaissance Literature: The Art of Making Knowledge, 1580–1670*. Cambridge: Cambridge University Press, 2004.

———. *Seventeenth-Century English Recipe Books: Cooking, Physic and Chirurgery in the Works of Elizabeth Grey and Aletheia Talbot*. Burlington, VT: Ashgate, 2008.

Stagl, Justin. *A History of Curiosity: The Theory of Travel, 1500–1800*. Chur, Switzerland: Harwood Academic, 1995.

Stallybrass, Peter. "Macbeth and Witchcraft." In *Focus on Macbeth*, ed. John Russell Brown, 189–209. Boston: Routledge, 1982.

Stephens, James. "Bacon's New English Rhetoric and the Debt to Aristotle." *Speech Monographs* 39, no. 4 (1972): 248–59.

———. "Science and the Aphorism: Bacon's Theory of the Philosophical Style." *Speech Monographs* 37, no. 3 (1970): 157–71.

Sugimura, N. K. *"Matter of Glorious Trial": Spiritual and Material Substance in Paradise Lost.* New Haven, CT: Yale University Press, 2009.
Svendsen, Kester. *Milton and Science.* Cambridge, MA: Harvard University Press, 1956.
Teskey, Gordon. *Allegory and Violence.* Ithaca, NY: Cornell University Press, 1996.
Thomas, Keith. *Religion and the Decline of Magic.* New York: Scribner, 1971.
Thompson, Ayanna. *Performing Race and Torture on the Early Modern Stage.* New York: Routledge, 2008.
Thornton, Tim. *Prophecy, Politics and the People in Early Modern England.* Rochester: Boydell, 2006.
Trubowitz, Rachel J. "Body Politics in *Paradise Lost.*" *PMLA* 121, no. 2 (2006): 388–404.
———. "The Reenchantment of Utopia and the Female Monarchical Self: Margaret Cavendish's *Blazing World.*" *Tulsa Studies in Women's Literature* 11, no. 2 (1992): 229–45.
Turner, Henry S. *The Corporate Commonwealth: Pluralism and Political Fictions in England, 1516–1651.* Chicago: University of Chicago Press, 2016.
———. *The English Renaissance Stage: Geometry, Poetics, and the Practical Spatial Arts, 1580–1630.* Oxford: Oxford University Press, 2006.
Ulreich, John C. "Two Great World Systems: Galileo, Milton, and the Problem of Truth." *Cithara: Essays in the Judaeo-Christian Tradition* 43, no. 1 (2003): 25–36.
Vaihinger, H. *The Philosophy of "As If."* London: K. Paul, Trench, Trubner, 1924.
Vickers, Brian. "Bacon and Rhetoric." In *The Cambridge Companion to Bacon*, ed. Markku Peltonen, 200–231. Cambridge: Cambridge University Press, 1996.
———. "Francis Bacon, Feminist Historiography, and the Dominion of Nature." *Journal of the History of Ideas* 69, no. 1 (2008): 117–41.
Wagner, Geraldine. "Romancing Multiplicity: Female Subjectivity and the Body Divisible in Margaret Cavendish's *Blazing World.*" *Early Modern Literary Studies* 9, no. 1 (2003). http://purl.oclc.org/emls/09-1/wagnblaz.htm.
Walker, William. "Milton's Dualistic Theory of Religious Toleration in *A Treatise of Civil Power, Of Christian Doctrine*, and *Paradise Lost.*" *Modern Philology* 99, no. 2 (2001): 201–30.
———. "Typology and *Paradise Lost*, Books XI and XII." *Milton Studies* 25 (1989): 245–64.
Wall, Wendy. *Recipes for Thought: Knowledge and Taste in the Early Modern English Kitchen.* Philadelphia: University of Pennsylvania Press, 2015.
Wall-Randell, Sarah. *The Immaterial Book: Reading and Romance in Early Modern England.* Ann Arbor: University of Michigan Press, 2013.
Walsh, Brian. "'Deep Prescience': Succession and the Politics of Prophecy in *Friar Bacon and Friar Bungay.*" *Medieval and Renaissance Drama in England* 23 (2010): 63–85.
———. *Shakespeare, the Queen's Men, and the Elizabethan Performance of History.* Cambridge: Cambridge University Press, 2009.
Walters, Lisa. *Margaret Cavendish: Gender, Science and Politics.* Cambridge: Cambridge University Press, 2014.
Webster, Charles. *The Great Instauration: Science, Medicine, and Reform, 1626–1660.* 2nd ed. New York: Peter Lang, 2002.
Webster, Erin. "Milton's Pandæmonium and the Infinitesimal Calculus." *English Literary Renaissance* 45, no. 3 (2015): 425–58.
Weinberg, Bernard. *A History of Literary Criticism in the Italian Renaissance.* Chicago: University of Chicago Press, 1961.
Wiggins, Martin, with Catherine Richardson. *British Drama, 1533–1642: A Catalogue.* 8 vols. Oxford: Oxford University Press, 2015.

Williams, Katherine Schaap. "'Strange Virtue': Staging Acts of Cure." In *Disability, Health, and Happiness in the Shakespearean Body*, ed. Sujata Iyengar, 93–108. New York: Routledge, 2014.

———. *Unfixable Forms: Disability, Performance, and the Early Modern English Theater.* Ithaca, NY: Cornell University Press, 2021.

Williams, Travis D. "Unspeakable Creation: Writing in *Paradise Lost* and Early Modern Mathematics." *Philological Quarterly* 98, nos. 1–2 (2019): 181–200.

Witmore, Michael. *Culture of Accidents: Unexpected Knowledges in Early Modern England.* Stanford, CA: Stanford University Press, 2001.

Wittreich, Joseph Anthony, Jr. *Visionary Poetics: Milton's Tradition and His Legacy.* San Marino, CA: Huntington Library, 1979.

Wofford, Susanne L. "Britomart's Petrarchan Lament: Allegory and Narrative in *The Faerie Queene* III, iv." *Comparative Literature* 39, no. 1 (1987): 28–57.

———. *The Choice of Achilles: The Ideology of Figure in the Epic.* Stanford, CA: Stanford University Press, 1992.

Woolf, Virginia. *A Room of One's Own.* In *The Norton Anthology of English Literature*, gen. ed. M. H. Abrams 7th ed. 2 vols. New York: W. W. Norton, 2000.

Yeats, William Butler. *The Cutting of an Agate.* New York: Macmillan, 1912.

Yusoff, Kathryn. *A Billion Black Anthropocenes or None.* Minneapolis: University of Minnesota Press, 2019.

Zavala, Silvio. *Sir Thomas More in New Spain: A Utopian Adventure of the Renaissance.* London: Hispanic and Luso-Brazilian Councils, 1955.

Zurcher, Amelia. "Serious Extravagance: Romance Writing in Seventeenth-Century England." *Literature Compass* 8, no. 6 (2011): 376–89.

INDEX

ACKNOWLEDGMENTS

At its heart, this is a book about the imaginative leaps that enable us to envision what we think is possible. I am grateful to the mentors, colleagues, friends, and family who encouraged me to take my own imaginative leaps, whose advice has reshaped my thinking, and whose care has sustained me through the process of completing this book.

Heather Dubrow transformed everything. She welcomed a questioning engineering student into her office in Madison and shared so much of her time and expertise as I grappled with what felt like a radical decision. Her guidance made possible my journey from the applied sciences into literary studies. Also at Madison, Henry Turner helped me realize that what then felt like a radical change was, in reality, a continuation: our early conversations attuned me to the fact that I was more interested in asking how scientific knowledge comes into being rather than what "science" does today. This insight shaped this project from its inception, and I consider myself extremely fortunate that I could continue these discussions with Henry in the course of my graduate career at Rutgers. Henry modeled for me how to think capaciously, and his enthusiasm for my ideas invigorated my own thinking. I cannot fathom this project existing without the time and energy he devoted to it.

My teachers in the English Department at Rutgers transformed my perceptions about what literature does, and what literary studies can do on its behalf. Ann Coiro told me which texts to read, Jacqueline Miller taught me how to read those texts, and Emily Bartels showed me that writing could help me see them in fresh ways. Yet, their brilliance and intellectual generosity tell only part of the story. Jackie's critical acumen and skepticism have made me a better scholar; her enthusiasm and care have nurtured me in immeasurable ways. I am forever indebted to Ann for her infectious enthusiasm and for never letting me lose sight of the joys of our intellectual pursuits. Emily's wit, humor, and encouragement as I navigated new cultural and professional spaces continue to guide and inspire me.

The Medieval-Renaissance Colloquium at Rutgers was a formative setting, and I am grateful to Ronald Levao, Thomas Fulton, Sarah Novacich, Stacy Klein, and Larry Scanlon for their advice. I also owe a debt of gratitude to the faculty at Rutgers whose guidance has informed this book and my orientation to the profession: Alastair Bellany, Lynn Festa, Christopher Iannini, Colin Jager, Ann Jurecic, David Kurnick, Meredith McGill, Michael McKeon, Sonali Perera, Stéphane Robolin, Margaret Ronda, Evie Shockley, Rebecca Walkowitz, and Cheryl Wall. I am immensely grateful to James Delbourgo for indulging my many questions on histories of science, for his deep investment in my ideas, and for his confidence in my ability to enter multiple critical discourses.

The graduate program at Rutgers was a vibrant intellectual space and social community, and I especially thank Sarah Balkin, Courtney Borack, Tyler Bradway, Mark DiGiacomo, Greg Ellermann, Joshua Fesi, Octavio González, Stephanie Hunt, Erin Kelly, Naomi Levine, Philip Longo, Lizzie Oldfather, Brian Pietras, Cheryl Robinson, John Savarese, and Scott Trudell. I am immensely grateful for the friendship of Aditi Gupta, Jennifer Raterman, and Sarah Rodgers, whose brilliance inspired me during graduate school and whose care continues to sustain me.

I thank my colleagues at Hendrix College for making me grow as a writer, thinker, and teacher. I am especially grateful to Kristi McKim and Hope Coulter for their profound care and compassion, to Tyrone Jaeger for his humor and perspective, to Jessica Jacobs for conversations that sustained me in ways that are impossible to document, to Toni Wall Jaudon, Giffen Maupin, and Josh Glick for being such generous interlocutors. I thank Andrea Duina, Alice Hines, Dorian Stuber, and Alex Vernon, for their continual guidance as I navigated a new institution. I am indebted to Anne Goldberg, Sasha Pfau, Marjorie Swann, Henryetta Vanaman, Carol West, and Sarah Engeler-Young for their support, and to Mark Barr, Peter Kett, Liz Lundeen, Mario Muscedere, Julie Nicol, and Mel White for their friendship. To Sarah Grant, Nickole Brown, and Kris McAbee, thanks for instilling such joy by your companionship!

At the University of Connecticut, I am grateful for the warmth and generosity of colleagues who have supported me in the final stages of writing this book. I am especially grateful to Gregory Kneidel, Gregory Semenza, and Evelyn Tribble, who read the manuscript in its entirety and have been continuous sources of support. I am deeply indebted to my Avery Point English colleagues, Pamela Bedore and Mary K. Bercaw Edwards, for all of their help—their guidance has meant the world, and I feel extremely fortunate to work alongside them. I thank Pamela Allen Brown, Jeffrey Shoulson, and Brendan Kane for welcoming me into the vibrant early modern community at UConn, and to Mary Cygan and Annamaria Csizmadia for welcoming me into the community at Stamford. For their collegiality and mentorship, I extend my gratitude to Margaret Breen, Katherine Capshaw, Dwight Codr, Eleni Coundouriotis, Anna Mae Duane, Wayne Franklin, Serkan Görkemli, Kathy Knapp, Charles Mahoney, Jean Marsden, Penelope Pelizzon, Thomas Recchio, Frederick Roden, Victoria Ford Smith, Fiona Somerset, and Sarah Winter; a special thanks to Yohei Igarashi, Grégory Pierrot, and Bhakti Shringarpure for their moral support, and for sharing generously of their time and experiences. I thank Robert Hasenfratz and Clare Costley King'oo for their support and advocacy in their roles as department head, and to Melanie Hepburn and Peter Carcia for their guidance and unfailing cheer. My Maritime Studies colleagues have modeled for me what it means to work in—indeed thrive in—an interdisciplinary community, and I thank in particular Syma Ebbin, Matthew McKenzie, Nathaniel Trumball, Michele Baggio, and Kroum Batchvarov. At Avery Point, I thank Daniel Mercier, Noemi Maldonado Picardi, Cynthia Bernardo, Laurie Wolfley, Mark Bond, Janene Vandi, and Annemarie Seifert. My gratitude to Helen Rozwadowski and Alexis Boylan knows no bounds—they welcomed me into varied communities at UConn, transformed how I approach cross-disciplinary research, and offered unflinching support.

While so much of research and writing can be isolated and isolating, the process of completing this book has been sustained by various communities. I thank mentors and friends who have given selflessly of their time; the more time I spend in this profession, the more I marvel at the generosity of spirit and the incalculable labor that have made my own work—and existence—in this field possible. Kathryn Vomero Santos wrote virtually with me (at any and all hours); Wendy Beth Hyman offered invaluable insight into early versions of the chapters; Vanessa Corredera's virtual companionship kept me writing in what often felt like impossible circumstances. Karen Raber's ardent support and firm belief gave me much-needed confidence, and her reminders to let go have always been perfectly timed. Mary Thomas Crane has shared her expertise for many years, and her clarity about the scope of my research has always helped me perceive it in fresh ways. Mary Baine Campbell offered incisive interventions into the project in its early stages, and I am beyond delighted this work took shape in dialogue with a thinker whose scholarship first modeled for me what literature/science studies could be. Ayanna Thompson has been a source of wisdom, and her reminders about how to navigate the seemingly impossible balance of personal and professional life have provided much-needed perspective. Being part of a writing group with Caralyn Bialo, Adhaar Desai, David Hershinow, Laura Kolb, Lauren Robertson, and Steven Swarbrick for the last several years has energized my thinking and sustained my writing. For their active engagement with the book's argument, I thank J. K. Barret, Liza Blake, Stephanie Elsky, Samuel Fallon, Timothy Harrison, Natasha Korda, Rebecca Laroche, Tara Lyons, Jennifer Munroe, Tessie Prakas, Benedict Robinson, Marjorie Rubright, Elizabeth Spiller, Jacqueline Wernimont, and Tiffany Werth; I am especially grateful to Brandi K. Adams and Ambereen Dadabhoy for their keen insight and enthusiasm—thinking with them about the broader implications of this project has made the book feel new again.

I owe my deepest thanks to four friends who continue to astonish me with their absolute care for me and for my work: Corey McEleney read and reread pieces at a moment's notice and always knew how to make my writing better; Colleen Rosenfeld engaged in uncountable conversations and read many drafts, and her brilliance inflects every aspect of my thinking on literary methodologies; Hillary Eklund has been an exceptional intellectual interlocutor and transformed how I want to pursue research; and Katherine Schaap Williams has for over a decade known exactly how to give form to my tangled ideas and is always available to think with me at the level of word, sentence, paragraph, and chapter.

This book has been shaped in immeasurable ways by the engagement of colleagues and friends who raised questions, shared resources, and offered encouragement. I thank in particular Patricia Akhimie, Pavneet Aulakh, Chris Barrett, Lara Bovilsky, Dennis Britton, Rebecca Bushnell, Joseph Campana, Urvashi Chakravarty, Jeffrey J. Cohen, Sara Coodin, Julie Crawford, Kevin Curran, Jane Hwang Degenhardt, Lara Dodds, Katherine Eggert, Ruben Espinosa, Nahyan Fancy, Mary Floyd-Wilson, Miles Grier, Kim F. Hall, Matthew Harrison, Stephanie Hershinow, Gavin Hollis, Phebe Jensen, Mira Assaf Kafantaris, Farah Karim-Cooper, Carol Mejia LaPerle, Caroline Levine, Erika Lin, Howard Marchitello, Carla Mazzio, Nedda Mehdizadeh, Steve Mentz, Erin Murphy, Noémie Ndiaye, Scott Newstok, Jennifer Park, Ayesha Ramachandran, Shankar Raman, Jessica Rosenberg, Anita Sherman, Ann Thompson, Samuel Truett, Maggie Vinter, Jennifer Waldron, Sarah Wall-Randell, Geoffrey Way, Travis Williams, Susanne Wofford, Julian Yates, Lehua Yim, and Adam Zucker. My students at Rutgers, Hendrix, and UConn have left indelible marks through their questions, enthusiasm, and even their skepticism—I especially thank participants of "The Early Modern Scientific Imagination" graduate seminar at UConn and the "Early Modern Possible Worlds" seminar at Hendrix for expanding my understanding on topics I thought I knew so well.

It is a pleasure to express my thanks to the institutions and organizations that have supported my work with material resources, access to archives, and time and space that made researching, writing, and thinking possible. These institutional spaces have also provided the intangible gifts of community that made thinking pleasurable. The earliest versions of the project bear the imprint of stimulating conversations at the Folger Institute's Researching the Archives seminar (led by James Siemon and Keith Wrightson), the Center for Cultural Analysis at Rutgers University (in the seminar "Public Knowledge" led by Meredith McGill and Henry Turner), the Rutgers Center for Historical Analysis (in the seminar "Networks of Exchange" led by James Delbourgo and Toby Jones), and the Rutgers English Department's Dissertation Writing Seminar (led by David Kurnick). A Mellon/ACLS Dissertation Completion Fellowship provided me the time to first conceive of this project in its entirety.

I am immensely grateful to the Folger Shakespeare Library for a National Endowment for the Humanities/Folger Shakespeare Library Long-Term Fellowship . I thank Amanda Herbert, Kathleen Lynch, and Owen Williams for their unfailing support of the project, Michael Witmore for his enthusiastic engagement with my ideas, and (the late) Betsy Walsh, LuEllen DeHaven, Caroline Duroselle-Melish, Rosalind Larry, Camille Seerattan, Abbie Weinberg, and Heather Wolfe for their help with and in the archives. I extend my deepest gratitude to my brilliant cohort of fellows whose questions gave direction to a project that seemed newly amorphous: Anston Bosman, Michelle DiMeo, Derek Dunne, Jessica Goethals, Marissa Greenberg, Megan Heffernan, Carmen Nocentelli, Joseph Ortiz, and Holly Crawford-Pickett. I am thrilled that this book is being published with the Folger imprint!

A yearlong fellowship at the University of Connecticut Humanities Institute (UCHI) enabled me to think about the interdisciplinary implications of the project, the Felberbaum Family Faculty Award allowed me to complete required research, and the Humanities Book Support Award provided support at the final stages of publication." My thanks, in particular, to Michael Lynch, Nasya Al-Saidy, Jo-Ann Waide, Morgne Cramer, Joseph Ulatowski, Kornel Chang, Andrea Celli, Nathan Braccio, Hayley Stefan, and Laura Godfrey. I thank the English Department at UConn for its generous support of a manuscript workshop, and I am immeasurably grateful to Mary Thomas Crane and Melissa Sanchez for reading the entire manuscript. Their incisive comments inflect all aspects of this book, and their interventions enabled me to see the final shape of my argument.

There could not have been a more idyllic place to complete the book than the Huntington, and I am grateful for the Fletcher Jones Foundation Research

Fellowship that provided me the time I needed to fully synthesize my ideas. I thank Steve Hindle for his enthusiastic support of my work and Juan Gomez, Natalie Serrano, and Catherine Wehrey-Miller for all their help. My deepest thanks to my fellowship cohort whose questions made the project stronger and whose reassurances bolstered me during the final stages of revision. I am especially grateful to Wendy Wall for helping me parse complicated ideas at this stage; to Alyssa Collins, Ardeta Gjikola, Aminah Hasan-Birdwell, and Nydia Pineda de Ávila for a year of conversations on histories of science and knowledge; to Lydia Barnett, Anthony Grafton, Joel Harrington, and Eileen Reeves for their wisdom; and to James Chandler, Raúl Coronado, Kathleen Donegan, Jonathan Koch, Lisa Mendelman, Stevie Ruiz, Elena Schneider, and Ula Taylor for their intellectual camaraderie and moral support.

Portions of this research have been presented at the Department of English Brown Bag Series at UConn, the Early Modern Working Group at UConn, the Folger, the Humanities Studio at Pomona College, the Huntington, the Medieval-Renaissance Colloquium at Rutgers, Scientiae, the UCHI, the University of Oklahoma's History of Science Colloquium, the University of Texas at Austin's Early Modern Temporalities Series, and the Wesleyan Renaissance Seminar. I thank the organizers of these events for inviting me to share my research and to the audiences for their feedback. I also extend my thanks to organizers, panelists, seminar members, and audiences at the American Comparative Literature Association, Conference on John Milton, Group for Early Modern Cultural Studies, the International Margaret Cavendish Society Biennial Meeting, Modern Language Association, Renaissance Society of America, Shakespeare Association of America, and Society for Literature, Science and the Arts Conference, who offered insights at various stages of development. A portion of Chapter 2 appeared as "'To crown my thoughts with acts': Prophecy and Prescription in *Macbeth*," in *Macbeth: The State of Play, Arden Shakespeare* (an imprint of Bloomsbury Publishing), ed. Ann Thompson (London: Bloomsbury Publishing, April 2014), 83–106; the material is reproduced by permission of Bloomsbury. Chapter 5 is derived from an article titled "'Sad Experiment' in *Paradise Lost*: Epic Knowledge and Evental Poetics," published in *Exemplaria: A Journal of Theory in Medieval & Renaissance Studies* 26, no. 4 (Winter 2014), copyright Taylor and Francis, available online: https://www.tandfonline.com/ and https://doi.org/10.1179/1041257314Z.00000000059.

At Penn Press, I am immensely grateful to Jenny Tan for her belief in the project from the beginning and for her encouragement and cheer throughout. I thank Lily Palladino, Kristen Bettcher, and Jennifer Backer for their support during the production process. I am indebted to the inimitable Puck Fletcher for their keen attention in the final stages of editing, to Christopher Bolster for his research assistance, and to Kavita Mudan Finn for her brilliant and perceptive indexing. I am delighted to express my thanks to Jenny C. Mann and Vin Nardizzi, who read the book for the press and revealed their identities afterward. Their comments got to the heart of the argument and helped me see the book's fullest potential, and I am especially thankful for their willingness to serve as generous interlocutors and their continued guidance.

My deepest thanks to my friends and family for their unqualified emotional support, for their constant care about my well-being, and for making sure I never felt the distance from home even continents away. My work as a literary scholar would have been unimaginable to my extended family, yet they have continued to be towers of strength, offering the vociferous encouragement as only (I am biased, I realize) Bengali families can offer. I am forever indebted to Rathindranath Sarkar, Sandhya Sarkar, Purnendu Datta, Shikha Dutta, Chandra Bhanu Sarkar, Susmita Sarkar, Ashish Sarkar, Kaberi Sarkar, Susmita Mallick, and Udita Biswas. I am grateful to Nivedita Budhalakoti and Shailendra Budhalakoti for their unabashed pride at every stage of my professional journey. Biju Muduli has been both friend and family as we made new lives in the United States, and Sailaja Gorti and Suratna Budalakoti have offered the warmth and welcome of their home over the years. Thanks to my niece, Nishka, for a decade of reminders about the joys of encountering new, imagined worlds. My warm gratitude to Jill Jemella and Karen King for all the camaraderie and for carrying me through the years of pandemic parenting.

To my brother Debopam, thanks for all the love and care, for the optimism, and for the interminable reminders that everything will be fine. My most insufficient thanks to my parents for their selfless love. My late father, Malay Kumar Sarkar, encouraged me to dream. I so wish he could hold this book today. And my mother, Sujata Sarkar, under extraordinarily hard circumstances, let me pursue those dreams. Her spirit, courage, and capacity for happiness take my breath away. I dedicate this book to them.

And finally, to the two people who have brought unlimited joy and laughter to my life. My warmest love to Takshak, who has made finishing this book unbelievably easy and impossibly hard, and whose presence has filled my life (and my office!) with cheer and color. And to Suvrat, thanks for everything—for finding me in the library, for sharing this life, for letting me be, and for being there, always.

CPSIA information can be obtained
at www.ICGtesting.com
Printed in the USA
JSHW080919080223
37451JS00002B/3